Frontiers in Clinical Drug Research – Diabetes and Obesity

(Volume 4)

Edited by

Atta-ur-Rahman, *FRS*
Honorary Life Fellow, Kings College, University of Cambridge, Cambridge, UK

Frontiers in Clinical Drug Research – Diabetes and Obesity

Volume # 4

Editor: Atta-ur-Rahman

ISSN (Online): 2352-3220

ISSN (Print): 2467-9607

ISBN (Online): 978-1-68108-445-9

ISBN (Print): 978-1-68108-446-6

©2019, Bentham eBooks imprint.

Published by Bentham Science Publishers – Sharjah, UAE. All Rights Reserved.

BENTHAM SCIENCE PUBLISHERS LTD.
End User License Agreement (for non-institutional, personal use)

This is an agreement between you and Bentham Science Publishers Ltd. Please read this License Agreement carefully before using the ebook/echapter/ejournal (**"Work"**). Your use of the Work constitutes your agreement to the terms and conditions set forth in this License Agreement. If you do not agree to these terms and conditions then you should not use the Work.

Bentham Science Publishers agrees to grant you a non-exclusive, non-transferable limited license to use the Work subject to and in accordance with the following terms and conditions. This License Agreement is for non-library, personal use only. For a library / institutional / multi user license in respect of the Work, please contact: permission@benthamscience.org.

Usage Rules:

1. All rights reserved: The Work is the subject of copyright and Bentham Science Publishers either owns the Work (and the copyright in it) or is licensed to distribute the Work. You shall not copy, reproduce, modify, remove, delete, augment, add to, publish, transmit, sell, resell, create derivative works from, or in any way exploit the Work or make the Work available for others to do any of the same, in any form or by any means, in whole or in part, in each case without the prior written permission of Bentham Science Publishers, unless stated otherwise in this License Agreement.
2. You may download a copy of the Work on one occasion to one personal computer (including tablet, laptop, desktop, or other such devices). You may make one back-up copy of the Work to avoid losing it. The following DRM (Digital Rights Management) policy may also be applicable to the Work at Bentham Science Publishers' election, acting in its sole discretion:

- 25 'copy' commands can be executed every 7 days in respect of the Work. The text selected for copying cannot extend to more than a single page. Each time a text 'copy' command is executed, irrespective of whether the text selection is made from within one page or from separate pages, it will be considered as a separate / individual 'copy' command.
- 25 pages only from the Work can be printed every 7 days.

3. The unauthorised use or distribution of copyrighted or other proprietary content is illegal and could subject you to liability for substantial money damages. You will be liable for any damage resulting from your misuse of the Work or any violation of this License Agreement, including any infringement by you of copyrights or proprietary rights.

Disclaimer:

Bentham Science Publishers does not guarantee that the information in the Work is error-free, or warrant that it will meet your requirements or that access to the Work will be uninterrupted or error-free. The Work is provided "as is" without warranty of any kind, either express or implied or statutory, including, without limitation, implied warranties of merchantability and fitness for a particular purpose. The entire risk as to the results and performance of the Work is assumed by you. No responsibility is assumed by Bentham Science Publishers, its staff, editors and/or authors for any injury and/or damage to persons or property as a matter of products liability, negligence or otherwise, or from any use or operation of any methods, products instruction, advertisements or ideas contained in the Work.

Limitation of Liability:

In no event will Bentham Science Publishers, its staff, editors and/or authors, be liable for any damages, including, without limitation, special, incidental and/or consequential damages and/or damages for lost data and/or profits arising out of (whether directly or indirectly) the use or inability to use the Work. The entire liability of Bentham Science Publishers shall be limited to the amount actually paid by you for the Work.

General:

1. Any dispute or claim arising out of or in connection with this License Agreement or the Work (including non-contractual disputes or claims) will be governed by and construed in accordance with the laws of the U.A.E. as applied in the Emirate of Dubai. Each party agrees that the courts of the Emirate of Dubai shall have exclusive jurisdiction to settle any dispute or claim arising out of or in connection with this License Agreement or the Work (including non-contractual disputes or claims).
2. Your rights under this License Agreement will automatically terminate without notice and without the need for a court order if at any point you breach any terms of this License Agreement. In no event will any delay or failure by Bentham Science Publishers in enforcing your compliance with this License Agreement constitute a waiver of any of its rights.
3. You acknowledge that you have read this License Agreement, and agree to be bound by its terms and conditions. To the extent that any other terms and conditions presented on any website of Bentham Science Publishers conflict with, or are inconsistent with, the terms and conditions set out in this License Agreement, you acknowledge that the terms and conditions set out in this License Agreement shall prevail.

Bentham Science Publishers Ltd.
Executive Suite Y - 2
PO Box 7917, Saif Zone
Sharjah, U.A.E.
Email: subscriptions@benthamscience.net

CONTENTS

PREFACE	i
LIST OF CONTRIBUTORS	ii

CHAPTER 1 PHARMACOLOGIC OBESITY TREATMENT – EMPHASIS ON EFFICACY AND CARDIOMETABOLIC MARKERS 1
Fariha Salman, Ivan Gerling and *Helmut O. Steinberg*
- INTRODUCTION 2
  - Weight-Loss Drug Decreasing Food Absorption 3
  - Weight-Loss Drugs Decreasing Food Intake 4
  - Combination of Topiramate and Phentermine 7
  - Combination of Naltrexone and Bupropion 10
  - Liraglutide 13
  - Summary 15
- CONSENT FOR PUBLICATION 16
- CONFLICT OF INTEREST 16
- ACKNOWLEDGEMENTS 16
- REFERENCES 16

CHAPTER 2 INTERPLAY BETWEEN BILE ACID AND GLP-1 RECEPTOR AGONIST SIGNALING INFORMS THE DESIGN OF DRUGS TO COMBAT OBESITY AND ITS METABOLIC COMPLICATIONS 20
Jessica Felton and *Jean-Pierre Raufman*
- INTRODUCTION AND OVERVIEW 21
  - Bile Acid Synthesis & Circulation 22
  - Bile Acid Feedback Mechanisms 22
- TAKEDA G PROTEIN-COUPLED RECEPTOR-5 (TGR5) 25
  - TGR5 & Immunity 25
  - TGR5 in the Liver and Gallbladder 26
  - TGR5 & Thyroid Hormone Activation 27
- GLUCAGON-LIKE PEPTIDE 1 (GLP-1) 27
  - Physiology of GLP-1 27
  - Effects of GLP-1 28
  - Pathophysiology of GLP-1 29
- THE ROLE OF TGR5 IN GLP-1 SECRETION 30
- THE ROLE OF FXR IN GLUCOSE HOMEOSTASIS 32
- MANIPULATING THE BILE ACID POOL FOR GLUCOSE HOMEOSTASIS 33
- BILE ACIDS & GLP-1 AFTER BARIATRIC SURGERY 35
  - Glucose Homeostasis following Gastric Bypass Surgery 36
  - Increased Bile Acid Levels after Gastric Bypass Surgery 38
- PHARMACOLOGY 40
  - GLP-1 Agonists 40
  - DPP4 Inhibitors 44
  - TGR5 Agonists 47
- FUTURE DIRECTIONS 48
- CONCLUSIONS 49
- CONSENT FOR PUBLICATION 50
- CONFLICT OF INTEREST 50
- ACKNOWLEDGEMENTS 50
- REFERENCES 50

CHAPTER 3 SODIUM–GLUCOSE CO-TRANSPORTERS INHIBITORS FOR TYPE 2 DIABETES MELLITUS: THE 'NEW KIDS ON THE BLOCK' IN THE ERA OF EVIDENCE-BASED MEDICINE 58
Cheow Peng Ooi, Norlaila Mustafa, Nor Azmi Kamaruddin and *Munn Sann Lye*
    INTRODUCTION 59
        Epidemiology 59
        Pathophysiology of T2DM 59
        Glycaemic Control 60
    NEW TREATMENT OPTION 60
        Role of SGLTs in Glucose Homoeostasis 61
        Drug Development Status of SGLT Inhibitors 62
        Efficacy of SGLT2 Inhibitors 63
        Safety of SGLT2 Inhibitors 69
        Post-regulatory Approval and Long-term Safety Concerns 71
    EVIDENCE-BASED MEDICINE 71
        Evidence-based Medicine at the Crossroads 72
        Dilemmas in Evidence-based Medicine 73
            Restricted View of Evidence 73
            Limitations of Clinical Trials 73
            Misleading Results 78
        Issues in Translating RCTs to the Clinical Care Arena 78
            The Efficacy-Efficiency Gap 79
            Secondary Research of Evidence Synthesis 79
            The Harm of Poor-quality Evidence 82
    CURRENT EFFORTS IN BRIDGING THE EFFICACY-EFFICIENCY GAP 83
    FUTURE RESEARCH 87
    CONCLUSION 89
    CONSENT FOR PUBLICATION 90
    CONFLICT OF INTEREST 90
    ACKNOWLEDGEMENTS 90
    REFERENCES 90

CHAPTER 4 THE EFFECTS OF TRADITIONAL CHINESE MEDICINE ON INFLAMMATORY CYTOKINES IN DIABETIC NEPHROPATHY: THE PROGRESS IN THE PAST DECADES 108
Xiang Tu, YuanPing Deng, Ming Chen, James B. Jordan and *Sen Zhong*
    INTRODUCTION 109
    MAJOR INFLAMMATORY CYTOKINES INVOLVED IN DN 110
        Transforming Growth Factor (TGF)-β1 110
        Tumor Necrosis Factor (TNF)-α 110
        Monocyte Chemoattractant Protein-1 (MCP-1) 111
        Interleukin (IL)-1 111
        IL-6 111
        IL-18 112
    THE EFFECTS OF TRADITIONAL CHINESE MEDICINE (TCM) ON INFLAMMATORY CYTOKINES IN DN 112
        Single TCM and Their Extracts 112
        Astragalus [Plant name: Astragalus membranaceus (Fisch.) Bunge] 112
        Danshen Root (Plant name: Salvia miltiorrhiza Bge.) 114
        Szechuan Lovage Rhizome (Plant name: Ligusticum chuanxiong Hort.) 116
        Kudzuvine Root (Plant name: Pueraria thomsonii Benth.) 117

Sanchi [Plant name: Panax notoginseng (Burkill) F.H.Chen]	117
Common Threewingnut Root (Plant Name: Tripterygium wilfordii Hook. F.)	118
Formulae	120
Buyang Huanwu Decoction (BHD)	120
Fufang Danshen Diwan (FDD)	122
Bushen Tongluo Formula (BTF)	122
QidanYishenJiangtang Capsule (QYJC)	123
CONCLUSIONS	124
ABBREVIATIONS	124
CONSENT FOR PUBLICATION	126
CONFLICT OF INTEREST	126
ACKNOWLEDGEMENTS	127
REFERENCES	127
CHAPTER 5 THROUGH THE PERSPECTIVE OF HISTOLOGY - THE ALZHEIMER'S DISEASE PROMOTION BY OBESITY AND GLUCOSE METABOLISM: TYPE 3 DIABETES	135
Cigdem Elmas and *Cemile Merve Seymen*	
THE PATHOPHYSIOLOGY OF ALZHEIMER'S DISEASE	136
Neuronal Loss	138
Synapse Loss	139
INTERACTIONS BETWEEN DIABETES MELLITUS AND ALZHEIMER'S DISEASE	139
OXIDATIVE STRESS	140
Diabetes and Oxidative Stress	140
Advanced Glycation Products (AGEs)	141
AGEs (Advanced Glycation End-Products) Formation Mechanism	142
GLUCOSE/ENERGY METABOLISM IN ALZHEIMER'S DISEASE	144
Glucose Autoxidation	145
Glycation	146
Poliol Pathway	146
Changes in Hypothalamic Neurons	147
Changes in Hippocampal Neurons	147
Changes in Neruronal and Glial Cells in Occipital and Frontal Cortex	148
INSULIN AND ALZHEIMER'S DISEASE	148
Insulin Resistance	149
Insulin Like Growth Factors (IGFs)	150
Insulin Receptors	151
INFLAMMATORY RESPONSE	152
Cytokines and Other Bioactive Substances	152
Ceramides	154
Endoplasmic Reticulum (ER) Stress	154
Toll-like Receptors (TLR)	155
NEURONAL CALCIUM (Ca^{2+}) DYSREGULATION	156
MITOCHONDRIAL DYSFUNCTION	157
TAU HYPERPHOSPHORYLATION	160
Neurofibrillary Tangles (NFT)	161
Tau Hyperphosphorylation in Diabetes and Alzheimer's Disease	162
Glucose Transporters (GLUTs)	164
Glycogen-Synthase Kinase-3 (GSK-3)	165
UBIQUITIN / PROTEOSOME SYSTEM (UPS)	166
Ubiquitin/Proteosome System (UPS) in Diabetes and Alzheimer's Disease	167
AMYLOID BETA (AB) DEPOSITION	168

Amyloid β-Derived Diffusible Ligands (ADDLs)	170
Amyloid Beta (Aβ) Deposition in Diabetes and Alzheimer's Disease	171
Insulin-Degrading Enzyme (IDE)	173
OBESITY	174
THE APOLIPOPROTEIN-E (APOE4)	175
CONCLUSION	176
CONSENT FOR PUBLICATION	176
CONFLICT OF INTEREST	177
ACKNOWLEDGEMENTS	177
REFERENCES	177

CHAPTER 6 PHARMACOLOGICAL MECHANISM OF PPARΓ RATIO IN DIABETES AND OBESITY 199

José Roberto Santin, Marina Jagielski Goss and *Nara Lins Meira Quintão*

INTRODUCTION	200
PHARMACOLOGY OF PPARΓ	200
OBESITY AND DIABETES	205
Adipose Tissue: Obesity and Diabetes	208
Role of PPARγ in Obesity and Insulin Resistance	210
PRE-CLINICAL RESEARCH IN OBESITY AND DIABETES X PPARΓ	211
CLINICAL TRIALS	213
Synthetic Ligands	213
Pioglitazone	213
INT131 Besylate	214
Aleglitazar	214
Saroglitazar	214
Lobeglitazone	215
Natural Ligands	215
Essential Fatty Acids	216
Bofutsushosan	216
Doenjang	217
CONCLUSION	221
CONSENT FOR PUBLICATION	221
CONFLICT OF INTEREST	221
ACKNOWLEDGMENTS	221
REFERENCES	221

CHAPTER 7 HYDROGEN SULFIDE AND CARBOHYDRATE METABOLISM 226

Zahra Bahadoran, Parvin Mirmiran and *Asghar Ghasemi*

INTRODUCTION	227
AN OVERVIEW ON H_2S BIOSYNTHESIS AND METABOLISM	228
H_2S METABOLISM IN DIABETES	231
Animal Studies	231
Human Evidence	232
H_2S AND GLUCOSE/INSULIN HOMEOSTASIS	232
H_2S and Glucose Output and Utilization in Hepatocytes	232
H_2S and Glucose Metabolism in Adipose Tissue	234
H_2S and Glucose Metabolism in Skeletal Muscle	236
H_2S and Insulin Release in Pancreatic β-cells	237
H_2S and Triggering Pathway of Insulin Secretion	239
H_2S and Amplifying Pathways of Insulin Secretion	241
H_2S and Insulin Synthesis	244

  Effects of H$_2$S on Pancreatic β-cell Differentiation and Survival 244
H$_2$S AND DIABETES COMPLICATIONS 246
  Effects of H$_2$S on Diabetic Cardiovascular Complications 246
  Effects of H$_2$S on Renal Complications of Diabetic Models 247
  H$_2$S and Other Diabetes Complications 248
CONCLUSION AND FUTURE PERSPECTIVE 248
CONSENT FOR PUBLICATION 249
CONFLICT OF INTEREST 249
ACKNOWLEDGEMENT 249
REFERENCES 249

SUBJECT INDEX 259

PREFACE

This volume of the eBook series entitled: *Frontiers in Clinical Drug Research – Diabetes and Obesity* presents some recent exciting developments in clinical trials and formulation of research plans in the field of diabetes and obesity. The book should prove to be a valuable resource for pharmaceutical scientists and postgraduate students seeking updated and critically important information for developing clinical trials.

The chapters are written by authorities in the field and are mainly focused on obesity treatment. They include contributions on cardiometabolic markers, bile acid and GLP-1 receptor agonists to combat obesity and its metabolic complications, sodium-glucose co-transporter inhibitors for type 2 diabetes mellitus, the effects of traditional Chinese medicine on inflammatory cytokines in diabetic nephropathy, Alzheimer's disease promotion by obesity and glucose metabolism, pharmacological mechanism of PPARγ ratio in diabetes and obesity, and the metabolism of hydrogen sulfide and carbohydrate.

I hope that the readers will find these reviews valuable and thought-provoking so that they may trigger further research in the quest for new developments in the field. I am thankful to the efficient team of Bentham Science Publishers especially Dr. Faryal Sami (Manager Publications), Mr. Shehzad Iqbal Naqvi (Editorial Manager Publications) and Mr. Mahmood Alam (Director Publications).

Atta-ur-Rahman, *FRS*
Honorary Life Fellow
Kings College
University of Cambridge
Cambridge
UK

List of Contributors

Asghar Ghasemi	Endocrine Physiology Research Center, Research Institute for Endocrine Sciences, Shahid Beheshti University of Medical Sciences, Tehran, Iran
Cemile Merve Seymen	Department of Histology and Embryology, Faculty of Medicine, Gazi University, Ankara, Turkey
Charmaine Gentles	North Shore University Hospital, Manhassett, NY, USA
Cheow Peng Ooi	Endocrine Unit, Department of Medicine, Faculty of Medicine and Health Sciences, Universiti Putra, Malaysia
Cigdem Elmas	Department of Histology and Embryology, Faculty of Medicine, Gazi University, Ankara, Turkey
Dominick Gadaleta	North Shore University Hospital, Manhassett, NY, USA
Fariha Salman	Department of Internal Medicine, University of Tennessee Health Science Center, Memphis, TN, USA
Helmut O. Steinberg	Department of Medicine, Division of Endocrinology, Diabetes and Metabolism, University of Tennessee Health Science Center, Memphis TN, USA
Ivan Gerling	Department of Medicine, Division of Endocrinology, Diabetes and Metabolism, University of Tennessee Health Science Center, Memphis TN, USA
James B. Jordan	National Traditional Chinese Medicine Clinical Research Base for Diabetes Mellitus/Teaching Hospital of Chengdu University of Traditional Chinese Medicine, Sichuan Province, China
Jean-Pierre Raufman	Department of Medicine, University of Maryland School of Medicine, Baltimore, MD, USA
Jessica Felton	Department of Surgery, University of Maryland School of Medicine, Baltimore, MD, USA
José Roberto Santin	Postgraduate Program of Pharmaceutical Science, Universidade do Vale do Itajaí, Itajaí, Santa Catarina, Brazil
Larry Gellman	North Shore University Hospital, Manhassett, NY, USA
Marina Jagielski Goss	Postgraduate Program of Pharmaceutical Science, Universidade do Vale do Itajaí, Itajaí, Santa Catarina, Brazil
Ming Chen	Department of Nephrology, Teaching Hospital of Chengdu University of Traditional Chinese Medicine, Sichuan Province, China
Munn Sann Lye	Department of Community Health, Faculty of Medicine and Health Sciences, Universiti Putra, Malaysia
Nara Lins Meira Quintão	Postgraduate Program of Pharmaceutical Science, Universidade do Vale do Itajaí, Itajaí, Santa Catarina, Brazil
Nor Azmi Kamaruddin	Endocrine Unit, Department of Medicine, Faculty of Medicine, National University of Malaysia, Malaysia
Norlaila Mustafa	Endocrine Unit, Department of Medicine, Faculty of Medicine, National University of Malaysia, Malaysia

Parvin Mirmiran	Nutrition and Endocrine Research Center, Research Institute for Endocrine Sciences, Shahid Beheshti University of Medical Sciences, Tehran, Iran
Sen Zhong	National Traditional Chinese Medicine Clinical Research Base for Diabetes Mellitus/Teaching Hospital of Chengdu University of Traditional Chinese Medicine, Sichuan Province, China
Xiang Tu	National Traditional Chinese Medicine Clinical Research Base for Diabetes Mellitus/Teaching Hospital of Chengdu University of Traditional Chinese Medicine, Sichuan Province, China
YuanPing Deng	Department of Internal Medicine, Traditional Chinese Medicine Hospital of Fushun County, Sichuan Province, China
Zahra Bahadoran	Nutrition and Endocrine Research Center, Research Institute for Endocrine Sciences, Shahid Beheshti University of Medical Sciences, Tehran, Iran

CHAPTER 1

Pharmacologic Obesity Treatment – Emphasis on Efficacy and Cardiometabolic Markers

Fariha Salman[1], **Ivan Gerling**[2] **and Helmut O. Steinberg**[*,2]

[1] Department of Internal Medicine, University of Tennessee Health Science Center, Memphis, TN, USA

[2] Department of Medicine, Division of Endocrinology, Diabetes and Metabolism, University of Tennessee Health Science Center, Memphis, TN, USA

Abstract: After a thirteen-year hiatus, the FDA approved two new anti-obesity drugs in 2012, lorcaserin (brand-name Belviq®) and a fixed-combination of topiramide and phentermine (brand-name Qsymia®); one new anti-obesity drug was approved in 2014, a fixed-combination of naltrexone and bupropion (brand-name Contrave®), and in 2015 the "high dose" liraglutide (brand-name Saxenda®) was approved for weight loss. During this time, the marketed anti-obesity drug sibutramine was withdrawn due to increase in non-fatal myocardial infarction and stroke incidence [1], two drugs targeting cannabinoid receptors were not approvable in the United States and the European Union due to concerns regarding suicidality and leptin at pharmacologic doses was not marketed due to disappointing efficacy. All approved drugs, in conjunction with diet and exercise, achieve more weight gain as compared to placebo. Efficacy between different drugs cannot be directly compared since no head-to-head studies have been performed; however, some drugs/drug combinations appear to provide substantially bigger reductions in weight than others or the respective monotherapies. Cardiovascular parameters, especially systolic blood pressure, triglycerides and HDL-cholesterol respond positively to even small amounts of weight loss; the same holds true for insulin and insulin resistance. Uric acid, an emerging risk factor for type 2 diabetes and cardiovascular disease, also tends to improve in response to weight loss although there is an increased short-term risk of gout. All drugs have specific side effects and several drugs do have black-box warnings; for example, for female patients, pregnancy needs to be ruled-out before starting topiramate and a negative pregnancy test is required every 4 weeks while on treatment with Qsymia, and buproprion needs to be tapered off slowly and not discontinued abruptly to decrease the risk of seizures. Treating overweight/obese subjects presents an opportunity and a challenge to physicians and patients. To achieve optimal weight loss with least complications, patients need to work on hypocaloric diets and exercise

---

[*] **Corresponding Author Helmut O. Steinberg:** Department of Medicine, Division of Endocrinology, Diabetes and Metabolism, University of Tennessee Health Science Center, Memphis, TN, USA, Tel: 9014481008, Fax: (901)4485332; E-mail: hsteinb1@uthsc.edu

Atta-ur-Rahman (Ed.)
All rights reserved-© 2019 Bentham Science Publishers

and physicians need to know the prescribing information of the prescribed weight-loss drugs.

Keywords: Blood Pressure, Combination-Therapy, Glucose, Lipids, Mono-Therapy, Weight Loss.

INTRODUCTION

The ability to store excess calories has developed in many species to counteract periods of decreased food availability even for prolonged times of famine and to ensure that progression to fecundity and fertility does not result in excessive mortality for mother and offspring. Excess calories are mostly stored as fat in adipose tissue and to a much smaller degree as protein in muscle. However, the body has few, if any, means to defend it against prolonged excessive oversupply of calories. The extent to which it can generate and maintain adipose tissue is genetically determined. After expanding adipose tissue mass and filling all the available storage space in adipose tissue, fat will be deposited in other tissues resulting in lipotoxicity. Depending on the organ affected by lipotoxicity, fatty liver and hepatic insulin resistance, skeletal muscle insulin resistance and pancreatic beta-cell dysfunction can occur; it is likely that the metabolic consequences of lipotoxicity contribute to the development of diabetes and hypertension, renal hyperfiltration and congestive heart failure. In addition, mechanical issues due to excessive adipose tissue mass may cause problems such as degenerative joint disease and sleep apnea.

Approximately one third of the American population is obese and approximately ten percent of the American population is diabetic with another approximately 27% percent being pre-diabetic. Even small amounts of weight loss can prevent or delay the onset of type 2 diabetes. The diabetes prevention program (DPP) and the Look AHEAD studies [2, 3] showed that diet with exercise can achieve up to ten percent of weight loss in a good proportion of subjects. The DPP demonstrated that weight loss achieved by life-style modification decreased the onset of diabetes by nearly 60% when compared to standard of care. In subjects with diabetes, weight loss achieved by diet and exercise can effectively lower glucose levels and reduce the need for medications for lipids, blood pressure and sugar control. However, many overweight and obese subjects struggle to achieve meaningful weight loss and, when weight loss is achieved, to maintain the new lower level of body weight. While bariatric surgery has been demonstrated to lead to large and prolonged reductions in body weight, up to 30% weight loss, and cause remission of diabetes and hypertension and reduce the incidence of cancer and cardiovascular disease [4, 5], it is not available to most subjects.

A first step in weight loss is to have a caloric intake that is lower than the energy needed to maintain body weight. Decreasing hunger, increasing satiety in response to a meal, reduced intestinal food absorption and increased metabolism could all support the goal of weight loss; none of the currently approved drugs increases metabolism and drugs that increase metabolism had serious safety issues and were not developed for clinical use (Fig. **1**). High-quality clinical studies of weight loss medications have provided us with good estimates of efficacy and safety including improvements in cardiovascular and metabolic parameters. Most common side effects are related to either the central nervous system and/or the gastro-intestinal tract. Long-term results for cardiovascular safety are only available for lorcaserin [6].

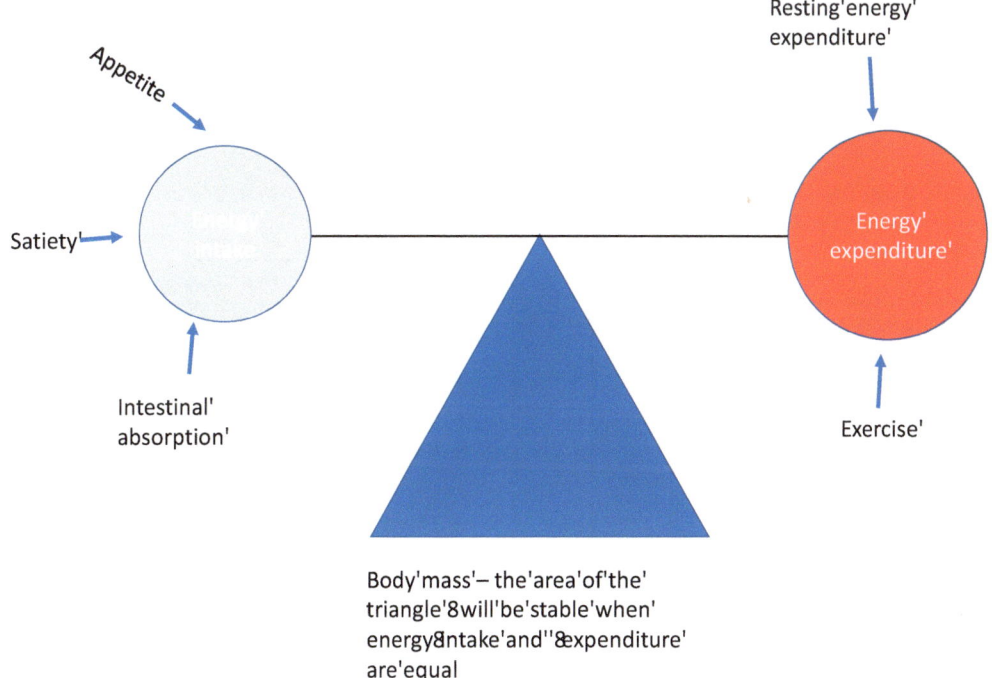

Fig. (1). Schematic display of variables that affect weight. Current approved weight loss drugs work on the left side of the triangle. Exercise is currently the only accepted and recommended obesity treatment that increases energy expenditure.

Weight-Loss Drug Decreasing Food Absorption

Orlistat (brand-names Xenical® and Alli®) is the only available drug for weight loss that interferes with absorption of food resulting in decreased caloric intake; it is taken at 120 mg three times a day with meals. Orlistat reduces fat absorption *via* inhibition of pancreatic lipase; it is minimally absorbed and does not accumulate in the circulation. Weight loss above that observed with placebo while on a diet and exercise regimen is in the 2-3 kilogram range over a one year time period (for review Hutton&Ferguson [6]). Compared to placebo, roughly twice as many subjects maintained a weight loss of >5.0% (range 35.5-54.8%) and >10.0% (range 16.4-24.8%) after one year (Table **1a**). Orlistat treatment was associated with small beneficial reductions in total and LDL-cholesterol and blood pressure (Table **1b**). Orlistat reduced the incidence of new onset diabetes by 42% in subjects with impaired glucose tolerance but only by 8% in subjects with normal glucose tolerance [7, 8]. In overweight subjects with diabetes, orlistat treatment led to better glycemic control; however, there was a higher incidence of hypoglycemia in the treatment group [9, 10]. The main adverse events to orlistat are gastro-intestinal and related to its mechanism of action; oily spotting and fecal incontinence have been reported to occur at a frequency of 26.6% and 7.7%, respectively during the first year of treatment. The frequency of these side effects appears to decrease over time. Orlistat is generally safe but rare serious adverse events have been reported in post marketing surveys and the prescribing physicians should be familiar with the package insert of this drug.

Table 1a. Proportion (%) of subjects achieving weight loss > 5.0% and >10.0% in response to orlistat or placebo in the Xendos study after 4 years.

	>5.0% weight loss	>10.0% weight loss
Orlistat	52.8	26.2
Placebo	37.3	15.6

Table 1b. Mean changes from baseline in cardio-metabolic parameters in response to orlistat or placebo in the Xendos study after 4 years.

	Δ systolic blood pressure	Δ diastolic blood pressure	Δ LDL-cholesterol (%)	Δ glucose (mg/dl)	Δ insulin (μU/ml)	Δ HbA1c
Orlstat	-4.9	-2.6	-12.8	1.801	-4.6	
Placebo	-3.4	-1.9	-5.1	3.6	-2.9	

Weight-Loss Drugs Decreasing Food Intake

Phentermine (brand-names Adipex® and Fastin®) has been used for weight management for almost 50 years. It acts at the trace amine-associated receptor 1 (TAAR1); this receptor is intracellular and located at the presynaptic end of the neuron. Phentermine enters the presynaptic neuron *via* a transport protein and activates TAAR1 which causes the release of norepinephrine; no effects on serotonin have been found. Increased CNS epinephrine levels account for the effects and side effects of this drug. It has been approved as monotherapy for short-term treatment (a few weeks) of obesity as adjunct to diet, exercise and behavioral modification. The reasons for the indication for only short term use include potential long-term adverse effects on the cardiovascular system such as tachycardia, hypertension, primary pulmonary hypertension, valvular heart disease and possible myocardial infarction and stroke. In addition, there is tachyphylaxis over time and the risk of drug dependence. Until recently, few high-quality clinical studies had been conducted with phentermine monotherapy and no dose response data for weight loss or appetite were available; and most of the older studies had small patient populations. Phentermine used to be combined with fenfluramine or dexfenfluramine when these drugs were available. The best data for phentermine monotherapy come from the development program for the combination of topiramate (see below) and phentermine [11]. In this study, all patients received counseling to decrease caloric intake by 500 cal/day and to increase physical activity as tolerated.

Phentermine monotherapy at doses of 7.5 and 15.0 mg/day over 28 weeks lead to a weight reduction of 6.7% and 7.4% respectively; weight loss in response to placebo was 1.7%. Systolic blood pressure (SBP) decreased slightly by ~3.5 mmHg in either treatment group; SBP decreased by ~2 mmHg in the placebo group. Changes in diastolic blood pressure (DBP) were small and, compared to placebo, not clinically meaningful. Heart rate did not change significantly from baseline. There was no change from baseline in fasting glucose and HbA1c in either treatment group.

No cardiovascular outcome studies have been performed to assess the long-term safety of phentermine.

Topiramate (brand-name Topomax®) is approved for the treatment of epileptic seizures and for migraine prophylaxis. The mechanism(s) by which topiramate works are not fully elucidated and may involve action on the GABA and the glutamate receptors as well as inhibition of carbonic anhydrase. The effect of topiramate on weight was first reported in 1996. Further dedicated weight loss studies in obese subjects without epilepsy studies showed robust weight loss; doses up to 384 mg per day were evaluated [12]. In this study by Bray *et al.* [12], topiramate was uptitrated every other week until the final target doses were achieved; all participants also participated in a commercially available weight loss program and were provided with a diet that was 600 calories lower per day than the estimated energy expenditure. The average weight loss (intention to treat, last observation carried forward analysis) at 24 weeks was 2.6% for placebo and 5.0%, 4.8%, 6.3%, and 6.3%, respectively, for groups treated with 64, 96, 192, and 384 mg/d topiramate (corresponding to mean decreases of 2.8 kg for placebo and of 5.2, 5.0, 6.4, and 6.6 kg, respectively, for the 64-, 96-, 192-, and 384-mg/d TPM groups). For subjects who completed the 24-week study, those receiving placebo experienced 3.6% weight loss compared with losses of 5.8%, 6.5%, 8.2%, and 8.5%, respectively, for those receiving 64, 96, 192, and 384 mg/d topiramate. There were no significant changes in glucose, insulin and lipids compared to placebo. However, all topiramate groups exhibited significantly greater decreases in SBP as compared to placebo. Due to a very high frequency of side effects leading to drop outs of the study, the highest dose was dropped in studies evaluating the effects of topiramate on weight-loss associated changes in glucose control [13, 14] and blood pressure [15]. The study by Stenlöf et al. showed a reduction in body weight of 2.5, 6.6 and 9.1%, respectively, in the placebo and the 96 mg and 192 mg topiramate groups [14]. In the study by Eliasson et al., diabetics on placebo lost no weight whereas patients on topiramate 192 mg per day lost on average 7.2 kg (baseline weight 108.4 kg) [13].

The study by Stenlöf et al. showed that hemoglobin A1c (HbA1c) decreased in the topiramate treatment groups (96mg and 192 mg) by ~0.6% (baseline ~6.8%) while HbA1c decreased by 0.2% in the placebo group [14]. Fasting plasma glucose (FPG) decreased by ~18 mg/dL in the topiramate groups while no change was observed in the placebo group. Insulin levels decreased in all groups but there were no differences between placebo and topiramate groups. No significant changes in lipid profiles were found in either group. SBP decreased significantly more in the topiramate treatment groups (-2.0, -6.3 and -7.6 mmHg) in the placebo and the 96 mg and 192 mg topiramate groups, respectively. Albumin excretion was also found to significantly decrease in the topiramate groups but not in the placebo group.

The study by Eliason et al. also showed a more pronounced decrease in HbA1c in the topiramate 192 mg group (1.1% vs. 0.3% at a baseline of ~%) [13]. FPG also was lowered more robustly with topiramate (~ 23mg/dL vs. 10 mg/dL). No between-group differences were observed for lipids and insulin levels. In addition, no differences in insulin sensitivity were found. And no significant changes in blood pressure were observed.

Assessment of topiramate's effect on blood pressure in a group of obese hypertensive subjects [13] showed that patients on drug (either 96 mg or 196 mg per day) lost significantly more weight (1.9%, 5.9% and 6.5%, respectively, for placebo or topiramate 96 mg or 196 mg per day). While SBP decreased in all groups, the reduction was not statistically significantly different as compared to placebo. DBP decreased also slightly in all groups and the reduction in the topiramate groups was significantly greater than placebo (2.1, 5.5 and 6.3 mmHg). In subjects who had lost >5.0% body weight induced by a low-calorie diet, topiramate was associated with additional reduction in weight with an efficacy similar to that described above [16]. All topiramate studies showed robust weight loss above placebo and it was speculated that the drug might increase the metabolic rate in addition to affecting food intake. However, a dedicated study by Tremblay *et al.* was unable to substantiate any effect on metabolic rate [17]; therefore, all weight loss in response to topiramate appear to be the result of its effect on appetite and/or satiety.

While topiramate was very effective, the side-effect profile was unfavorable showing mainly central nervous system site effects, especially at high doses. A dedicated dose-response study on topiramate's effect on cognition showed that these became significantly more frequent and pronounced at studied [18] doses of greater than 96 mg per day. At the end, due to the side-effect profile, the topiramate monotherapy program for a weight loss indication was discontinued.

No cardiovascular outcome studies have been performed to assess the long-term safety of topiramate.

Combination of Topiramate and Phentermine

The effect of the combination of topiramate and phentermine (Qsymia®) on weight has been studied in several studies [11, 19, 20]. A pooled analysis of results [21] showed that, in non-diabetic obese subjects, the combination leads to weight loss at all available doses. Mean weight loss was 4.7%, 8.2% and 10.4% at doses of 3.75mg/23mg for Phentermine/Topiramide Extended Release (Phen/TpmER), 7.5mg/46mg (Phen/TpmER) and 15mg/92mg (Phen/TpmER), respectively *vs.* 1.5% in response to placebo. In obese diabetic subjects, the 15mg/92mg(Phen/TpmER) dose lead to a 9.4% weight loss *vs.* 2.7% with placebo. Combination therapy achieved clinically meaningful weight loss (>5.0% body weight) in approximately 45% of the subjects treated with the lowest dose 3.75mg/23mg (Phen/TpmER), and in approximately 60-70% of subjects treated with the higher doses *vs.* approximately 20% in the placebo groups (Table **2a**).

Table 2a. Proportion of subjects achieving weight loss > 5.0% and >10.0% in response to the highest marketed dose of Qsymia (combination of phentermine 15 mg and topiramate 92 mg) after 56 weeks (CONQUER and EQUIP studies) or 108 weeks (SEQEL study).

Study		>5.0% weight loss	>10.0% weight loss
CONQUER	Qsymia	70	48
	Placebo	21	7
EQUIP	Qsymia	66.7	47.2
	Placebo	17.3	7.4
SEQEL	Qsymia	79.3	53.9

Study		>5.0% weight loss	>10.0% weight loss
	Placebo	30	11.5

SBP decreased from baseline in all the combination therapy groups which was significantly different from the placebo group where a lesser drop in blood pressure was observed; greater weight loss was associated with greater reductions in pressure. DBP, compared to placebo, decreased more with the two higher doses of the combination. Fasting insulin levels decreased more in the combination therapy groups which was even slightly more pronounced in a subgroup of pre-diabetic subjects; no overall changes were seen in glucose levels except in pre-diabetic and diabetic subgroups. In the pre-diabetic subgroup, glucose decreased by approximately 5-6 mg/dL in the combination therapy groups *vs.* approximately 2.5 mg/dL in the placebo group; the annualized incidence of progression from prediabetes to diabetes was lower in the combination therapy groups and this decrease in incidence was related to the degree of weight loss. In the diabetic subgroup, glucose decreased by approximately 5-10 mg/dL in the combination therapy groups *vs.* approximately 5 mg/dL in the placebo group. Furthermore, in the diabetic subgroup [13, 22], glycated hemoglobin decreased by 0.4% *vs.* 0.1% in the combination *vs.* the placebo groups, respectively. Triglycerides and HDL-cholesterol improved slightly with combination therapy compared to placebo. LDL-cholesterol may improve slightly as compared to placebo (Table **2b**).

Table 2b. Mean changes from baseline in cardio-metabolic parameters in response to the highest marketed dose of Qsymia (combination of phentermine 15 mg and topiramate 92 mg) after 56 weeks (CONQUER and EQUIP studies) or 108 weeks (SEQEL study).

Study		Δ systolic blood pressure	Δ diastolic blood pressure	Δ LDL-cholesterol (%)	Δ glucose (mg/dl)	Δ insulin (μU/ml)	Δ HbA1c
CONQUER	Qsymia	-5.6	-3.8	-6.9	-1.26	-3.97	-0.1
	Placebo	-2.4	-2.7	-4.1	2.13	0.734	0.1
EQUIP	Qsymia	-2.9	-1.5	-8.4	-0.6		
	Placebo	0.9	0.4	-5.5	1.9		
SEQEL	Qsymia	-4.3	-3.5	-5.6	-1.2	-5.2	0
	Placebo	-3.2	-3.9	-10.7	3.7	-2.6	0.2

A subgroup analysis [23] of subjects with prediabetes and/or the metabolic syndrome suggests clinical meaningful reduction in the rate of progression to diabetes of 80-90% with the highest dose of Phen/TpmER combination.

No cardiovascular outcome studies have been performed to assess the long-term safety of the combination therapy of phentermine and topiramate.

Lorcaserin (Belviq®) is a highly selective and potent 5-hydroxy tryptamine 2c (5-HT_{2c}) agonist. It works on receptors in the central nervous system, particularly in the hypothalamus. It activates 5-HT_{2c} receptors on the pro-opiomelanocortin (POMC) neurons in the arcuate nucleus which leads to the release of alpha-melanocortin-stimulating hormone (alpha-MSH). MSH acts on melanocortin-4 receptors in the paraventricular nucleus to suppress appetite. The mode of action is similar to fenfluramine and dexfenfluramine; however, the latter drugs had metabolites with activity on 5-HT_{2A} and 5-HT_{2B} receptors which mediate hallucinations and cardiovascular side effects such as valvulopathy and pulmonary hypertension. Lorcaserin, due to its selectivity for 5-HT_{2c} receptors, appears to be void of any effects on the 5-HT_{2A} and 5HT_{-2B} receptors. The FDA recommends to monitor for worsening and emergence of suicidal thoughts and behaviors and there is a black-box warning that refers to this potential side-effect.

A small 12-week dose-finding study [24] showed weight loss of 1.8 kg. 2.6 kg and 3.6 kg, respectively, at doses of 10 mg qd. 15 mg qd and 10 mg bid compared to a 0.3 kg weight loss with placebo. The BLOSSOM trial [25] evaluated the effect of lorcaserin at doses of 10 mg qd and 10 mg bid over one year; it showed a weight loss of 5.8 kg and 4.7 kg at doses of 10 mg qd and 10 mg bid, respectively, compared to 2.9 kg weight loss with placebo. The BLOOM trial [26] evaluated weight loss in response to 10 mg bid over two years; at one year, the weight loss was 5.8 kg with lorcaserin *versus* 2.2 kg with placebo. During year two, part of the subjects randomized to lorcaserin were changed to placebo as per protocol; around week 72, weight in the subjects changed to placebo became indistinguishable to that of subjects who had been on placebo from the beginning of the study. Subjects who continued on locarserin for the whole two-year period maintained more weight loss as compared to the placebo group(s) although the between-group difference in weight loss became smaller.

The BLOOM-DM [27] trial was a dedicated, one-year weight-loss study in a type 2 diabetic population that evaluated the response to lorcarserin 10 mg qd and 10 mg bid *versus* placebo. The two lorcaserin treatment groups experienced similar weight loss of ~5.5 kg while the placebo group lost ~2.0 kg over the one year period. Results for the proportion of subjects achieving clinically meaningful weight loss are in (Table **3a**).

BLOSSOM [25] showed no changes in HbA1c in response to either dose of lorcaserin and glucose and insulin levels were not reported. Small decreases in LDL-cholesterol were observed with either dose of lorcaserin; these changes were not different from placebo. Similarly, no changes in either SBP or DBP were observed. In the BLOOM [26] trial, significant but tiny reductions were observed in HbA1c, glucose and insulin levels. Reductions in blood SBP and DBP were small and not different from placebo Table (**3b**).

Table 3a. Proportion of subjects achieving weight loss > 5.0% and >10.0% in response to lorcaserin 10 mg twice a day or placebo in the BLOOM, BLOSSOM and BLOOM-DM studies study after 52 weeks.

Study		>5.0% weight loss	>10.0% weight loss
BLOOM	Lorcaserin	47.5	22.6
	Placebo	20.3	7.7
BLOSSOM	Lorcaserin	47.2	22.6
	Placebo	25	9.7
BLOOM-DM	Lorcaserin	37.5	16.3
	Placebo	16.1	4.4

Table 3b. Mean changes from baseline in cardio-metabolic parameters in response to lorcaserin 10 mg twice a day or placebo in the BLOOM, BLOSSOM and BLOOM-DM studies study after 52 weeks.

Study		Δ systolic blood pressure	Δ diastolic blood pressure	Δ LDL-cholesterol (%)	Δ glucose (mg/dl)	Δ insulin (μU/ml)	Δ HbA1c
BLOOM	Lorcaserin	-1.4	-1.1	2.87	-0.8	-3.33	-0.04
	Placebo	-0.8	-0.6	4.03	1.1	-1.28	0.03
BLOSSOM	Lorcaserin	-1.9	-1.9	0.3			-0.19
	Placebo	-1.2	-1.4	1.7			-0.14
BLOOM-DM	Lorcaserin	-0.8	-1.1	4.2	-27.4	-3	-0.9
	Placebo	-0.9	-0.7	5.0	-11.9	-1.6	-0.4

The results of the trial (CAMELIA-TIMI) to assess long-term cardiovascular safety with locaserin were recently published [6]; rates of cardiovascular events were not different from placebo.

Combination of Naltrexone and Bupropion

The combination of naltrexone and bupropion (Contrave®) was approved for the treatment of obesity in 2014. Interestingly, none of these drugs is FDA-approved as monotherapy for weight loss. Both drugs have been on the market for several decades with indications to treat depression and smoking cessation (bupropion) and as an antidote to opioid overdoses (naltrexone). Most monotherapy (early) studies for either component were small and limited to patients with depression or other behavioral problems. The FDA recommends to monitor for worsening and emergence of suicidal thoughts and behaviors and there is a black-box warning that refers to this potential side-effect.

Naltrexone is an opioid receptor antagonist with affinity to all three opioid receptor subtypes; it acts most potently on mu and delta receptor subtypes and works on central and peripheral opioid receptors. Its mechanism on weight is not well understood; it is thought to be mediated *via* receptors in the mesolimbic system that affect the dopaminergic reward- and learning center. It is thought that opioid antagonist remove an inhibitory feedback on pro-opiomelanocortin (POMC) neurons in the arcuate nucleus of the hypothalamus that are involved in energy balance and eating behavior; studies in mice do support this concept [28]. The effect of removing POMC neuron inhibition becomes evident mostly when measuring the frequency of action currents in isolated neurons (mouse-brain slice). If the proposed mechanism of naltrexone action is correct, naltrexone might also enhance the efficacy of lorcaserin; however, there are no studies to refute or support the combination of naltrexone and lorcaserin.

Studies with naltrexone showed little or no effect on weight [28, 29]. In normal subjects (not taking any opioids), administration of naltrexone has no significant cardio-vascular effects. It also has no effect on energy expenditure.

In the two studies [28, 29] with high quality data of monotherapy with naltrexone, naltrexone 50 mg qd showed an adjusted weight loss of 2% *vs.* 1.0% for placebo after 16 weeks of treatment; in the second study, naltrexone 48 mg qd (immediate release) showed 1.2% weight *vs.* a 0.8% weight loss with placebo. Weight loss of >5.0% was reported in 15.0% and 12.0% of subjects taking naltrexone 50 mg or placebo, respectively. No subject in the naltrexone group achieved a >10.0% weight loss. Similarly, subjects taking the immediate release formulation of naltrexone 48 mg did not achieve better weight loss outcomes compared to placebo. Changes from baseline in blood pressure, fasting insulin and glucose and lipids were small and not different between naltrexone and placebo groups.

Bupropion acts centrally on norepinephrine transporters, on the vesicular monoamine transporter and possibly also on dopamine transporters. It inhibits norepinephrine and dopamine reuptake. In subjects with depression, it had been recognized to induce mild weight loss, which was in contrast to other classes of antidepressants that were associated with weight gain. Weight loss for bupropion monotherapy 300 mg qd was 3.6% *vs.* 1.0% in the placebo group and 400 mg daily resulted in 2.7% *vs.* 0.8% weight loss. Changes from baseline in blood pressure, fasting insulin and glucose and lipids were small and not different between bupropion and placebo groups.

For the combination therapy with naltrexone and bupropion, most results have been obtained with the 32 mg daily dose sustained release formulation of naltrexone and the 360 mg daily dose of sustained release bupropion; different daily doses of naltrexone (16 mg or 48 mg) were studied in combination with bupropion 360 mg daily dose and the results were overall not meaningfully different from the 32 mg naltrexone dose [11, 30]. In obese, non-diabetic subjects, weight loss in response to the combination therapy was slightly more than 6.0% *versus* approximately 1.5% in the placebo groups. Combination therapy achieved clinically meaningful weight loss (>5.0% body weight) in approximately 50%the treated subjects *vs.* approximately 15% in the placebo groups.

Blood pressure did not change from baseline in the combination therapy groups which was significantly different from the placebo group where, despite less weight loss, blood pressure slightly decreased. Fasting insulin levels decreased more in the combination therapy groups; no overall changes were seen in glucose levels. Triglycerides and HDL-cholesterol improved slightly with combination therapy compared to placebo. LDL-cholesterol may improve slightly as compared to placebo.

In obese subjects with type 2 diabetes [31] treatment resulted in a 5.0% *vs.* a 1.8% weight loss in the combination therapy group and the placebo group, respectively. Combination therapy achieved clinically meaningful weight loss (>5.0% body weight) in approximately 45%the treated subjects *vs.* approximately 20% in the placebo group (Table **4a**).

Table 4a. Proportion of subjects achieving weight loss > 5.0% and >10.0% in response the highest marketed dose of Contrave (combination of naltrexone 32 mg and buproprion 360 mg) after 56 weeks.

Study		>5.0% weight loss	>10.0% weight loss
COR-I	Contrave	48	25
	Placebo	16	7
COR-II	Contrave	50.5	28.3
	Placebo	17.1	5.7
COR-BMOD	Contrave	66.4	41.5
	Placebo	42.5	20.2
COR-Diab	Contrave	44.5	18.5
	Placebo	18.9	5.7

Blood pressure tended to be slightly decreased from baseline in both groups with no significant between-group differences. HbA1c decreased significantly more in the combination group (0.6% *vs.* 0.1%). Fasting glucose levels tended to improve more in the combination group but the difference missed statistical significance.

Fasting insulin levels decreased similarly in both groups. Triglycerides and HDL-cholesterol improved slightly with combination therapy compared to placebo. LDL-cholesterol did not change in either group (Table **4b**).

Table **4b**. Mean changes from baseline in cardio-metabolic parameters in response the highest marketed dose of Contrave (combination of naltrexone 32 mg and buproprion mg) after 56 weeks.

Study		Δ systolic blood pressure	Δ diastolic blood pressure	Δ LDL-cholesterol (%)	Δ glucose (mg/dl)	Δ insulin (μU/ml)	Δ HbA1c
COR-I	Contrave	-0.1	0	-2	-3.24	-2.4	
	Placebo	-1.9	-0.9	-0.5	-1.26	-0.66	
COR-II	Contrave	0.6	0.4	-6.2	-2.8	-11.4	
	Placebo	-0.5	0.3	-2.1	-1.3	-3.5	
COR-BMOD	Contrave	-1.3	-1.4	7.1	-2.4	-3.5	
	Placebo	-3.9	-2.8	10	-1.1	-2.2	
COR-Diab	Contrave	0	-1.1	-1.4	-11.9	-13.5	-0.6
	Placebo	-1.1	-1.5	0	-4	-10.4	-0.1

A cardiovascular study was under way but was discontinued due to inappropriately disclosed interim analyses. At the time the study was closed, there did not appear to be an increased cardiovascular risk with the combination therapy.

Liraglutide

Liraglutide (brand-name Victoza® for the treatment of diabetes and Saxenda® for weight management) is a glucagon-like peptide-1 (GLP-1) analog that is resistant to degradation by the enzyme dipeptidyl-peptidase-4 (DPP-4) and acts on several organ systems including the brain where is decreases appetite, the stomach where it delays gastric emptying and the pancreas where it increases insulin secretion in a glucose-dependent fashion. In humans, the effect of liraglutide on weight is due to decreasing food intake and increasing satiety [32, 33]. In rodents, it may increase energy expenditure but this observation appears to be dependent on which obesity model is used. Liraglutide is administered daily *via* subcutaneous injection. Liraglutide levels are in steady state approximately 12 hours after injection and more than 10-fold higher than native GLP-1. A dose-response study [34] demonstrated significant weight loss *vs.* placebo at doses of 1.8 mg per per day and higher; the dose-response study and pharmacokinetic modeling [35] indicated that the 3.0 mg dose would be optimal for weight loss. Liraglutide needs to be slowly up-titrated over a four-week period to administer the full daily dose of 3.0 mg which reduces the gastro-intestinal side-effects. There is a black box warning that liraglutide is contraindicated in subjects with a personal or family history of medullary thyroid carcinoma or in patients with multiple endocrine neoplasia syndrome 2; this black box warning is based on animal data. In addition, there is concern about a potential association between incretin-based therapies and pancreatitis; however, the current data are inconclusive and a recently completed large, placebo-controlled and randomized study using 1.8 mg of liraglutide did not demonstrate any increased risk.

In obese, non-diabetic subjects, weight loss in response to liraglutide was 8.0% *versus* approximately 2.6% in the placebo group [36]. Liraglutide therapy achieved clinically meaningful weight loss (>5.0% body weight) in approximately 60% of the treated subjects *vs.* approximately 27% in the placebo group (Table **5a**). Subgroup analyses suggested that liraglutide may be less effective is subjects with body mass index >=40. A study by Wadden *et al.* [37], showed that the effect of liraglutide on weight loss was additive to that achieved after a 12-week low calorie (1200-1400 cal/day) diet; in addition to the 5.9% of body weight lost due to diet, subjects on liraglutide lost another 6.2% *vs.* a 0.2% loss in the placebo group. In diabetic subjects [38], liraglutide 3.0 mg qd reduced weight by 6.0% *vs.* 2.0% observed in the placebo group. Clinically meaningful weight loss (>5.0% body weight) was observed in approximately 55% of the treated subjects in the liraglutide group *vs.* approximately 20% in the placebo group (Table **5a**).

Table **5a**. Proportion of subjects achieving weight loss > 5.0% and >10.0% in response liraglutide 3 mg s.c. after 56 weeks.

Study		>5.0% weight loss	>10.0% weight loss
Scale	Liraglutide	63.2	33.1
	Placebo	27.1	10.6
Scale Maintanance	Liraglutide	50.5	26.1
	Placebo	21.8	6.3
Scale-Diab	Liraglutide	54.3	25.2
	Placebo	21.4	6.7

In type 2 diabetic subjects, the reduction in HbA1c was 1.3% *vs.* 0.3%, respectively, in the liraglutide and placebo group. SBP decreased slightly more in the liraglutide *vs.* the placebo group (2.8 mmHg *vs.* 0.4 mmHg); no between-group differences were seen for the decrease in DBP Table (**5b**).

SBP and DBP decreased from baseline in the liraglutide and the placebo groups; the decrease was more pronounced with liraglutide. Heart rate increased by approximately 2.5 beats per minute with liraglutide while there was no change in the placebo group.

Table 5b. Mean changes from baseline in cardio-metabolic parameters in response liraglutide 3 mg s.c. after 56 weeks.

Study		Δ systolic blood pressure	Δ diastolic blood pressure	Δ LDL (%)	Δ glucose (mg/dl)	Δ insulin (μU/ml)	Δ HbA1c
Scale	Liraglutide	-4.2	-2.6	-3	-7.1	-12.6	-0.30
	Placebo	-1.5	-1.9	-1	0.1	-4.4	-0.06
Scale Maintanance	Liraglutide	0.2	1.4	0.2	-9	0.40	-0.1
	Placebo	2.8	1.2	0.3	-3.6	2.33	0.1
Scale-Diab	Liraglutide	-2.8	-0.9	0.58	-34.3	6.87	-1.3
	Placebo	-0.4	-0.5	5.02	-0.2	1.94	-0.3

In non-diabetic obese subjects, fasting glucose and fasting insulin levels decreased more in response to liraglutide. In non-diabetic obese subjects, fasting glucose levels decreased by 7.1 mg/dL and 0.1 mg/dl, respectively, in the liraglutide and placebo groups, respectively Table (**5b**), and HbA1c decreased significantly more in the liraglutide group (0.6% *vs.* 0.1%). Fasting insulin levels decreased by 12.6% and 4.4%, respectively, in the liraglutide and placebo groups. Interestingly, the decrease in C-peptide was similar in both groups.

All lipid parameters (LDL, HDL-, VLDL-, non-HDL-, total cholesterol and triglycerides) improved in the liraglutide group and the changes were significantly more pronounced *vs.* placebo (Table **5b**).

No cardiovascular-outcomes results are available for the 3.0 mg dose of liraglutide in obese non-diabetic subjects; however, a study in high-risk, obese diabetic subjects [39] showed that subjects receiving liraglutide (~80% on the 1.8 mg dose) experienced ~13% lower incidence of the primary composite endpoint (consisting of first occurrence of death from any cardiovascular causes, nonfatal myocardial infarction, or nonfatal stroke).

Summary

Over the last 40 years, based on the work of Jules Hirsch, Rudolph Leibel, Jeffrey Friedman, Michael Nauck and Jens Holst and many others, progress has been made in unraveling mechanisms of control of appetite, satiety and feeding that appears to be involved in generating unhealthy amounts of adipose tissue. Their work has provided evidence for cross talk between adipocytes and the brain (*e.g.* leptin – not developed for commercial use) and for cross talk between the gut and the brain (*e.g.* GLP-1 – liraglutide which is marketed for weight loss). Their work has led to identification of neuronal pathways and neuro-transmitters that present potential targets for weight management drugs. Work is underway to coax (fat) cells into being metabolically more active and thus aide weight management.

Currently available weight-loss drugs are moderately effective, although not as effective as gastric sleeve or Roux-en-Y bariatric surgery. Neither surgery nor drug treatment work without diet; all patients can gain weight while on drug or after surgery when high-density, high-calorie food items are consumed.

Side effects often limit the dosing of the drugs; this problem can be overcome in part by combining lower doses of drugs that affect different targets. For example, one might combine and GLP-1 agonist with one or two drugs described in this chapter; this would be off-label use requiring vigilance for side effects and ideally conducted in randomized and blinded fashion as part of clinical research and drug development.

CONSENT FOR PUBLICATION

Not applicable.

CONFLICT OF INTEREST

The authors confirm that they have no conflict of interest to declare for this publication.

ACKNOWLEDGEMENTS

Declared none.

REFERENCES

[1] James WPT, Caterson ID, Coutinho W, *et al.* Effect of sibutramine on cardiovascular outcomes in overweight and obese subjects. N Engl J Med 2010; 363(10): 905-17.
[http://dx.doi.org/10.1056/NEJMoa1003114] [PMID: 20818901]

[2] Knowler WC, Barrett-Connor E, Fowler SE, *et al.* Reduction in the incidence of type 2 diabetes with lifestyle intervention or metformin. N Engl J Med 2002; 346(6): 393-403.
[http://dx.doi.org/10.1056/NEJMoa012512] [PMID: 11832527]

[3] Pi-Sunyer X, Blackburn G, Brancati FL, *et al.* Reduction in weight and cardiovascular disease risk factors in individuals with type 2 diabetes: one-year results of the look AHEAD trial. Diabetes Care 2007; 30(6): 1374-83.
[http://dx.doi.org/10.2337/dc07-0048] [PMID: 17363746]

[4] Sjöström L, Narbro K, Sjöström CD, *et al.* Effects of bariatric surgery on mortality in Swedish obese subjects. N Engl J Med 2007; 357(8): 741-52.

[http://dx.doi.org/10.1056/NEJMoa066254] [PMID: 17715408]

[5] Chang J, Wittert G. Effects of bariatric surgery on morbidity and mortality in severe obesity. Int J Evid-Based Healthc 2009; 7(1): 43-8.
[http://dx.doi.org/10.1111/j.1744-1609.2009.00123.x] [PMID: 21631845]

[6] Bohula EA, Wiviott SD, McGuire DK, et al. Cardiovascilar safty of Lorcaserin in overweight or obese patients. N Engl J Med 2018; 379(12): 1107-7.
[http://dx.doi.org/10.1056/NEJMoa1808721]

[7] Hutton B, Fergusson D. Changes in body weight and serum lipid profile in obese patients treated with orlistat in addition to a hypocaloric diet: a systematic review of randomized clinical trials. Am J Clin Nutr 2004; 80(6): 1461-8.
[http://dx.doi.org/10.1093/ajcn/80.6.1461] [PMID: 15585756]

[8] Heymsfield SB, Segal KR, Hauptman J, et al. Effects of weight loss with orlistat on glucose tolerance and progression to type 2 diabetes in obese adults. Arch Intern Med 2000; 160(9): 1321-6.
[http://dx.doi.org/10.1001/archinte.160.9.1321] [PMID: 10809036]

[9] Torgerson JS, Hauptman J, Boldrin MN, Sjöström L. XENical in the prevention of diabetes in obese subjects (XENDOS) study: a randomized study of orlistat as an adjunct to lifestyle changes for the prevention of type 2 diabetes in obese patients. Diabetes Care 2004; 27(1): 155-61.
[http://dx.doi.org/10.2337/diacare.27.1.155] [PMID: 14693982]

[10] Kelley DE, Bray GA, Pi-Sunyer FX, et al. Clinical efficacy of orlistat therapy in overweight and obese patients with insulin-treated type 2 diabetes: A 1-year randomized controlled trial. Diabetes Care 2002; 25(6): 1033-41.
[http://dx.doi.org/10.2337/diacare.25.6.1033] [PMID: 12032111]

[11] Miles JM, Leiter L, Hollander P, et al. Effect of orlistat in overweight and obese patients with type 2 diabetes treated with metformin. Diabetes Care 2002; 25(7): 1123-8.
[http://dx.doi.org/10.2337/diacare.25.7.1123] [PMID: 12087008]

[12] Aronne LJ, Wadden TA, Peterson C, Winslow D, Odeh S, Gadde KM. Evaluation of phentermine and topiramate *versus* phentermine/topiramate extended-release in obese adults. Obesity (Silver Spring) 2013; 21(11): 2163-71.
[http://dx.doi.org/10.1002/oby.20584] [PMID: 24136928]

[13] Bray GA, Hollander P, Klein S, et al. A 6-month randomized, placebo-controlled, dose-ranging trial of topiramate for weight loss in obesity. Obes Res 2003; 11(6): 722-33.
[http://dx.doi.org/10.1038/oby.2003.102] [PMID: 12805393]

[14] Eliasson B, Gudbjörnsdottir S, Cederholm J, Liang Y, Vercruysse F, Smith U. Weight loss and metabolic effects of topiramate in overweight and obese type 2 diabetic patients: randomized double-blind placebo-controlled trial. Int J Obes 2007; 31(7): 1140-7.
[http://dx.doi.org/10.1038/sj.ijo.0803548] [PMID: 17264849]

[15] Stenlöf K, Rössner S, Vercruysse F, Kumar A, Fitchet M, Sjöström L. Topiramate in the treatment of obese subjects with drug-naive type 2 diabetes. Diabetes Obes Metab 2007; 9(3): 360-8.
[http://dx.doi.org/10.1111/j.1463-1326.2006.00618.x] [PMID: 17391164]

[16] Tonstad S, Tykarski A, Weissgarten J, et al. Efficacy and safety of topiramate in the treatment of obese subjects with essential hypertension. Am J Cardiol 2005; 96(2): 243-51.
[http://dx.doi.org/10.1016/j.amjcard.2005.03.053] [PMID: 16018851]

[17] Astrup A, Caterson I, Zelissen P, et al. Topiramate: long-term maintenance of weight loss induced by a low-calorie diet in obese subjects. Obes Res 2004; 12(10): 1658-69.
[http://dx.doi.org/10.1038/oby.2004.206] [PMID: 15536230]

[18] Tremblay A, Chaput JP, Bérubé-Parent S, et al. The effect of topiramate on energy balance in obese men: a 6-month double-blind randomized placebo-controlled study with a 6-month open-label extension. Eur J Clin Pharmacol 2007; 63(2): 123-34.

[http://dx.doi.org/10.1007/s00228-006-0220-1] [PMID: 17200837]

[19] Loring DW, Williamson DJ, Meador KJ, Wiegand F, Hulihan J. Topiramate dose effects on cognition: a randomized double-blind study. Neurology 2011; 76(2): 131-7.
[http://dx.doi.org/10.1212/WNL.0b013e318206ca02] [PMID: 21148119]

[20] Gadde KM, Allison DB, Ryan DH, et al. Effects of low-dose, controlled-release, phentermine plus topiramate combination on weight and associated comorbidities in overweight and obese adults (CONQUER): a randomised, placebo-controlled, phase 3 trial. Lancet 2011; 377(9774): 1341-52.
[http://dx.doi.org/10.1016/S0140-6736(11)60205-5] [PMID: 21481449]

[21] Garvey WT, Ryan DH, Look M, et al. Two-year sustained weight loss and metabolic benefits with controlled-release phentermine/topiramate in obese and overweight adults (SEQUEL): a randomized, placebo-controlled, phase 3 extension study. Am J Clin Nutr 2012; 95(2): 297-308.
[http://dx.doi.org/10.3945/ajcn.111.024927] [PMID: 22158731]

[22] Jordan J, Astrup A, Engeli S, Narkiewicz K, Day WW, Finer N. Cardiovascular effects of phentermine and topiramate: a new drug combination for the treatment of obesity. J Hypertens 2014; 32(6): 1178-88.
[http://dx.doi.org/10.1097/HJH.0000000000000145] [PMID: 24621808]

[23] Garvey WT, Ryan DH, Bohannon NJ, et al. Weight-loss therapy in type 2 diabetes: effects of phentermine and topiramate extended release. Diabetes Care 2014; 37(12): 3309-16.
[http://dx.doi.org/10.2337/dc14-0930] [PMID: 25249652]

[24] Garvey WT, Ryan DH, Henry R, et al. Prevention of type 2 diabetes in subjects with prediabetes and metabolic syndrome treated with phentermine and topiramate extended release. Diabetes Care 2014; 37(4): 912-21.
[http://dx.doi.org/10.2337/dc13-1518] [PMID: 24103901]

[25] Smith SR, Prosser WA, Donahue DJ, Morgan ME, Anderson CM, Shanahan WR. Lorcaserin (APD356), a selective 5-HT(2C) agonist, reduces body weight in obese men and women. Obesity (Silver Spring) 2009; 17(3): 494-503.
[http://dx.doi.org/10.1038/oby.2008.537] [PMID: 19057523]

[26] Fidler MC, Sanchez M, Raether B, et al. A one-year randomized trial of lorcaserin for weight loss in obese and overweight adults: the BLOSSOM trial. J Clin Endocrinol Metab 2011; 96(10): 3067-77.
[http://dx.doi.org/10.1210/jc.2011-1256] [PMID: 21795446]

[27] Smith SR, Weissman NJ, Anderson CM, et al. Multicenter, placebo-controlled trial of lorcaserin for weight management. N Engl J Med 2010; 363(3): 245-56.
[http://dx.doi.org/10.1056/NEJMoa0909809] [PMID: 20647200]

[28] O'Neil PM, Smith SR, Weissman NJ, et al. Randomized placebo-controlled clinical trial of lorcaserin for weight loss in type 2 diabetes mellitus: the BLOOM-DM study. Obesity (Silver Spring) 2012; 20(7): 1426-36.
[http://dx.doi.org/10.1038/oby.2012.66] [PMID: 22421927]

[29] Greenway FL, Whitehouse MJ, Guttadauria M, et al. Rational design of a combination medication for the treatment of obesity. Obesity (Silver Spring) 2009; 17(1): 30-9.
[http://dx.doi.org/10.1038/oby.2008.461] [PMID: 18997675]

[30] Greenway FL, Dunayevich E, Tollefson G, et al. Comparison of combined bupropion and naltrexone therapy for obesity with monotherapy and placebo. J Clin Endocrinol Metab 2009; 94(12): 4898-906.
[http://dx.doi.org/10.1210/jc.2009-1350] [PMID: 19846734]

[31] Greenway FL, Fujioka K, Plodkowski RA, et al. Effect of naltrexone plus bupropion on weight loss in overweight and obese adults (COR-I): a multicentre, randomised, double-blind, placebo-controlled, phase 3 trial. Lancet 2010; 376(9741): 595-605.
[http://dx.doi.org/10.1016/S0140-6736(10)60888-4] [PMID: 20673995]

[32] Hollander P, Gupta AK, Plodkowski R, et al. Effects of naltrexone sustained-release/bupropion

sustained-release combination therapy on body weight and glycemic parameters in overweight and obese patients with type 2 diabetes. Diabetes Care 2013; 36(12): 4022-9.
[http://dx.doi.org/10.2337/dc13-0234] [PMID: 24144653]

[33] Holst JJ. Incretin hormones and the satiation signal. Int J Obes 2013; 37(9): 1161-8.
[http://dx.doi.org/10.1038/ijo.2012.208] [PMID: 23295502]

[34] van Can J, Sloth B, Jensen CB, Flint A, Blaak EE, Saris WH. Effects of the once-daily GLP-1 analog liraglutide on gastric emptying, glycemic parameters, appetite and energy metabolism in obese, non-diabetic adults. Int J Obes 2014; 38(6): 784-93.
[http://dx.doi.org/10.1038/ijo.2013.162] [PMID: 23999198]

[35] Astrup A, Carraro R, Finer N, et al. Safety, tolerability and sustained weight loss over 2 years with the once-daily human GLP-1 analog, liraglutide. Int J Obes 2012; 36(6): 843-54.
[http://dx.doi.org/10.1038/ijo.2011.158] [PMID: 21844879]

[36] Wilding JP, Overgaard RV, Jacobsen LV, Jensen CB, le Roux CW. Exposure-response analyses of liraglutide 3.0 mg for weight management. Diabetes Obes Metab 2016; 18(5): 491-9.
[http://dx.doi.org/10.1111/dom.12639] [PMID: 26833744]

[37] Pi-Sunyer X, Astrup A, Fujioka K, et al. A Randomized, Controlled Trial of 3.0 mg of Liraglutide in Weight Management. N Engl J Med 2015; 373(1): 11-22.
[http://dx.doi.org/10.1056/NEJMoa1411892] [PMID: 26132939]

[38] Wadden TA, Hollander P, Klein S, et al. Weight maintenance and additional weight loss with liraglutide after low-calorie-diet-induced weight loss: the SCALE Maintenance randomized study. Int J Obes 2013; 37(11): 1443-51.
[http://dx.doi.org/10.1038/ijo.2013.120] [PMID: 23812094]

[39] Davies MJ, Bergenstal R, Bode B, et al. Efficacy of Liraglutide for Weight Loss Among Patients With Type 2 Diabetes: The SCALE Diabetes Randomized Clinical Trial. JAMA 2015; 314(7): 687-99.
[http://dx.doi.org/10.1001/jama.2015.9676] [PMID: 26284720]

[40] Marso SP, Daniels GH, Brown-Frandsen K, et al. Liraglutide and Cardiovascular Outcomes in Type 2 Diabetes. N Engl J Med 2016; 375(4): 311-22.
[http://dx.doi.org/10.1056/NEJMoa1603827] [PMID: 27295427]

CHAPTER 2

Interplay Between Bile Acid and GLP-1 Receptor Agonist Signaling Informs the Design of Drugs to Combat Obesity and its Metabolic Complications

Jessica Felton[1] and Jean-Pierre Raufman[2],*

[1] Department of Surgery, University of Maryland School of Medicine, Baltimore, MD, USA

[2] Department of Medicine, University of Maryland School of Medicine, Baltimore, MD, USA

Abstract: For more than a century the physiological role of bile acids was considered limited to their actions in cholesterol metabolism and lipid absorption from the gastrointestinal tract. Evidence emerging over the past 20 years has greatly changed this perspective. It is now apparent that these complex molecules play an integral signaling function within the gut and have extra-intestinal hormonal actions. Bile acid interaction with plasma membrane G protein-coupled receptors (*e.g.* TGR5, M3R) and nuclear receptors (*e.g.* FXR) expressed on intestinal epithelial cells modulates post-receptor signaling and gene transcription. Herein, we review the fundamentals of how bile acid structure governs the interaction of these molecules with cell receptors and transport proteins (*e.g.* ASBT), and how these interactions are important for nutritional balance. We focus on bile acid interaction with TGR5, a receptor whose activation stimulates release of glucagon-like peptide-1 (GLP-1) from enteroendocrine L cells; GLP-1, an intestinal incretin, is important for glucose homeostasis. Drugs that mimic the actions of GLP-1 or retard its degradation are effective treatments for diabetes, obesity, and their metabolic complications (*e.g.* non-alcoholic fatty liver disease). Altered gut and plasma levels of bile acids and GLP-1 are important for the clinical benefits of bariatric surgery. Hence, there is great interest in developing novel pharmaceutical approaches to imitate these changes and, in particular, the beneficial actions of bile acids. We offer a critical analysis of these approaches and propose novel opportunities for drug design to combat the current obesity epidemic and its metabolic complications.

Keywords: Bariatric surgery, Bile acids, Bile acid sequestrants, DPP4, Enteroendocrine L cells, Exenatide, FXR, Gastric bypass, GLP-1, Incretin, Metabolic syndrome, Non-Alcoholic fatty liver disease, Non-Alcoholic steatohepatitis, Obesity, Roux-en-Y, Sitagliptin, TGR5, Type 2 diabetes.

---

* Corresponding author Jean-Pierre Raufman: Department of Medicine, University of Maryland School of Medicine, Baltimore, Maryland, USA; Tel: 410-328-8728. Fax: 410-328-8315; E-mail: jraufman@som.umaryland.edu

Atta-ur-Rahman (Ed.)
All rights reserved-© 2019 Bentham Science Publishers

INTRODUCTION AND OVERVIEW

Adult obesity, defined as body mass index (BMI) 30 or greater, has blossomed into a major worldwide epidemic; as of 2016, over one-third of adults in the United States (U.S.) met this criterion and the percentage continues to increase [1]. As the prevalence of obesity increases, so have obesity-related disorders – type 2 diabetes, hyperlipidemia, hypertension, obstructive sleep apnea, heart disease, stroke, asthma, depression, non-alcoholic fatty liver disease (NAFLD), and many types of cancer [1, 2]. In 2008 the U.S. medical cost of obesity was estimated at $ 147 billion; on average, yearly medical care costs for the obese were $1,429 greater than those for normal-weight persons [1]. Not only does the growing obesity epidemic add substantial burden to already strapped healthcare resources, but also it is a major cause of morbidity and mortality [3]; obesity-related conditions are leading causes of preventable death [1]. In comparison to normal-weight persons, a 25-year-old morbidly obese man faces a 22% reduction in expected lifespan [2]. Obesity, type 2 diabetes, and NAFLD are manageable but many currently used treatments are costly, invasive, and lack durability. As reviewed here, recent advances in bile acid research have expanded our understanding of the mechanistic underpinnings of obesity-related disorders and promise innovative safer, more cost-effective, and durable therapeutic alternatives.

Bile acids are well-established regulators of cholesterol homeostasis, and facilitate the uptake, digestion, and metabolism of lipids. However, in recent years, bile acids are receiving increased recognition as playing equally important roles as endocrine modulators, including actions on glucose metabolism and energy regulation. Here, we review the fundamentals of bile acid synthesis, structure, and transport, and how these features govern interaction of these molecules with cell receptors and transport proteins and why these interactions are important for nutritional balance. We focus on bile acid interactions with a G protein-coupled receptor, TGR5, which stimulates release of a key intestinal incretin, glucagon-like peptide-1 (GLP-1), from small intestinal enteroendocrine L cells. Drugs that mimic the actions of GLP-1 or prevent its degradation are effective treatments for diabetes, obesity, and perhaps most importantly their metabolic complications. Post-bariatric surgery changes in gut and plasma levels of bile acids and GLP-1 appear to be important for the clinical benefits of these procedures. Hence, there is great interest in developing novel pharmaceutical approaches that imitate these changes and, in particular, the beneficial actions of bile acids. We offer a critical analysis of these approaches and propose novel opportunities for drug design that can help combat the current obesity epidemic and its metabolic complications.

Bile Acid Synthesis & Circulation

Bile acids are synthesized in the liver from hepatic cholesterol through a series of enzymatic reactions [4]. Primary bile acids (cholic and chenodeoxycholic acids) derive from two major synthetic routes, the classical (neutral) and alternative (acidic) pathways [4, 5]. In the classical pathway, the first, rate-limiting step is hydroxylation of cholesterol, catalyzed by the cytochrome P450 enzyme cholesterol 7α-hydroxylase (CYP7A1). The classical pathway produces the majority of bile acids in circulation, with cholic and chenodeoxoycholic acids produced in nearly equal amounts. Secondary bile acids, deoxycholic (from cholic acid) and lithocholic (from chenodeoxycholic acid) acids, derive from intestinal bacterial enzyme actions on primary bile acids after they are partially 7α-dehydroxylated [5].

Bile acid synthesis can also derive from the alternative pathway, which involves the enzyme oxysterol 7α-hydroxylase CYP27A1 and conversion of oxysterols to bile acids [4, 5]. It is estimated that less than 10% of bile acid synthesis occurs *via* this pathway, but during fetal development and in chronic liver disease, the contribution of this pathway to the bile acid pool may increase substantially [4, 6].

Before primary bile acids are secreted into the bile canalicular lumen, they are conjugated with taurine or glycine and stored in the gallbladder as mixed micelles with phospholipids and cholesterol [4]. After ingestion of a meal, release of intestinal cholecystokinin stimulates gallbladder contraction and release of bile into the proximal small intestine, where bile acids facilitate digestion, absorption, and transport of dietary fats and fat-soluble vitamins [4, 5]. Approximately 95% of bile acids are reabsorbed by a very effective transport mechanism in the distal ileum and transported back to the liver *via* the portal circulation – the enterohepatic circulation [6]. Approximately 5% of intestinal bile acids, not reabsorbed from the distal ileum, is excreted in feces; these 'lost' bile acids are replenished through *de novo* bile acid synthesis in the liver [4, 5]. The recycling of this small pool of bile acids occurs 5-10 times per day, and the size of the bile acid pool is kept relatively constant by a variety of feedback mechanisms [5].

Bile Acid Feedback Mechanisms

To prevent cytotoxic accumulation of bile acids, a tight feedback mechanism governed by several nuclear receptors represses *de novo* bile acid synthesis in the liver [4] (Fig. **1**). Thus, bile acids act as signaling molecules that regulate their own biosynthesis through these negative feedback loops. Primarily, bile acids limit their own synthesis by inhibiting the production of CYP7A1, the enzyme that is the first and rate-limiting step of the aforementioned classical pathway [7], as well as inhibiting another important classical pathway enzyme, CYP8B1 [8].

These inhibitory circuits are regulated by several predominantly farnesoid X receptor (FXR)-dependent pathways.

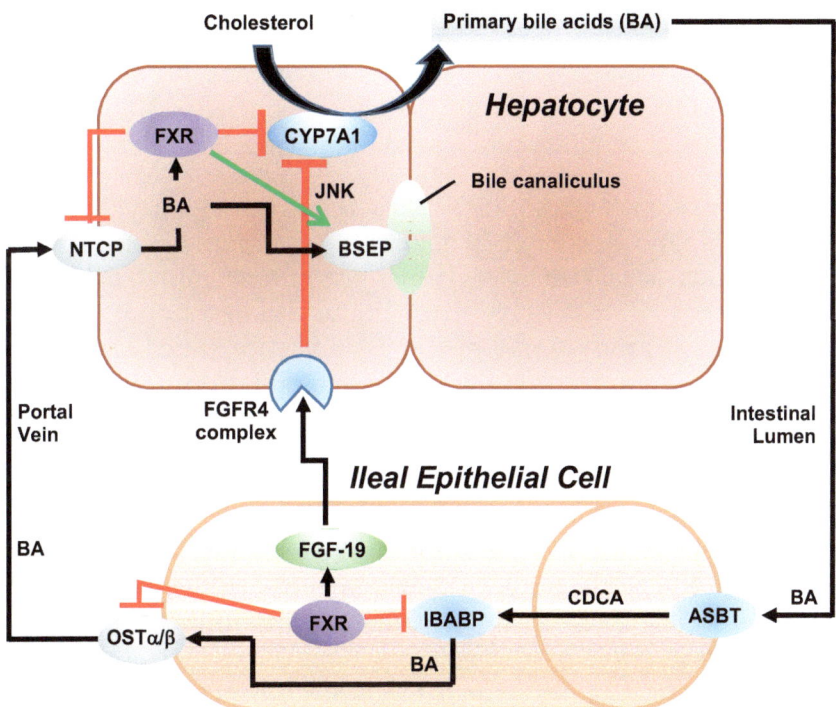

Fig. (1). Bile acids regulate their own synthesis via a tight feedback loop. In the classical pathway, CYP7A1 catalyzes primary bile acid synthesis from cholesterol. ASBT transports bile acids from the intestinal lumen into epithelial cells. IBABP shuttles bile acids across the enterocytes. Then, basolateral membrane proteins OSTα and β, both regulated by FXR, transport bile acids into the portal circulation. Activation of intestinal FXR stimulates FGF-19 expression and release; FGF-19 then circulates via the portal vein to hepatocytes where it binds to the FGF4 receptor complex. This represses *CYP7A1* transcription and bile acid biosynthesis. Bile acids travel from enterocytes to the liver via the portal vein where NTCP, also regulated by FXR, transports bile acids into hepatocytes – the enterohepatic circulation. In hepatocytes, bile acid-induced activation of FXR represses bile acid absorption and synthesis, while FXR-induced BSEP expression increases bile acid transport across the bile canalicular membrane into bile ducts.

FXR is a nuclear hormone receptor that acts as an endogenous bile acid sensor in both the liver and small intestine [7, 9]. Bile acid binding and activation of FXR upregulates a small heterodimer partner (SHP), an atypical nuclear receptor that acts as a co-repressor [10]. FXR is most potently activated by chenodeoxycholic acid, but can be activated by other primary and secondary bile acids [4]. SHP suppresses the activity of hepatocyte nuclear factor-4α (HNF-4α), liver X receptor (LXR), and liver receptor homolog-1 (LRH-1). Both LXR and LRH-1 modulate

CYP7A1 transcription by binding to bile acid response elements in the CYP7A1 promotor region [8, 11].

Activation of intestinal FXR by bile acids also induces enterocyte production of fibroblast growth factor 19/15 (FGF-19 in humans and Fgf-15 in rodents). FGF-19 is released by enterocytes and circulates *via* the portal vein to hepatocytes where it binds to the FGF receptor 4 (FGFR4) complex [4, 8]. This represses CYP7A1 transcription *via* the c-Jun N-terminal kinase (JNK) pathway and attenuates biosynthesis of bile acids from intrahepatic cholesterol [4, 8].

FXR also regulates bile acid export and reabsorption in hepatocytes by regulating expression of the ATP-dependent bile acid salt export pump BSEP/ABCB11 that transports bile acids across the canalicular membrane into the bile duct and gallbladder [12, 13]. FXR also upregulates expression of transporters responsible for hepatobiliary conveyance of phospholipids and hydrophilic organic anions like divalent bile acid conjugates and of a basolateral bile acid import pump, OATP1B3/SLC01B3 [12]. In contrast, FXR down-regulates expression of NTCP/SLC10A1, the sinusoidal/basolateral sodium-taurocholate co-transporting polypeptide, which returns approximately 80% of bile acids to the liver [4, 12].

FXR not only regulates hepatic bile acid transport expression, but also controls bile acid uptake by enterocytes, a critical step in bile acid metabolism. Expression of the murine apical sodium-dependent bile acid transporter (ASBT)/intestinal bile acid transporter (IBAT) is positively controlled by LRH-1. LRH-1 is repressed by FXR-induced SHP-mediated repression *via* the CYP7A1 pathway in hepatocytes. ASBT transports bile aids from the intestinal lumen into mucosal epithelial cells [13]. FXR partially controls expression of intestinal bile acid-binding protein (IBABP), which shuttles bile acids from the apical to basolateral membranes of enterocytes, in addition to the organic solute transporter α/β (OSTα/β), which transports bile acids into the portal circulation [4, 13]. Finally, FXR directly regulates transcription of fatty acid binding protein 6 (FABP6), whose likely role is to sequester excess intracellular bile acids, as well as to serve as a coactivator of FXR [12]. Thus, by modulating enterohepatic circulation and the *de novo* synthesis of bile acids, FXR plays a critical role in tightly regulating the size of the bile acid pool.

In addition to FXR, the G protein-coupled bile acid receptor-1 (GPABR1, commonly referred to as TGR5, the Takeda G-protein-coupled receptor-5) also regulates bile acid homeostasis, although the underlying mechanisms are not as clear. Unlike FXR, TGR5 is expressed in a wide variety of tissues, including liver, gallbladder, ileum, and skeletal muscle [13]. When activated by bile acids, TGR5 increases cellular levels of cyclic adenosine monophosphate (cAMP) and stimulates energy expenditure in both brown adipose tissue and skeletal muscle [8, 13].

Note that sphingosine-1-phosphate receptor 2 (S1PR2) may also play a role in bile acid-mediated hepatic glucose and lipid metabolism [14]. S1PRs expressed in a variety of tissues are coupled to specific G proteins that mediate unique functions, *e.g.* cell degeneration and chemotaxis towards foci of inflammation [14]. S1PR2 are highly expressed in liver and recently reported to play a role in bile acid-mediated hepatic lipid metabolism. Activation of S1PR2 by conjugated bile acids activates downstream ERK1/2 and AKT signaling which may regulate hepatic glucose and lipid metabolism [14, 15]. In rat hepatocytes, infusion of bile acids (taurocholic acid, TCA) activates AKT and glycogen synthase (GS) [14, 16]. This suggests that, in conjunction with insulin, bile acids may modulate post-prandial glucose storage in the liver and regulate plasma glucose homeostasis [16]. While there is no current evidence that S1PR2 plays a role in GLP-1 secretion, targeting S1PR2 activation and signaling may hold promise for treating diabetes, obesity, and other metabolic disorders.

TAKEDA G PROTEIN-COUPLED RECEPTOR-5 (TGR5)

TGR5, a G protein-coupled bile acid receptor, can be stimulated by all bile acids but the most potent natural agonists are lithocholic and taurine-conjugated lithocholic acids [17]. TGR5 activation promotes cAMP production and activation of protein kinase A; downstream effects depend on the cell type [13, 17] (Fig. **2**). TGR5 is highly expressed in the gallbladder and at lower levels in brown adipose tissue (BAT), liver, intestine, and areas of the central nervous system [18]. At present, the diverse metabolic effects of TGR5 activation remain incompletely understood.

TGR5 & Immunity

In 2002, when TGR5 was discovered it was shown to play a role in immune cell function [19, 20]. TGR5 mRNA is expressed abundantly in monocytes and macrophages, and when activated by bile acids suppresses the functions of these cells [20]. In rabbit models, bile acid-induced cAMP production in macrophages suppressed lipopolysaccharide (LPS)-stimulated cytokine production [20]. While this role of bile acids may appear unrelated to metabolism, it is now generally accepted that in many tissues chronic inflammation promotes metabolic dysfunction [18].

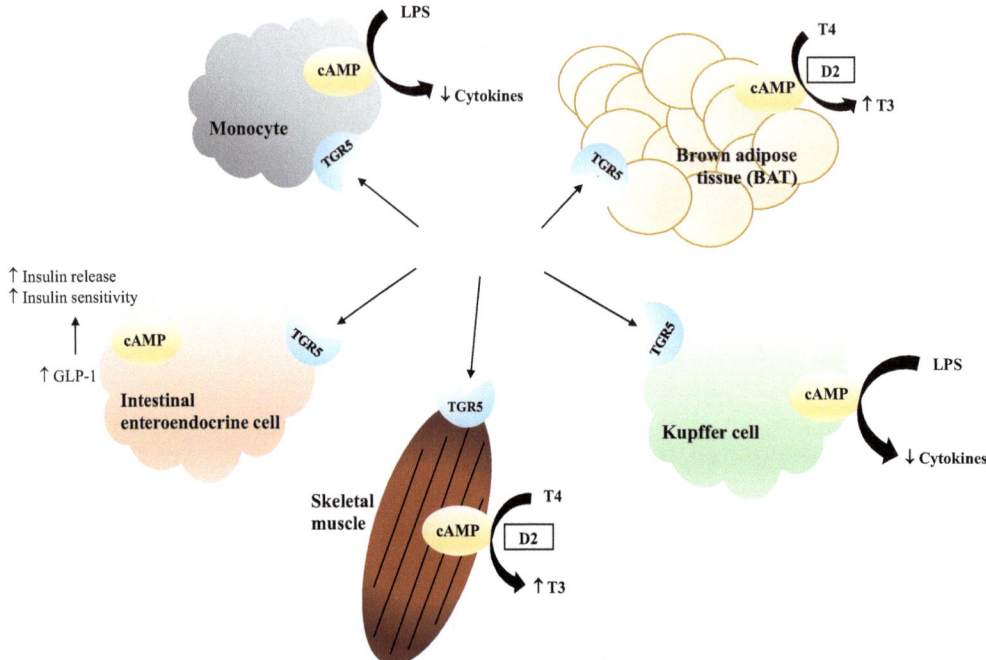

Fig. (2). Bile acids act on several tissues by activating TGR5. In both monocytes and hepatic Kupffer cells, bile acids suppress LPS-stimulated cytokine production. Bile acids also act on brown adipose tissue and skeletal muscle, increasing activity of the enzyme D2 that converts inactive thyroxine (T4) to active T3, thereby playing a critical role in energy homeostasis and metabolism. Bile acids stimulate GLP-1 release from intestinal enteroendocrine L cells, which increases insulin release and insulin sensitivity.

TGR5 in the Liver and Gallbladder

TGR5 is expressed in both liver sinusoidal endothelial and Kupffer cells; the latter act as liver macrophages [18, 21]. Sinusoidal endothelial cells, a permeable barrier between hepatocytes and blood, are routinely exposed to variable concentrations of nutrients and bile salts whose levels increase in the portal circulation after eating [21]. Changes in TGR5 expression in sinusoidal endothelial cells may allow these cells to adapt to changing concentrations of bile salts. In line with the downstream effects of TGR5 activation, primary and secondary bile salts increase cAMP levels in sinusoidal endothelial cells [21]. Although incompletely understood, this increase in cAMP levels may protect sinusoidal endothelial cells from cold storage/reperfusion injury, as well as from oxidative stress and lipid peroxidation *via* eNOS and rapid nitric oxide release [21].

TGR5 expression in Kupffer cells is consistent with the initial discovery of a role for TGR5 in mediating cellular immunity. Kupffer cells play a central role in the hepatic response to injury where they are a major source of inflammatory cytokines [22]. Similar to the aforementioned mechanisms, bile acids activate TGR5 in the Kupffer cell plasma membrane, resulting in increased cAMP production; this mechanism inhibits expression of several cytokines within the liver, possibly repressing cytokine production in obstructive cholestasis and minimizing hepatic injury [22].

In addition to liver cells, TGR5 is expressed in gallbladder epithelium, where bile acid-induced TGR5 activation increases cAMP levels. TGR5 also interacts with the cAMP-regulated chloride channel cystic fibrosis transmembrane conductance regulator (CFTR) and the apical sodium-dependent bile salt uptake transporter (ASBT), suggesting a relationship between TGR5 and bile acid uptake and chloride secretion [23]. TGR5 deficiency markedly decreased gallstone formation in TGR5 knockout mice fed a lithogenic diet [24]. The mechanisms underlying these actions require further elucidation.

TGR5 & Thyroid Hormone Activation

In mice, administration of bile acids increases energy expenditure in brown adipose tissue, thereby preventing obesity and insulin resistance [25]. These effects are mediated by the increased cAMP production resulting from bile acid binding to TGR5; cAMP activates type-2 iodothyronine deiodinase (D2), thereby stimulating conversion of inactive thyroxine (T4) to active 3,5,3'-tri-iodothyronine (T3). D2 plays a vital role in energy homeostasis. Ultimately, these actions increase oxygen consumption by brown adipocytes and skeletal myocytes [25]. While humans do not possess as much brown adipose tissue as mice, they express substantial D2 levels in skeletal muscle, suggesting that in humans bile acids contribute to increased energy expenditure primarily *via* their actions on skeletal muscle metabolism [25].

GLUCAGON-LIKE PEPTIDE 1 (GLP-1)

Before describing how bile acids and TGR5 play a role in mediating release of glucagon-like peptide 1 (GLP-1), it is essential to understand the physiology and actions of GLP-1. GLP-1, a 30-amino acid peptide produced from proglucagon in endocrine L cells in the intestinal epithelium, is an incretin, a hormone that decreases blood glucose levels [26, 27]. L cell density is greatest in the ileum, although L cells are also numerous in jejunum and colon [26, 28].

Physiology of GLP-1

GLP-1 secretion is modulated by eating; plasma concentrations are low in the fasting state [26]. Following meals, GLP-1 is rapidly secreted by endocrine L cells (within 10-15 minutes), most likely in response to interaction of nutrients in the gut lumen with microvilli on L cells [26, 27]; L cells face the gut lumen, allowing them to sense luminal concentrations of nutrients directly [28]. Ingestion of lipid- or carbohydrate-rich foods stimulates GLP-1 release (also modified by meal size) [26, 27].

The mechanisms whereby nutrients stimulate GLP-1 secretion are not completely understood. Work published in 2002 reported that GLP-1 secretion and electrical activity in GLUTag cells, an L cell model, were triggered by glucose concentrations ranging from 0.5-25 mmol/L [28]. Regulation of KATP channel activity plays a crucial role in glucose sensing by GLUTag cells [28]. GLP-1 secretion from GLUTag cells is stimulated by glucose through a mechanism involving KATP channel closure [28]. The same investigators found GLP-1 was also secreted from GLUTag cells in response to glucose exposure by way of electrogenic Na-coupled glucose uptake by sodium glucose cotransporters (SGLTs) [29, 30]. In a canine model, blocking luminal SGLT-1 inhibited GLP-1 secretion, implicating glucose as a key regulator [26]. Also in the GLUTag cell model, SGLT-1 was important for glucose-induced GLP-1 secretion [26, 27]. The role of acetylcholine as a transmitter in a neural stimulatory pathway for GLP-1 secretion is controversial. Conversely, norepinephrine-mediated sympathetic innervation of the gut inhibits GLP-1 secretion [26]. Finally, GLP-1 secretion from L cells is also under the paracrine control of neighboring somatostatin-producing D-cells. Without restraint from somatostatin, GLP-1 release is increased up to eight-fold [26].

GLP-1 binds the GLP-1 receptor, a class 2, G protein-coupled receptor expressed by cells in pancreatic islets, brain, heart, kidney, and the gastrointestinal tract [26]. GLP-1 is rapidly degraded following release from intestinal endocrine L cells, a process catalyzed by dipeptidyl peptidase IV (DPP4) expressed by the enterocyte brush border and endothelial cells lining capillaries in the lamina propria [26]. Only 10-15% of newly-secreted GLP-1 reaches the systemic circulation intact. As is true for most peptide hormones, GLP-1 has a short plasma half-life (1-2 minutes) and is rapidly metabolized in the kidneys [26].

Effects of GLP-1

GLP-1 has several systemic effects, most prominently as a member of the incretin family (Table **1**). This "incretin effect" is responsible for amplifying gastro-intestinal hormone-mediated insulin secretion. After eating, insulin release is augmented as a consequence of the insulinotropic actions of gut hormones, primarily GLP-1 and glucose-dependent insulinotropic peptide (GIP) [26]; these insulinotropic (incretin) actions are not observed when the same composition of nutrients is delivered parenterally.

Table 1. Actions of GLP-1.

Location	Action
Pancreatic beta cells	- Glucose sensitization - Insulin secretion - Upregulation of insulin gene transcription - Beta cell proliferation - Inhibition of beta cell apoptosis
Pancreatic alpha cells	- Inhibition of glucagon release at glucose levels at or above fasting levels (no increased risk of hypoglycemia)
Gastrointestinal tract	- Inhibition of meal-induced gastric motility and pancreatic secretion - Inhibition of gastrin-induced acid secretion - "Ileal break" effect with peptide YY (PYY) - Induction of satiety (possibly)

GLP-1 binding to its G protein-coupled receptor on the cell membrane of pancreatic islet beta cells activates adenylyl cyclase and generates cAMP, the key mediator of almost all its secondary actions [26]. GLP-1 also plays a putative role in activating beta cell KATP channels sensitive to glucose fluxes. Additional evidence suggests GLP-1 acts as a glucose sensitizer and may potentiate insulin gene transcription and upregulate genes coding for the cellular machinery involved in insulin secretion [26]. Finally, GLP-1 trophic actions stimulate beta cell proliferation, enhance differentiation of new beta cells from progenitor cells, and inhibit beta cell apoptosis [26].

GLP-1 is also a strong inhibitor of glucagon secretion, though the mechanism is not completely understood [26]. Thus, even in the absence of beta cell activity, *e.g.* type 1 diabetes, GLP-1 can still lower fasting plasma glucose concentrations. GLP-1 only inhibits glucagon at or above fasting glucose levels; hence, this action does not increase the risk of hypoglycemia [26].

GLP-1 also inhibits meal-induced gastric acid secretion, gastric emptying, and pancreatic secretion [26]. Together with peptide YY (PYY), released from the L cell in parallel with GLP-1, GLP-1 exerts an "ileal break" effect, the endocrine inhibition of upper gastrointestinal functions stimulated by unabsorbed nutrients within the ileum [26]. GLP-1 may also induce satiety, most likely through its interaction with sensory afferents though the mechanism remains uncertain [26, 27]. GLP-1's actions on other tissues are beyond the scope of this chapter.

Pathophysiology of GLP-1

GLP-1 plays an important role in obesity [26]. Compared with lean controls, obese subjects tend to have reduced L cell secretion of GLP-1 and diminished postprandial GLP-1 secretion. The mechanisms are uncertain but may be related to insulin resistance associated with weight gain [26]. Reduced postprandial GLP-1 release may contribute to obesity through inadequate postprandial satiation [26]. Notably, bariatric surgery immediately results in substantial elevation of GLP-1 levels that dramatically improves glucose tolerance. This is thought to be mediated by GLP-1 effects on insulin and glucagon secretion rather than weight loss per se, since the increase in GLP-1 occurs very soon postoperatively before major weight loss has occurred (discussed in further detail below) [26].

Subjects with type 2 diabetes have substantially reduced postprandial GLP-1 secretion. GLP-1 still exerts its insulinotropic effects in type 2 diabetes, but its efficacy is significantly reduced (by ~20%) in diabetics compared to healthy controls [26]. Recent evidence suggests this abnormal incretin effect is a consequence of diabetes and/or insulin resistance [26]. In hyperglycemic clamp experiments, infusing GLP-1 normalized glucose-induced insulin secretion [26]. Also, type 2 diabetics have reduced suppression of glucagon secretion by GLP-1 [26].

Type 2 diabetes and non-alcoholic fatty liver disease (NAFLD) are associated with insulin resistance and dyslipidemia. A recent study found higher fasting glucagon levels in NAFLD, with sustained alpha cell sensitivity to the inhibitory effect of GLP-1 [31]. Subjects and healthy controls had similar GLP-1-induced glucagon suppression when fasting and after meals. Furthermore, subjects with NAFLD had higher basal levels of insulin and increased insulin secretion with fasting levels of plasma glucose. At fasting glucose levels, these subjects also had higher rates of insulin secretion following infusion of GLP-1 [31]. This finding suggests insulin-dependent inhibition of glucagon release by GLP-1 [31]. Taken together, these mechanisms may help maintain normoglycemia in NAFLD with insulin resistance [31].

THE ROLE OF TGR5 IN GLP-1 SECRETION

As reviewed above, TGR5 activation by bile acids increases cAMP levels and protein kinase A (PKA) activation, resulting in important downstream effects regulating the functions of the immune system, liver, gallbladder, and thyroid hormone, and also GLP-1 secretion (Fig. **3**). This links bile acids, through their interactions with TGR5, to glucose homeostasis. Even before the discovery that bile acids promote GLP-1 secretion, it was observed that unsaturated free fatty acids, like linolenic acid, stimulate GLP-1 secretion [32]. Bile acids promote GLP-1 secretion by murine enteroendocrine STC-1 cells. In this model, lithocholic (LCA) and deoxycholic (DCA) acids potently promote dose-dependent secretion of GLP-1 [33]. The underlying mechanism is bile acid-induced TGR5 activation causing intracellular cAMP accumulation [33].

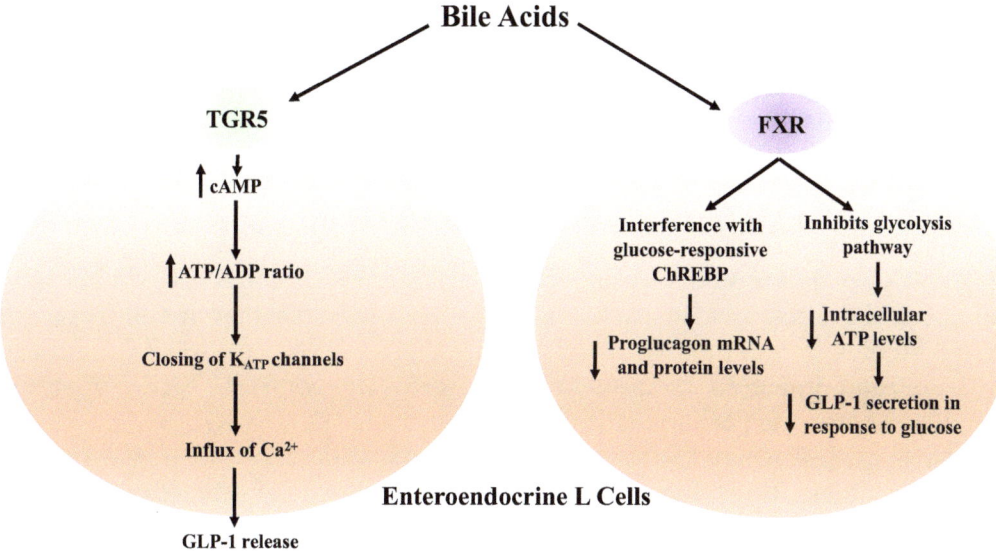

Fig. (3). Differences between bile acid activation of TGR5 and FXR in enteroendocrine L cells. Bile acid activation of TGR5 results in immediate increase in cAMP levels that can activate protein kinase A. More importantly, the increase in cAMP levels causes a similar increase in the ATP/ADP ratio that results in closing of KATP channels, an influx of calcium, and release of GLP-1. On the other hand, bile acid activation of L cell FXR can activate two different pathways. First, through interference with the glucose-responsive factor Carbohydrate-Responsible Element Binding Protein (ChREBP), FXR reduces proglucagon mRNA levels. Second, by inhibiting the glycolysis pathway, FXR activation reduces GLP-1 secretion in response to glucose, an action opposite to that of TGR5 activation.

An important 2009 study by Thomas, *et al.* established the physiologic impact of TGR5 activation on GLP-1 secretion and furthered our understanding of the relationship between TGR5 function and GLP-1 release. Their work showed that intestinal lipids, carbohydrates, and bile acids stimulate endocrine L cells to release GLP-1 by a cAMP-dependent increase in the ATP/ADP ratio which results in the closing of KATP channels and influx of calcium [27, 34]. This mechanism is similar to that underlying the release of insulin from pancreatic beta cells [34]. Furthermore, these investigators found that TGR5 overexpression markedly improved glucose tolerance in obese mice fed a high-fat diet [34]. Increased glucose tolerance associated with substantial postprandial GLP-1 and insulin release was more prominent after a test meal compared to a glucose challenge, most likely due to increased bile acids with the former [34].

Consistent with previous data showing GLP-1 promotes pancreatic islet survival and proliferation, TGR5 expression contributes to the maintenance of normal islet distribution profile, most likely secondary to higher GLP-1 levels; TGR5 overexpression in mice on a high-fat diet increases pancreatic islet insulin levels [34]. In contrast, TGR5-deficient mice had significant impairment of glucose tolerance when fed a high-fat diet for eight weeks; in mice fed standard chow no differences were observed. Plasma GLP-1 levels were blunted in TGR5-deficient mice stimulated with a semi-synthetic cholic acid derivative that is a potent TGR5 agonist [34]. This evidence strongly favors an important role for TGR5 in GLP-1 release that contributes to glucose homeostasis.

THE ROLE OF FXR IN GLUCOSE HOMEOSTASIS

While TGR5 is most commonly associated with glucose homeostasis *via* bile acid activation, the latest evidence shows that FXR may also play a role (Fig. 3). Recent studies using mouse models found FXR deficiency leads to insulin resistance and impaired glucose tolerance [35]. In contrast, FXR activation promotes insulin sensitivity. In diabetic animals, liver FXR expression is diminished [35]. A recent mouse study was the first to demonstrate bile acids alter pancreatic beta cell function *via* FXR [35]. FXR-mediated sodium taurochenodeoxycholate (TCDC) augmented glucose-induced insulin secretion, an effect replicated by an FXR agonist and suppressed by an FXR antagonist [35]. In murine beta cells, TCDC-induced activation of FXR-stimulated electrical activity and increased concentrations of calcium, which in turn triggers insulin secretion. Finally, in pancreatic beta cells, FXR activation by TCDC inhibited KATP channel activity, and stimulated calcium influx and insulin release [35].

Other researchers investigated FXR expression in enterocytes and L cells and its role in glucose homeostasis [36]. In several models, as well as mice and human intestines, FXR activation decreased proglucagon mRNA levels by interfering with the glucose-responsive factor Carbohydrate-Responsible Element Binding Protein (ChREBP) and GLP-1 secretion by inhibiting glycolysis [36, 37]. It is likely that FXR activation induces a more delayed response after food ingestion, compared with the immediate response mediated through TGR5. *In vivo*, FXR-deficiency was beneficial for glucose homeostasis because of increased GLP-1 gene expression and secretion in response to glucose [36].

A recent study adds additional support for the importance of FXR in glucose control and obesity. Treating mice with Gly-MCA, a selective high-affinity FXR inhibitor, reversed or prevented high-fat-diet-induced and genetic obesity, insulin resistance, and hepatic steatosis [38]. These improvements resulted from decreased intestinal ceramide synthesis after FXR inhibition [38]. Furthermore, compared to their lean counterparts, obese humans had increased intestinal FXR expression and signaling. While the evidence is conflicting regarding the exact role of FXR in glucose metabolism, it is clear it is important and targeting FXR has therapeutic potential in diabetes and obesity [37, 38].

MANIPULATING THE BILE ACID POOL FOR GLUCOSE HOMEOSTASIS

Bile acid sequestrants, or bile acid-binding resins (BABRs), are ion-exchange resins that bind negative-charge bile salts in the intestinal lumen and divert them from the enterohepatic circulation. Bile acid sequestrants lower circulating levels of LDL cholesterol by reducing the bile acid pool size, which stimulates bile acid synthesis, up-regulation of LDL receptors, and enhanced delivery of LDL cholesterol to the liver. In addition to lowering plasma levels of LDL, by manipulating the bile acid pool, BABRs can also enhance glucose metabolism (Fig. **4**).

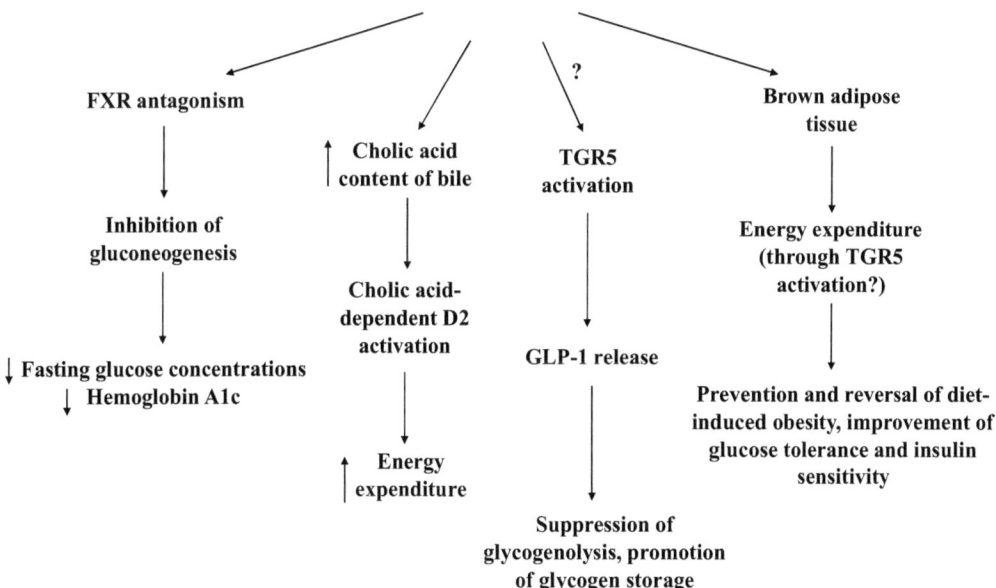

Fig. (4). Varying downstream effects of bile acid-binding resins (BABRs). Through these pathways, not all completely elucidated, BABRs improve glucose homeostasis, increase energy expenditure, cause GLP-1 release, and have the potential to prevent and reverse diet-induced obesity.

Bile acid pool composition is altered in diabetic humans and animals; this is manifested by bile acid pool enlargement and increased fecal excretion of bile acids, features normalized by insulin treatment [17]. Insulin suppresses hepatic *FXR* gene expression, which subsequently increases bile acid synthesis [17]. Compared to controls, type 2 diabetics have lower cholic acid and higher deoxycholic acid levels, suggesting increased conversion of cholic to deoxycholic acid [17]. While the mechanisms underlying this phenomenon have not been clearly elucidated, it may be associated with changes in the gut microbiome [17].

In 1994, Garg and Grundy discovered that a BABR (cholestyramine) not only reduced LDL levels in type 2 diabetics but also improved glycemic control. This randomized, double-blind study found that compared to placebo, cholestyramine lowered mean plasma glucose concentrations by 13%, decreased urinary glucose excretion, and reduced mean hemoglobin A1c (HbA1c) levels [39].

BABR effects on glucose metabolism were evaluated in an animal model of type 2 diabetes. Colestimide, a BABR, markedly improved hyperglycemia and dysplipidemia [40]. Mice had significantly lower fasting blood glucose after colestimide treatment and reductions in body weight that were not due to decreased food intake. Colestimide treatment also improved insulin sensitivity [40]. Expression of mRNA for CYP7A1, the rate-limiting step in cholesterol catabolism, was up-regulated four-fold in mice treated with colestimide. mRNA levels of SHP, which represses *CYP7A1* gene transcription, were significantly decreased [40]. This animal study provided additional evidence that BABRs enhance glucose metabolism in type 2 diabetes.

Human studies replicated these findings. In a study comparing colestimide and pravastatin treatment after three months in type 2 diabetics with hyperlipidemia, colestimide significantly improved glycemic control – lower fasting glucose concentrations and hemoglobin A1c; pravastatin had no effect [41]. The authors speculated that colestimide, an FXR antagonist, inhibits gluconeogenesis through a pathway involving inhibition of FXR. Furthermore, colestimide increases the cholic acid content of bile, and cholic acid-dependent D2 activation induces energy expenditure [25, 41]. A similar study, published concurrently, investigated the effect of colesevelam, another BABR, on type 2 diabetes. Compared to placebo, in type 2 diabetics 12 weeks of colesevalam significantly reduced hemoglobin A1c, fructosamine, and postprandial glucose levels [42]. The authors concluded type 2 diabetics develop improved glycemic control with BABRs because of dysregulated bile acid metabolism and "deactivation" of FXR and downstream pathways [42].

One group took this a step further and examined the effects of BABRs on GLP-1 levels. While colestimide was only administered for one rather than 12 weeks as in previous studies, 1-hour and 2-hour postprandial glucose levels were significantly decreased [43]. Also, after 1 week of colestimide, 2-hour postprandial plasma levels of GLP-1 were considerably increased; however, there were no significant changes in preprandial and 1-hour postprandial GLP-1 levels.

This was one of the first, if not the first, study to find colestimide administration increased postprandial levels of GLP-1 [43]. This study links changes in bile acid composition and concentration with GLP-1, an incretin that increases insulin secretion and lowers plasma glucose levels.

BABRs not only improve glycemic metabolism but also reduce obesity and ameliorate insulin resistance. In mice with diet-induced obesity, colestimide curbed weight gain, significantly reduced blood glucose concentrations, and improved insulin sensitivity [10]. Colestimide also prevented high fat diet-induced changes in liver and adipose mass. Substantial weight loss without a change in diet was probably explained by colestimide's actions on basal metabolic rate [10]. Colestimide's actions were similar to those of cholic acid when fed to mice with diet-induced obesity. Investigators confirmed brown adipose tissue and liver are primary targets contributing to positive effects on energy, lipid, and glucose metabolism [10]. BABRs stimulated energy expenditure in brown adipose tissue that prevented and reversed diet-induced obesity in mice and improved glucose tolerance and insulin sensitivity, possibly by TGR5 activation [10].

Finally, recent work confirmed the relationship between BABRs, TGR5, and GLP-1. Colesevelam, a BABR, suppressed hepatic glycogenolysis in diet-induced obese mice, improving hyperglycemia and hyperinsulinemia [44]. This effect was mediated in part by TGR5 *via* GLP-1 release. In TGR5-deficient mice, the ability of colesevelam to induce GLP-1, lower plasma glucose levels, and spare hepatic glycogen content was compromised [44]. In mice receiving the BABR, hepatic glycogen levels were elevated, indicating colesevelam suppresses glycogen breakdown and promotes glycogen storage [44]. Consistent with previous studies, GLP-1 signaling was required for suppression of glycogenolysis, and mice treated with colesevelam had higher portal GLP-1 concentrations. Finally, colesevelam failed to induce GLP-1 secretion in TGR5-deficient mice, suggesting TGR5 is required for colesevelam-mediated induction of GLP-1 and glycogen sparing [44]. While these studies showed repeatedly that BABRs improve glucose homeostasis in type 2 diabetics, effects mediated by TGR5 and GLP-1, it is still not understood how bile acid sequestration in the gut increases TGR5 activation [17]. Regardless, it is evident that increasing serum bile acid levels increases GLP-1 secretion *via* TGR5.

BILE ACIDS & GLP-1 AFTER BARIATRIC SURGERY

GLP-1 levels rise dramatically after gastric bypass surgery; GLP-1 plays an important role in restoring glycemic control even before there is substantial weight loss. Consequently, in addition to promoting weight loss, gastric bypass surgery has proved highly effective in correcting manifestations of the metabolic syndrome (Table **2**). There is broad interest in understanding the mechanistic underpinnings of these benefits as such would guide the development of non-surgical treatments that could achieve similar efficacy.

Table 2. Effects of gastric bypass surgery.

Mechanisms of weight loss	- Gastric restriction from the small gastric pouch - Early satiety - Mild malabsorption - Altered neural signals - Duodenal exclusion - Early delivery of nutrients to the mid and distal small intestine
Altered glucose homeostasis	- Increased postprandial levels of GLP-1 augment insulin release - Increased levels of PYY reduce hunger and impart satiety - Decreased insulin resistance - Improved beta-cell function
Changes in bile acids and their receptors	- Increased total circulating conjugated bile acid levels - Increased *TGR5* gene expression

As rates of obesity rise worldwide and, more specifically, in the U.S., there has been a concordant expansion in the use of bariatric surgery to treat obesity and its attendant comorbid conditions. The most common form of bariatric surgery in the U.S. is the Roux-en-Y gastric bypass (RYGB), which is very effective in producing sustained weight loss [45]. Several mechanisms may explain how RYGB achieves weight loss; these include gastric volume restriction, early satiety, mild malabsorption, and more rapid postprandial delivery of nutrients to the distal small intestine [45]. There is evidence of altered levels of metabolically-active hormones, including GLP-1, which appear to be associated with an increased bile acid pool that in turn stimulates TGR5 activation. However, the mechanisms underlying the improved metabolic homeostasis and long-term weight loss following gastric bypass surgery remain incompletely understood.

Glucose Homeostasis following Gastric Bypass Surgery

It was first discovered in 1995 that RYGB provided not only long-term control for obesity, but also durable control of type 2 diabetes mellitus [46]. Glucose metabolism normalized rapidly after gastric bypass in diabetics even before significant weight loss [46]. It was hypothesized this was a consequence of the immediate reduction in caloric intake, concluding that the lesion in type 2 diabetes was related to food intake and the signaling mechanisms stimulated by food. The investigators reported altered incretin release in the gut as well as deficient insulin signaling in muscle, which caused insulin resistance [46]. While these findings did not completely explain remission of diabetes following bariatric gastric bypass surgery, this group was the first to suggest that gastric bypass was valuable not only for weight loss but also to correct the associated metabolic disorders. In 2004, a large meta-analysis confirmed these findings and reported more than three-quarters of those undergoing bariatric surgery had complete resolution of type 2 diabetes; of those without complete resolution, more than half had significant improvement [2].

Since these initial discoveries more benefits of gastric bypass were found and mechanisms underlying these effects further elucidated. A 2004 study found significant decreases in blood glucose, insulin, IGF-1, and leptin with increased ACTH levels three weeks after RYGB, with minimal change in mean BMI [47]. Unfortunately, the lack of significant changes in other hormones [GIP, GLP-1, CCK, corticosterone, or neuropeptide Y (NPY)] shed no further light on the mechanistic underpinnings of the metabolic benefits [47, 48]. In 2007, an animal study revealed reduced fasting glucose levels and improved glucose tolerance 1 week and 1 month after gastrojejunal bypass in non-obese diabetic Goto-Kakizaki rats [48, 49], again without changes in GLP-1, GIP, or glucagon levels [49].

Despite these initial studies, more recent work indicates GLP-1 levels are significantly elevated after RYGB; a finding consistent across multiple studies [50]. The earlier studies sampled fasting GLP-1, whereas the later studies found elevations in postprandial GLP-1 levels – a finding of greater physiological importance [50]. Investigators speculated increased levels of postprandial GLP-1 after RYGB resulted from surgical anatomy that promoted more rapid delivery of nutrients to the distal gut [50]. In addition to GLP-1, most studies report increased postprandial levels of PYY after RYGB; PYY reduces hunger and imparts satiety [50].

Holst, et al. sought to elucidate the potential mechanisms responsible for improved glucose control in type 2 diabetes only days after RYGB [51]. Consistent with previous studies, in their subjects they found a dramatic change in glucose metabolism immediately post-RYGB. In subjects with both type 2 diabetes and normal glucose tolerance, there was decreased insulin resistance, decreased levels of insulin, improved beta-cell function, and increased GLP-1 secretion [51]. They could not identify the mechanism underlying the early increase in GLP-1 following RYGB, but hypothesized it could be due to earlier exposure of small intestinal L cells to nutrients, which would lead to GLP-1 secretion, possibly mediated by TGR5 [51]. The current consensus is that increased GLP-1 and the subsequent decrease in plasma glucose levels and insulin resistance after gastric bypass surgery is not simply a consequence of weight loss, since it precedes meaningful change in BMI; the underlying mechanism remains to be fully elucidated.

Recently, moderate caloric restriction was hypothesized to explain early post-RYGB improvements in glucose homeostasis [52]. Caloric restriction, about 600-800 kcal/day for 3-10 days, replicated many improvements seen following RYGB. This was also shown to improve glycemic control and insulin sensitivity without concurrent weight loss [52]. Benefits of caloric restriction appear to be a consequence of reduced fasting and post-prandial hepatic glucose production, in addition to suppressed glycogenolysis. In addition to the concept that earlier delivery of nutrients to the intestine strongly stimulates GLP-1 release from L-cells, this hypothesis may explain early, rapid normalization of glucose levels and remission of type 2 diabetes after RYGB [52].

Increased Bile Acid Levels after Gastric Bypass Surgery

The mechanisms underlying immediate changes in plasma glucose levels and insulin resistance following RYGB remain to be completely elucidated. Increased GLP-1 secretion from enteroendocrine L cells may play a role, given the earlier delivery of nutrients to the small bowel; this could be mediated by TGR5 activation due to increased levels of bile acids, but bile acid levels have not been shown to increase in the immediate post-operative period. There is evidence that increased GLP-1 levels one month following RYGB coincide with increases in total bile acids [52]. Increased levels of bile acids may play a role in sustained and long-term glucose control following gastric bypass surgery, but the dots between the quick post-prandial rise in bile acid and GLP-1 levels following surgery have yet to be connected.

By activating TGR5, bile acids stimulate dose-dependent GLP-1 release that enhances insulin secretion. Thus, it is not surprising that circulating bile acid levels correlate with measures of insulin sensitivity [45]. This may explain improved glucose homeostasis following gastric bypass surgery, a procedure that results in prompt postprandial release of bile acids into the small intestine, activation of L cell TGR5, followed by GLP-1 release. Two to four years after gastric bypass surgery, total serum bile acid concentrations were two-fold higher compared to matched obese persons who did not have surgery [45]. The investigators did not elucidate the mechanisms responsible for increased serum bile acid levels, but suggested this was due to increased postprandial intestinal uptake of bile acids [45]. Elevated bile acid levels correlated with lipid and glucose metabolism. A robust inverse relationship was discovered between bile acid and serum glucose levels two hours after a mixed meal, most likely due to improved insulin sensitivity and increased incretin (GLP-1)-mediated insulin secretion [45]. Previous animal studies showed improved glucose tolerance following surgical procedures, such as gastric bypass, that result in 'premature' exposure of the distal intestine to bile acids [45].

Additional studies described similar rises in bile acid levels following gastric bypass surgery. One group observed obesity was associated with a blunted postprandial rise in circulating conjugated bile acids; however, these subjects had fasting bile acid levels similar to those in lean subjects [53]. In this longitudinal study, 40 weeks after RYGB the blunted response normalized and there was a significant increase in postprandial circulating conjugated bile acids. As early as four weeks after RYGB the postprandial bile acid response was accelerated [53]. The investigators postulated the postprandial rise in GLP-1 and energy expenditure were due to the affinity of conjugated bile acids for TGR5. Increased levels of postprandial bile acids are likely a consequence of accelerated nutrient delivery to the ileum; as this was a late observation, it may reflect time-dependent mechanisms that alter gut response to nutrient ingestion [53].

Investigators performed RYGB or sham surgery in Zucker diabetic fatty (ZDF) and normoglycemic Sprague Dawley (SD) rats and measured plasma and fecal bile acid levels, and plasma glucose, insulin, GLP-1, and PYY levels before and after surgery [54]. Starting the first week post-RYGB, body weight decreased in both SD and ZDF rats and plasma GLP-1 increased, while plasma insulin and glucose decreased in ZDF rats. In both SD-RYGB and ZDF-RYGB compared to sham controls, total plasma bile acids were increased by the end of the second post-operative week [54]. Finally, bile acid reabsorption increased in the proximal small intestine; however, this did not explain increased circulating bile acid levels [54]. Similar to other studies, the investigators found increased bile acids after RYGB, although not in the immediate post-operative period, and the mechanisms underlying post-RYGB metabolic changes remain uncertain.

A recent study compared obese type 2 diabetics randomized to either RYGB or a hypocaloric diet to obese subjects with and without type 2 diabetes undergoing vertical sleeve gastrectomy (VSG) [55]. There was a trend toward increased levels of mostly conjugated total fasting serum bile acids after RYGB in addition to VSG. Levels of *TGR5* gene expression in subcutaneous white adipose tissue decreased markedly after both RYGB and VSG, but since levels of TGR5 protein did not change, the significance of the correlation between *TGR5* gene expression and obesity is uncertain [55]. Thus, similar to other studies, the mechanisms responsible for increased levels of bile acids following surgery remain unclear. These data confirm increased bile acid levels after gastric bypass surgery likely drive increased postprandial GLP-1 levels that improve glucose homeostasis – yet many mechanistic dots remain to be connected.

Circulating bile acid levels and TGR5 signaling increase after RYGB; this does not occur with laparoscopic adjustable gastric banding (LAGB). In a study comparing outcomes after either RYGB or LAGB, both fasting and postprandial total plasma bile acid concentrations more than doubled after RYGB but tended to decrease following LAGB, a finding independent of weight loss [56]. Similar to the studies previously described, the change in postprandial bile acid concentrations correlated with an increase in GLP-1. Interestingly, plasma TSH also decreased after weight loss induced by RYGB but not after LAGB, but there was no difference between resting energy expenditure, insulin response, or skeletal muscle sensitivity between the two groups [56]. Therefore, while increased bile acid levels following RYGB alter GLP-1 secretion and thyroid hormone activity, this does not predict changes in the major factors that regulate glucose or energy metabolism.

PHARMACOLOGY

Over the past decade, to exploit these recently discovered mechanisms, several new classes of agents to manage diabetes were developed (Table **3**); these include bile acid sequestrants (BABRs), agents that mimic or extend the actions of incretins (*e.g.* GLP-1), and TGR5 agonists – these agents improve plasma glucose levels and insulin sensitivity in type 2 diabetes. In 2008, BABRs were FDA-approved to treat type 2 diabetes [8]. Incretin-based therapy was shown to improve glycemic control modestly in type 2 diabetes, reducing postprandial glucose and hemoglobin A1c levels [57]. Incretin-based medications were found non-inferior compared with non-incretin based pharmacotherapies [57]. Exenatide (exendin-4), a GLP-1 receptor agonist resistant to DPP4 degradation, was approved in April 2005 as adjunctive treatment for type 2 diabetes [57]. The first selective DPP4 inhibitor, sitagliptin, was FDA-approved in October 2006 for use as monotherapy or in combination with metformin or thiazolidinedione for type 2 diabetes [57]. Other drugs, including TGR5 agonists that target bile acid signaling, are currently under investigation for type 2 diabetes, NAFLD, and metabolic syndrome.

GLP-1 Agonists

The first GLP-1 agonist, exenatide (exendin-4), was developed in response to the need for long-acting anti-diabetic drugs with potential superiority to insulin [58]. Before exenatide, therapies for type 2 diabetes were limited by adverse effects, such as weight gain, edema, or hypoglycemia, and did not ameliorate the decline in pancreatic endocrine function [57]. This spurred interest in developing agents that mimicked the properties of GLP-1 but resisted degradation by DPP4. Consequently, exenatide was developed for long-term administration by injection and approved as adjunctive therapy to improve glycemic control in type 2 diabetics taking metformin and sulfonylurea, alone or in combination, with inadequate glycemic control [57, 58].

Table 3. Comparison of currently available incretin-based pharmacotherapies.

Drug Class	Mechanism of Action	Current Uses	Side Effects	Examples
GLP-1 Receptor Agonist	- GLP-1 receptor analogue - Resistant to degradation by DPP4 - Stimulates insulin release from the pancreas	- Not initial therapy for T2DM - Usually used in combination with an oral agent - Can result in weight loss - Injectable, short- and long-acting forms available	- Nausea, vomiting, diarrhea - Small risk of hypoglycemia - Questionable risk of pancreatitis - Contraindicated if creatinine clearance <30 mL/min	- Exenatide, lixisenatide - Liraglutide, extended release exenatide, albiglutide, dulaglutide
DPP4 Inhibitor	- Prevents GLP-1 degradation by inhibiting DPP4 enzyme that hydrolyzes GLP-1 - Modestly increases GLP-1 levels	- Not initial therapy for T2DM - Considered as monotherapy in patients with contraindications to metformin, sulfonylureas, or thiazolidinediones - Usually considered add-on therapy in patients inadequately controlled with metformin, sulfonylureas, or thiazolidinediones	- Headache, nasopharyngitis, upper respiratory tract infection - No effects on body weight - No risk of hypoglycemia - Dose adjustments required for patients with chronic kidney disease (except for linagliptin) - Questionable risk of pancreatitis - Some risk of hypersensitivity skin reactions - Some associated with severe joint pain	- Sitagliptin, saxagliptin, linagliptin, alogliptin - Vildagliptin is available in several countries but not the U.S.

Shortly after exenatide was approved for use with metformin or sulfonylureas, based on several phase 3, 30-week controlled clinical trials, it was tested against insulin glargine [58]. This 26-week multicenter, open-label, randomized, controlled trial evaluated the effects of exenatide and insulin glargine on glycemic control in type 2 diabetics with suboptimal glucose control with metformin and a sulfonylurea. Exenatide and insulin glargine achieved similar improvements in overall glycemic control; compared to insulin glargine, exenatide was associated with greater weight loss but more gastrointestinal adverse effects [58]. Exenatide achieved better postprandial glucose control and nocturnal hypoglycemia while insulin glargine resulted in lower fasting glucose levels, more subjects achieving target fasting glucose levels, and fewer daytime hypoglycemic episodes [58].

In addition to beneficial effects on glucose metabolism, GLP-1 agonists may play a therapeutic role in non-alcoholic fatty liver disease (NAFLD) [59]. A central problem in NAFLD appears to be insulin resistance – NAFLD is clearly associated with the metabolic syndrome. By reducing hepatocyte triglyceride accumulation, which increases hepatocyte vulnerability to oxidative stress, attenuated insulin resistance can prevent or slow NAFLD progression [59]. Around the same time that exenatide was approved for type 2 diabetes, it was shown to ameliorate NAFLD in obese mice with hepatic steatosis [59]. Compared to placebo, exendin-4 improved liver transaminase and insulin levels, reduced plasma glucose, and improved homeostatic model assessment (HOMA) scores, a measure of insulin resistance and pancreatic beta-cell function. Exendin-4-treated mice had markedly reduced net weight gain and hepatic lipid content. These effects mimicked those of GLP-1 on hepatocytes.

Data in humans are scarce, however a recent study examined the effects of a GLP-1 receptor agonist liraglutide on liver fat content in uncontrolled type 2 diabetics [60]. After undergoing treatment with liraglutide for 6 months, these subjects had significantly decreased body weight, hemoglobin A1c, and liver fat. Reduced liver fat was mainly driven by reduced body weight [60]. In the parallel group, intensely treated with insulin for 6 months, there was no significant change in liver fat. Therefore, GLP-1 receptor agonists improve glucose homeostasis and also have the potential to treat NAFLD.

A novel GLP-1/glucagon receptor dual agonist, designated G49, improved steatohepatitis and liver regeneration in mice [61]. In mice fed a methionine and choline-deficient (MCD) diet, G49 ameliorated non-alcoholic steatohepatitis (NASH) and animal survival after partial hepatectomy [61]. Also, when G49 was administered to diet-induced obese mice, body weight, plasma insulin, and glucagon levels decreased, along with improved glucose tolerance and insulin sensitivity [61]. This preclinical evidence supports the use of GLP-1/glucagon receptor dual agonists to treat NASH and improve glucose homeostasis in obesity.

A large, multi-center, randomized, double-blinded controlled trial published in 2015 showed beneficial effects of liraglutide, a GLP-1 analogue, on weight loss [62]. This study investigated subjects without type 2 diabetes with BMI greater or equal to 30, or BMI greater or equal to 27 with treated or untreated dyslipidemia or hypertension. Compared to placebo, administering 3 mg liraglutide as a supplement to diet and exercise was associated with a statistically significant decrease in body weight [62]. After an oral glucose tolerance test, liraglutide caused a greater reduction in hemoglobin A1c as well as fasting glucose and insulin levels than placebo. However, compared to placebo, liraglutide was associated with mild nausea and vomiting, and more episodes of cholelithiasis and cholecystitis [62]. This trial confirmed previous findings that when used in conjunction with a healthful diet and regular physical activity, once-daily liraglutide increased weight loss in overweight and obese adults without type 2 diabetes.

In 2016, a systematic review compared the clinical profiles of GLP-1 receptor agonists and recommended their use as second-line therapy for type 2 diabetes when glucose control with metformin is insufficient [63]. As anticipated from the published literature, compared to placebo, all GLP-1 receptor agonists improved hemoglobin A1c and fasting plasma glucose levels. However, compared to short-acting agonists, long-acting agents achieved significantly greater hemoglobin A1c reduction. Compared to placebo, GLP-1 receptor agonists resulted in greater reduction of body weight but there were no significant differences between the various agonists [63]. Some GLP-1 agonists were associated with hypoglycemia but again, there were no significant differences between them. Finally, gastrointestinal symptoms, well-known side effects of GLP-1 receptor agonists, were more frequent compared to placebo or other glucose-lowering agents, and some GLP-1 receptor agonists were more frequent offenders [63].

While GLP-1 receptor agonists improve glycemic control and weight control, differences in cardiovascular outcomes among these agents could be due to different drug properties or study populations. The ELIXA trial (Evaluation of Lixisenatide in Acute Coronary Syndrome), a multicenter, randomized, double-blind, placebo-controlled trial involving more than 6,000 subjects, investigated the effects of lixisenatide, a once-daily GLP-1 receptor agonist, on cardiovascular outcomes in type 2 diabetics with a recent coronary event [64]. The investigators found that over a three-year study period, adding lixisenatide to conventional therapy did not alter the rate of major cardiovascular events or other serious adverse events [64].

Other studies published in the past year investigated cardiovascular outcomes associated with GLP-1 receptor agonists. A large, multicenter, international, randomized, double-blind, placebo-controlled parallel-group trial was designed to assess semaglutide, a GLP-1 analogue with a half-life of approximately one week, in terms of cardiovascular safety in over 3,000 type 2 diabetics with cardiovascular disease, chronic kidney disease, or both [65]. These high-risk patients had significantly lower rates of cardiovascular death, nonfatal myocardial infarction, or nonfatal stroke after receiving semaglutide compared to placebo [65]. A similar trial evaluated a different GLP-1 receptor analogue, liraglutide, and its effects on cardiovascular outcomes. This multicenter, double-blind, placebo-controlled trial at hundreds of international sites involved over 9,000 type 2 diabetics with high cardiovascular risk [66]. After a median follow up of 3.8 years, the rate of death from cardiovascular causes, nonfatal myocardial infarction, or nonfatal stroke was lower with liraglutide than placebo [66].

A large survey and FDA database suggested an increased risk of pancreatitis in persons treated with GLP-1 receptor agonists [67]. However, a large meta-analysis of 41 randomized controlled trials did not support the hypothesis that GLP-1 receptor agonist treatment increased the risk of pancreatitis [67]. It should be noted only pancreatitis classified as a serious adverse event was considered; less severe pancreatitis may not have been reported. Nonetheless, a large systematic review conducted the same year including both randomized controlled trials and observational studies also concluded there was insufficient evidence to support the claim that treatment with GLP-1 receptor analogues increases the risk of pancreatitis [68]. While both studies came to similar conclusions, there is insufficient evidence to consider this conclusion definitive.

Treatment with GLP-1 agonists can result in adverse gastrointestinal side effects that limit adherence. Nausea is the predominant side effect reported with injectable GLP-1 receptor agonists [69]. A large meta-analysis of 29 randomized, controlled trials found dose-dependent nausea, vomiting, and diarrhea were the most frequently reported adverse events with exenatide [57]; nausea tended to decline after two months of use.

Although GLP-1 receptor agonists were approved for use in the U.S., their use has not induced remission of type 2 diabetes [17]. Currently, the American Diabetes Association (ADA)/European Association for the Study of Diabetes (EASD) guidelines recommend using GLP-1 agonists as supplemental therapy to metformin and lifestyle modification [63]. Partly this may be because GLP-1 effects are elicited proximate to its site of secretion in the small intestine, and synthetic drugs may not be able to fully exploit this mode of action [17]. Thus, targeting GLP-1 secretion and resistance to degradation may be more beneficial.

While GLP-1 receptor agonists are effective in improving glucose homeostasis, lowering hemoglobin A1c, and promoting weight loss in type 2 diabetics, the exact role for GLP-1 agonists in managing type 2 diabetes and other conditions remains unclear. Additional research must determine the long-term benefits and safety of these agents, and their role as adjuncts to other forms of therapy.

DPP4 Inhibitors

In October 2006, the first dipeptidyl peptidase-4 (DPP4) inhibitor, sitagliptin, was FDA-approved for use in type 2 diabetes, either as monotherapy or in combination with metformin or thiazolidinedione [57]. DPP4 is an enzyme that deactivates GLP-1 very rapidly, resulting in GLP-1 half-life of 1-2 minutes. To extend the plasma life of GLP-1 and its beneficial physiologic effects, DPP4 inhibitors were developed as adjuncts to the treatment of type 2 diabetes.

An early study investigated whether sitagliptin would be safe and effective as monotherapy in type 2 diabetes [70]. This multinational, randomized, double-blind control study found that after 18 weeks of treatment, sitagliptin significantly reduced hemoglobin A1c and fasting glucose levels relative to placebo in subjects with mild to moderate hyperglycemia. This DPP4 inhibitor was well-tolerated with a low incidence of hypoglycemia and gastrointestinal adverse experiences [70].

However, DPP4 inhibitors are uncommonly used as monotherapy for type 2 diabetes. Few patients achieve glycemic control with just one anti-hyperglycemic medication, and initial combination therapy has emerged as an alternative approach [71]. Another early study evaluated initial combination therapy with sitagliptin and metformin on glycemic control in patients with type 2 diabetes [71]. This multinational, randomized, double-blind, placebo-controlled study found significant and clinically meaningful reductions in hemoglobin A1c, fasting plasma glucose, and two-hour postprandial glucose levels compared with placebo; combination therapy provided greater reductions compared to monotherapy [71]. Combined sitaglipitin and high-dose metformin resulted in about two-thirds of subjects achieving the hemoglobin A1c goal <7%. The incidence of adverse events was moderate, with a low incidence of hypoglycemia and gastrointestinal events, more commonly in those receiving high-dose metformin (monotherapy and in combination) [71].

The same year, another study compared sitagliptin with glipizide, a sulfonylurea, in combination with metformin [72]. The goal of this multinational, randomized, parallel-group non-inferiority study with an active-controlled, double-blind treatment period was to assess whether sitalgliptin was non-inferior to glipizide regarding hemoglobin A1c levels. Over a 52-week period in more than 1,000 subjects adding sitaglipitin or glipizide to metformin provided similar hemoglobin A1c-lowering effects [72]. Both treatments were well-tolerated, although glipizide was associated with more hypoglycemia and weight gain, while those on sitagliptin lost a clinically meaningful amount of weight and had decreased waist circumference [72].

Adding a DPP4 inhibitor to metformin or insulin improves glycemic control in type 2 diabetes. In a randomized, double-blind, placebo-controlled trial, adding alogliptin, a DPP4 inhibitor, to insulin significantly decreased hemoglobin A1c levels [73]. The full effect of alogliptin was observed by week 8 and lasted throughout the entire 26-week treatment period. There were no differences in hypoglycemic events between placebo and alogliptin; overall adverse events were similar in all groups and adding alogliptin to insulin therapy had an acceptable safety profile [73].

Most DPP4 inhibitors require dose adjustment for chronic kidney disease but are effective and safe. A 54-week, international, multi-center, randomized, double-blind controlled trial evaluated the efficacy and safety of sitagliptin, a DPP4 inhibitor, and glipizide monotherapy in type 2 diabetes and end-stage renal disease on dialysis [74]. Sitagliptin or glipizide monotherapy effectively lowered hemoglobin A1c levels and was well-tolerated [74]. A similar study by the same investigators compared the efficacy and safety of sitagliptin with glipizide in type 2 diabetes and moderate-to-severe chronic renal insufficiency (eGFR <50mL/min/1.73 m2) [75]. In this 54-week, randomized controlled trial sitagliptin was non-inferior to glipizide and provided similar A1c-lowering efficacy. Furthermore, sitagliptin was well-tolerated with less symptomatic hypoglycemia than glipizide, decreased body weight, and few gastrointestinal adverse events [75].

DPP4 inhibitors are not free of side effects. As with GLP-1 receptor agonists, several recent studies examined the cardiovascular effects of DPP4 inhibitors. A recently published systematic review and meta-analysis of 38 randomized and observational studies investigated the relationship between treatment with DPP4 inhibitors and the risk of heart failure in type 2 diabetes [76]. The analysis suggested a small increase in heart failure with DPP4 inhibitors; however, results were only borderline significant and low-quality evidence. Twelve observational studies also provided low-quality evidence consistent with trial findings [76]. This does suggest a small increase in the risk of hospital admission for heart failure in those with or at high risk of cardiovascular disease. The relatively short follow-up and low quality evidence make it difficult to draw definite conclusions [76].

After DPP4 inhibitors were FDA-approved, observational studies suggested an increased risk of pancreatitis [77]. However, a meta-analysis of over 100 randomized controlled trials found no increased risk of pancreatitis during treatment with DPP4 inhibitors. There have also been reports of infectious complications. DPP4 acts on several different substrates and is not specific to GLP-1. DPP4 is expressed on many cell types, including lymphocytes, but studies have shown that DPP4 inhibitors do not substantially inhibit cell proliferation *in vitro* in human lymphocytes [69]. Regardless, a large meta-analysis of 29 randomized controlled trials found an increased risk of nasopharyngitis, more evident with sitagliptin use, and urinary tract infection [57]. Headache is also commonly reported, especially with vildagliptin [57].

Like GLP-1 receptor agonists, the role of DPP4 inhibitors in the long-term management of type 2 diabetes remains unclear. Few long-term studies have assessed the drug class' overall glycemic efficacy, health outcomes, and safety. Typically, DPP4 inhibitors are part of combination therapy for type 2 diabetics with inadequate glycemic control and not used as monotherapy. DPP4 is a ubiquitous enzyme expressed in many cell types and not a specific deactivator of GLP-1. Thus, inhibitors may have unexpected side effects.

TGR5 Agonists

Targeting TGR5 is attractive because its ligands can decrease blood glucose level to treat type 2 diabetes and increase energy expenditure to treat obesity [78]. Oleanolic acid, isolated from olive tree (*Olea europeaea*) leaves, was one of the first natural non-bile acid TGR5 agonists identified [79]. Extracts from these olive tree leaves, long known to improve metabolic disorders, including diabetes, activated TGR5 with similar efficacy as lithocholic acid, a positive control [79]. In a rodent model of metabolic syndrome treatment with oleanolic acid decreased body weight and epidydimal fat pad mass [79]. Furthermore, treatment with oleanolic acid decreased plasma glucose and insulin levels in these high-fat diet fed mice, thus improving metabolic homeostasis [79].

Attempts at designing potent TGR5 agonists have focused on derivatives of chenodeoxycholic acid. Targeted methylation of natural bile acids can direct their binding selectively to TGR5 rather than FXR [80]. Analysis of the structure-activity relationships of TGR5 agonists (naturally-occurring bile acids, semisynthetic bile acid derivatives, and steroid hormones) resulted in the discovery of three potent and selective TGR5 agonists [80]. These approaches identified the important structural components that play key roles in binding and activating TGR5 [81].

Using similar strategies, investigators built on the scaffold of cholic acid to develop 6α-ethyl-23(*S*)-methylcholic acid (*S*-EMCA, INT-777), a potent and selective TGR5 agonist [82]. INT-777 is stable to intestinal bacteria and resistant to conjugation; relatively strong albumin binding facilitates systemic circulation. Consistent with its *in vivo* actions, INT-777 strongly induces GLP-1 release *ex vivo* and stimulates D2 activity in brown adipose and muscle tissue [82, 83].

Other, more recently developed, TGR5 agonists lower blood glucose levels. One study comparing INT-777 to a new small molecule non-bile acid TGR5 agonist, compound 18, showed the latter caused robust GLP-1 secretion from mouse enteroendocrine cells but only modest GLP-1 secretion from human enteroendocrine cells [84]. When compared to INT-777, administering compound 18 to fasted mice significantly increased GLP-1 and PYY secretion. Also, compared to both placebo and the DPP4 inhibitor sitagliptin, compound 18 prolonged the increase in GLP-1 after an oral glucose tolerance test in mice [84]. Diet-induced overweight mice treated with compound 18 demonstrated a small but significant reduction of weight gain and a trend towards reduced gain in fat mass. Finally, compound 18 induced dose-dependent increases in bile weight in addition to elevated GLP-1 levels, whereas INT-777 increased gallbladder filling but did not increase GLP-1 [84].

Another small molecule TGR5 agonist, WB403, modestly activated TGR5 but potently stimulated GLP-1 activity in mice [85]. In obese diabetic mice, one week of WB403 treatment improved blood glucose tolerance and lowered HbA1c levels without inducing hypoglycemia. WB403 increased beta-cell mass in pancreatic islets of diabetic mice and did not alter gallbladder filling [85]. In summary, creating TGR5 agonists appears to be a promising line of drug development.

FUTURE DIRECTIONS

Currently, more than 300 registered clinical trials are evaluating the potential therapeutic role of bile acids [86]. Several are investigating the role of bile acids in obesity and diabetes, and their relationship to GLP-1, gastric bypass surgery, and energy expenditure. One is focused on the impact of gallbladder emptying and bile acids on human GLP-1 secretion; the investigators hypothesize that CCK-induced gallbladder emptying in healthy subjects stimulates GLP-1 release that could be enhanced by metformin or a BABR [87]. Another study is examining how and if bile acid sequestration influences postprandial GLP-1 secretion, gut microbiota, and glucose homeostasis in type 2 diabetics and healthy controls [88]; they hypothesize higher luminal bile acid levels in the distal gut change bacterial composition and postprandial hormone secretion. Another group studies effects of bariatric surgery on bile acid homeostasis [89] with the aims of evaluating bile acid synthesis and changes in levels of bile acids, GLP-1, TSH, and other hormones.

To test the hypothesis that humans respond to bile acids by increasing energy expenditure *via* the production of active thyroid hormone (T3), investigators are examining the relationship between stimulated bile acid release and changes in thyroid hormone homeostasis and energy metabolism [90]. One study is evaluating the serum bile acid profile in type 2 diabetics and healthy controls, and the association between changes in bile acid species and serum GLP-1 and FGF-19 levels [91]. They hypothesize modulating bile acid composition may have potential for treating type 2 diabetes [91]. A similar study is looking at the effects of bile acids and bile acid sequestrants on GLP-1 secretion following RYGB [92];

they will compare the effects of chenodeoxycholic acid to chenodeoxycholic acid plus colesevelam, a bile acid sequestrant.

A group is investigating the influence of metabolic syndrome and circulating bile acid levels on development of hepatocellular carcinoma and biliary tract cancers to determine whether pre-surgical bile acid profiles predict the efficacy of vertical sleeve gastrectomy, and reflect liver and metabolic dysfunction [93]. This handful of ongoing clinical trials reflects great interest and innovation but also highlight the need for more research to understand the complex relationships between bile acids, glucose homeostasis, and energy expenditure.

Several TGR5 agonists are in pre-clinical and early clinical trials. There remains a paucity of data evaluating the use of TGR5 agonists in humans. Fully exploiting the physiological connections between bile acid signaling, TGR5 activation, and GLP-1 release and stability will require an integrated systems biology approach with combination therapy. Just last year, investigators developed and studied OL3, a combination of linagliptin, a DPP4 inhibitor, and MN6, a TGR5 agonist [94]. The promise was to design an effective TGR5 agonist with fewer side effects, since others reported that systemic TGR5 activation increased gallbladder filling, impaired immune responses, and altered heart rate [94]. In testing OL3 in mice, the investigators found it potently activated TGR5 *in vitro*, and weakly inhibited DPP4 activity. In diabetic mice, OL3 improved glucose tolerance; in ICR mice, OL3 did not induce gallbladder filling but, as anticipated, increased GLP-1 plasma levels [94].

To date only one human trial involving a specific TGR5 agonist was published [95, 96]. The investigators examined the effects of the TGR5 agonist SB-756050 in type 2 diabetes. Subjects tolerated the drug well during the 6-day treatment but effects were highly variable between different treatment groups [96]. Effects on glucose levels with SB-756050 plus sitagliptin, a DPP4 inhibitor administered on day 6, were comparable to those with sitagliptin alone. These pre-clinical results were equivocal and this compound has yet to be tested in clinical trials [95, 96].

Undoubtedly, similar approaches that take into account the varied biological actions of these agents and their structure-function relationships will result in a new generation of potent and safe drugs to treat diabetes, obesity, and the metabolic syndrome.

CONCLUSIONS

Our understanding of the physiological actions of bile acids has expanded greatly beyond their classical role in regulating the metabolism, digestion, and absorption of lipids. Bile acids are complex signaling molecules that alter cell function by interacting with plasma membrane and nuclear receptors. They also act as hormones in a variety of tissues; by interacting with TGR5, bile acids modulate regional and circulating levels of GLP-1. Exploiting these actions has therapeutic potential for type 2 diabetes, obesity, non-alcoholic fatty liver disease, and other manifestations of the metabolic syndrome. Several studies have shown that agents stimulating GLP-1 release or extending its duration of action improve glycemic control and attenuate insulin resistance. Changes in the bile acid pool after gastric bypass and the ramifications on GLP-1 release and function have yet to be fully explored, understood, and exploited for therapeutic benefit. Nonetheless, strong evidence of improved glucose homeostasis and metabolic balance following bariatric surgery, most likely because the altered gastrointestinal anatomy allows earlier delivery of nutrients and bile acids to the mid- and distal small intestine, portends a dramatic change in our ability to manage diabetes, obesity, and the metabolic syndrome. Now, investigators are exploiting these regulatory pathways by developing GLP-1 receptor agonists, DPP4 inhibitors, and TGR5 agonists for use alone and in combination as alternative pharmacotherapies. These incretin approaches have great potential to safely normalize hemoglobin A1c levels, body weight, and visceral fat distribution. Clearly, although much has yet to be learned and considerable work remains, the future is promising.

CONSENT FOR PUBLICATION

Not applicable.

CONFLICT OF INTEREST

The author (editor) declares no conflict of interest, financial or otherwise.

ACKNOWLEDGEMENTS

Jessica Felton was supported by T32 training grant DK067872. This work was supported in part by Merit Review Award # BX002129 from the United States (U.S.) Department of Veterans Affairs Biomedical Laboratory Research and Development Program. The contents do not represent the views of the U.S. Department of Veterans Affairs or the United States Government.

REFERENCES

[1] Division of Nutrition, Physical Activity, and Obesity, National Center for Chronic Disease Prevention and Health Promotion (US). Overweight & Obesity. Atlanta (GA): Centers for Disease Control and Prevention. 2017. Available at: https://www.cdc.gov/obesity/index.html

[2] Buchwald H, Avidor Y, Braunwald E, et al. Bariatric surgery: a systematic review and meta-analysis. JAMA 2004; 292(14): 1724-37.
[http://dx.doi.org/10.1001/jama.292.14.1724] [PMID: 15479938]

[3] le Roux CW, Aylwin SJ, Batterham RL, et al. Gut hormone profiles following bariatric surgery favor

an anorectic state, facilitate weight loss, and improve metabolic parameters. Ann Surg 2006; 243(1): 108-14.
[http://dx.doi.org/10.1097/01.sla.0000183349.16877.84] [PMID: 16371744]

[4] Staels B, Fonseca VA. Bile acids and metabolic regulation: mechanisms and clinical responses to bile acid sequestration. Diabetes Care 2009; 32 (Suppl. 2): S237-45.
[http://dx.doi.org/10.2337/dc09-S355] [PMID: 19875558]

[5] Insull W Jr. Clinical utility of bile acid sequestrants in the treatment of dyslipidemia: a scientific review. South Med J 2006; 99(3): 257-73.
[http://dx.doi.org/10.1097/01.smj.0000208120.73327.db] [PMID: 16553100]

[6] Camilleri M, Gores GJ. Therapeutic targeting of bile acids. Am J Physiol Gastrointest Liver Physiol 2015; 309(4): G209-15.
[http://dx.doi.org/10.1152/ajpgi.00121.2015] [PMID: 26138466]

[7] Wang H, Chen J, Hollister K, Sowers LC, Forman BM. Endogenous bile acids are ligands for the nuclear receptor FXR/BAR. Mol Cell 1999; 3(5): 543-53.
[http://dx.doi.org/10.1016/S1097-2765(00)80348-2] [PMID: 10360171]

[8] Taoka H, Yokoyama Y, Morimoto K, et al. Role of bile acids in the regulation of the metabolic pathways. World J Diabetes 2016; 7(13): 260-70.
[http://dx.doi.org/10.4239/wjd.v7.i13.260] [PMID: 27433295]

[9] Forman BM, Goode E, Chen J, et al. Identification of a nuclear receptor that is activated by farnesol metabolites. Cell 1995; 81(5): 687-93.
[http://dx.doi.org/10.1016/0092-8674(95)90530-8] [PMID: 7774010]

[10] Watanabe M, Morimoto K, Houten SM, et al. Bile acid binding resin improves metabolic control through the induction of energy expenditure. PLoS One 2012; 7(8): e38286.
[http://dx.doi.org/10.1371/journal.pone.0038286] [PMID: 22952571]

[11] Hashimoto M, Kobayashi K, Watanabe M, et al. Knockout of mouse Cyp3a gene enhances synthesis of cholesterol and bile acid in the liver. J Lipid Res 2013; 54(8): 2060-8.
[http://dx.doi.org/10.1194/jlr.M033464] [PMID: 23709690]

[12] Lefebvre P, Cariou B, Lien F, Kuipers F, Staels B. Role of bile acids and bile acid receptors in metabolic regulation. Physiol Rev 2009; 89(1): 147-91.
[http://dx.doi.org/10.1152/physrev.00010.2008] [PMID: 19126757]

[13] Gonzalez FJ, Jiang C, Patterson AD. An Intestinal Microbiota-Farnesoid X Receptor Axis Modulates Metabolic Disease. Gastroenterology 2016; 151(5): 845-59.
[http://dx.doi.org/10.1053/j.gastro.2016.08.057] [PMID: 27639801]

[14] Kwong E, Li Y, Hylemon PB, Zhou H. Bile acids and sphingosine-1-phosphate receptor 2 in hepatic lipid metabolism. Acta Pharm Sin B 2015; 5(2): 151-7.
[http://dx.doi.org/10.1016/j.apsb.2014.12.009] [PMID: 26579441]

[15] Cao R, Cronk ZX, Zha W, et al. Bile acids regulate hepatic gluconeogenic genes and farnesoid X receptor via G(alpha)i-protein-coupled receptors and the AKT pathway. J Lipid Res 2010; 51(8): 2234-44.
[http://dx.doi.org/10.1194/jlr.M004929] [PMID: 20305288]

[16] Fang Y, Studer E, Mitchell C, et al. Conjugated bile acids regulate hepatocyte glycogen synthase activity in vitro and in vivo via Galphai signaling. Mol Pharmacol 2007; 71(4): 1122-8.
[http://dx.doi.org/10.1124/mol.106.032060] [PMID: 17200418]

[17] Sonne DP, Hansen M, Knop FK. Bile acid sequestrants in type 2 diabetes: potential effects on GLP1 secretion. Eur J Endocrinol 2014; 171(2): R47-65.
[http://dx.doi.org/10.1530/EJE-14-0154] [PMID: 24760535]

[18] Thomas C, Pellicciari R, Pruzanski M, Auwerx J, Schoonjans K. Targeting bile-acid signalling for metabolic diseases. Nat Rev Drug Discov 2008; 7(8): 678-93.

[http://dx.doi.org/10.1038/nrd2619] [PMID: 18670431]

[19] Maruyama T, Miyamoto Y, Nakamura T, *et al.* Identification of membrane-type receptor for bile acids (M-BAR). Biochem Biophys Res Commun 2002; 298(5): 714-9.
[http://dx.doi.org/10.1016/S0006-291X(02)02550-0] [PMID: 12419312]

[20] Kawamata Y, Fujii R, Hosoya M, *et al.* A G protein-coupled receptor responsive to bile acids. J Biol Chem 2003; 278(11): 9435-40.
[http://dx.doi.org/10.1074/jbc.M209706200] [PMID: 12524422]

[21] Keitel V, Reinehr R, Gatsios P, *et al.* The G-protein coupled bile salt receptor TGR5 is expressed in liver sinusoidal endothelial cells. Hepatology 2007; 45(3): 695-704.
[http://dx.doi.org/10.1002/hep.21458] [PMID: 17326144]

[22] Keitel V, Donner M, Winandy S, Kubitz R, Häussinger D. Expression and function of the bile acid receptor TGR5 in Kupffer cells. Biochem Biophys Res Commun 2008; 372(1): 78-84.
[http://dx.doi.org/10.1016/j.bbrc.2008.04.171] [PMID: 18468513]

[23] Keitel V, Cupisti K, Ullmer C, Knoefel WT, Kubitz R, Häussinger D. The membrane-bound bile acid receptor TGR5 is localized in the epithelium of human gallbladders. Hepatology 2009; 50(3): 861-70.
[http://dx.doi.org/10.1002/hep.23032] [PMID: 19582812]

[24] Vassileva G, Golovko A, Markowitz L, *et al.* Targeted deletion of Gpbar1 protects mice from cholesterol gallstone formation. Biochem J 2006; 398(3): 423-30.
[http://dx.doi.org/10.1042/BJ20060537] [PMID: 16724960]

[25] Watanabe M, Houten SM, Mataki C, *et al.* Bile acids induce energy expenditure by promoting intracellular thyroid hormone activation. Nature 2006; 439(7075): 484-9.
[http://dx.doi.org/10.1038/nature04330] [PMID: 16400329]

[26] Holst JJ. The physiology of glucagon-like peptide 1. Physiol Rev 2007; 87(4): 1409-39.
[http://dx.doi.org/10.1152/physrev.00034.2006] [PMID: 17928588]

[27] Knop FK. Bile-induced secretion of glucagon-like peptide-1: pathophysiological implications in type 2 diabetes? Am J Physiol Endocrinol Metab 2010; 299(1): E10-3.
[http://dx.doi.org/10.1152/ajpendo.00137.2010] [PMID: 20424139]

[28] Reimann F, Gribble FM. Glucose-sensing in glucagon-like peptide-1-secreting cells. Diabetes 2002; 51(9): 2757-63.
[http://dx.doi.org/10.2337/diabetes.51.9.2757] [PMID: 12196469]

[29] Reimann F, Habib AM, Tolhurst G, Parker HE, Rogers GJ, Gribble FM. Glucose sensing in L cells: a primary cell study. Cell Metab 2008; 8(6): 532-9.
[http://dx.doi.org/10.1016/j.cmet.2008.11.002] [PMID: 19041768]

[30] Gribble FM, Williams L, Simpson AK, Reimann F. A novel glucose-sensing mechanism contributing to glucagon-like peptide-1 secretion from the GLUTag cell line. Diabetes 2003; 52(5): 1147-54.
[http://dx.doi.org/10.2337/diabetes.52.5.1147] [PMID: 12716745]

[31] Junker AE, Gluud LL, van Hall G, Holst JJ, Knop FK, Vilsbøll T. Effects of glucagon-like peptide-1 on glucagon secretion in patients with non-alcoholic fatty liver disease. J Hepatol 2016; 64(4): 908-15.
[http://dx.doi.org/10.1016/j.jhep.2015.11.014] [PMID: 26626496]

[32] Hirasawa A, Tsumaya K, Awaji T, *et al.* Free fatty acids regulate gut incretin glucagon-like peptide-1 secretion through GPR120. Nat Med 2005; 11(1): 90-4.
[http://dx.doi.org/10.1038/nm1168] [PMID: 15619630]

[33] Katsuma S, Hirasawa A, Tsujimoto G. Bile acids promote glucagon-like peptide-1 secretion through TGR5 in a murine enteroendocrine cell line STC-1. Biochem Biophys Res Commun 2005; 329(1): 386-90.
[http://dx.doi.org/10.1016/j.bbrc.2005.01.139] [PMID: 15721318]

[34] Thomas C, Gioiello A, Noriega L, *et al.* TGR5-mediated bile acid sensing controls glucose

homeostasis. Cell Metab 2009; 10(3): 167-77.
[http://dx.doi.org/10.1016/j.cmet.2009.08.001] [PMID: 19723493]

[35] Düfer M, Hörth K, Wagner R, *et al.* Bile acids acutely stimulate insulin secretion of mouse β-cells *via* farnesoid X receptor activation and K(ATP) channel inhibition. Diabetes 2012; 61(6): 1479-89.
[http://dx.doi.org/10.2337/db11-0815] [PMID: 22492528]

[36] Trabelsi MS, Daoudi M, Prawitt J, *et al.* Farnesoid X receptor inhibits glucagon-like peptide-1 production by enteroendocrine L cells. Nat Commun 2015; 6: 7629.
[http://dx.doi.org/10.1038/ncomms8629] [PMID: 26134028]

[37] Trabelsi MS, Lestavel S, Staels B, Collet X. Intestinal bile acid receptors are key regulators of glucose homeostasis. Proc Nutr Soc 2016; 1-11.
[http://dx.doi.org/10.1017/S0029665116002834] [PMID: 27846919]

[38] Jiang C, Xie C, Lv Y, *et al.* Intestine-selective farnesoid X receptor inhibition improves obesity-related metabolic dysfunction. Nat Commun 2015; 6: 10166.
[http://dx.doi.org/10.1038/ncomms10166] [PMID: 26670557]

[39] Garg A, Grundy SM. Cholestyramine therapy for dyslipidemia in non-insulin-dependent diabetes mellitus. A short-term, double-blind, crossover trial. Ann Intern Med 1994; 121(6): 416-22.
[http://dx.doi.org/10.7326/0003-4819-121-6-199409150-00004] [PMID: 8053615]

[40] Kobayashi M, Ikegami H, Fujisawa T, *et al.* Prevention and treatment of obesity, insulin resistance, and diabetes by bile acid-binding resin. Diabetes 2007; 56(1): 239-47.
[http://dx.doi.org/10.2337/db06-0353] [PMID: 17192488]

[41] Yamakawa T, Takano T, Utsunomiya H, Kadonosono K, Okamura A. Effect of colestimide therapy for glycemic control in type 2 diabetes mellitus with hypercholesterolemia. Endocr J 2007; 54(1): 53--.
[http://dx.doi.org/10.1507/endocrj.K05-098] [PMID: 17102570]

[42] Zieve FJ, Kalin MF, Schwartz SL, Jones MR, Bailey WL. Results of the glucose-lowering effect of WelChol study (GLOWS): a randomized, double-blind, placebo-controlled pilot study evaluating the effect of colesevelam hydrochloride on glycemic control in subjects with type 2 diabetes. Clin Ther 2007; 29(1): 74-83.
[http://dx.doi.org/10.1016/j.clinthera.2007.01.003] [PMID: 17379048]

[43] Suzuki T, Oba K, Igari Y, *et al.* Colestimide lowers plasma glucose levels and increases plasma glucagon-like PEPTIDE-1 (7-36) levels in patients with type 2 diabetes mellitus complicated by hypercholesterolemia. J Nippon Med Sch 2007; 74(5): 338-43.
[http://dx.doi.org/10.1272/jnms.74.338] [PMID: 17965527]

[44] Potthoff MJ, Potts A, He T, *et al.* Colesevelam suppresses hepatic glycogenolysis by TGR5-mediated induction of GLP-1 action in DIO mice. Am J Physiol Gastrointest Liver Physiol 2013; 304(4): G371-80.
[http://dx.doi.org/10.1152/ajpgi.00400.2012] [PMID: 23257920]

[45] Patti ME, Houten SM, Bianco AC, *et al.* Serum bile acids are higher in humans with prior gastric bypass: potential contribution to improved glucose and lipid metabolism. Obesity (Silver Spring) 2009; 17(9): 1671-7.
[http://dx.doi.org/10.1038/oby.2009.102] [PMID: 19360006]

[46] Pories WJ, Swanson MS, MacDonald KG, *et al.* Who would have thought it? An operation proves to be the most effective therapy for adult-onset diabetes mellitus. Ann Surg 1995; 222(3): 339-50.
[http://dx.doi.org/10.1097/00000658-199509000-00011] [PMID: 7677463]

[47] Rubino F, Gagner M, Gentileschi P, *et al.* The early effect of the Roux-en-Y gastric bypass on hormones involved in body weight regulation and glucose metabolism. Ann Surg 2004; 240(2): 236-42.
[http://dx.doi.org/10.1097/01.sla.0000133117.12646.48] [PMID: 15273546]

[48] Mingrone G. Role of the incretin system in the remission of type 2 diabetes following bariatric surgery.

Nutr Metab Cardiovasc Dis 2008; 18(8): 574-9.
[http://dx.doi.org/10.1016/j.numecd.2008.07.004] [PMID: 18790374]

[49] Pacheco D, de Luis DA, Romero A, *et al.* The effects of duodenal-jejunal exclusion on hormonal regulation of glucose metabolism in Goto-Kakizaki rats. Am J Surg 2007; 194(2): 221-4.
[http://dx.doi.org/10.1016/j.amjsurg.2006.11.015] [PMID: 17618808]

[50] Beckman LM, Beckman TR, Earthman CP. Changes in gastrointestinal hormones and leptin after Roux-en-Y gastric bypass procedure: a review. J Am Diet Assoc 2010; 110(4): 571-84.
[http://dx.doi.org/10.1016/j.jada.2009.12.023] [PMID: 20338283]

[51] Jørgensen NB, Jacobsen SH, Dirksen C, *et al.* Acute and long-term effects of Roux-en-Y gastric bypass on glucose metabolism in subjects with Type 2 diabetes and normal glucose tolerance. Am J Physiol Endocrinol Metab 2012; 303(1): E122-31.
[http://dx.doi.org/10.1152/ajpendo.00073.2012] [PMID: 22535748]

[52] Albaugh VL, Banan B, Ajouz H, Abumrad NN, Flynn CR. Bile acids and bariatric surgery. Mol Aspects Med 2017; S0098-2997(16)30118-2.
[PMID: 28390813]

[53] Ahmad NN, Pfalzer A, Kaplan LM. Roux-en-Y gastric bypass normalizes the blunted postprandial bile acid excursion associated with obesity. Int J Obes 2013; 37(12): 1553-9.
[http://dx.doi.org/10.1038/ijo.2013.38] [PMID: 23567924]

[54] Bhutta HY, Rajpal N, White W, *et al.* Effect of Roux-en-Y gastric bypass surgery on bile acid metabolism in normal and obese diabetic rats. PLoS One 2015; 10(3): e0122273.
[http://dx.doi.org/10.1371/journal.pone.0122273] [PMID: 25798945]

[55] Jahansouz C, Xu H, Hertzel AV, *et al.* Bile Acids Increase Independently From Hypocaloric Restriction After Bariatric Surgery. Ann Surg 2016; 264(6): 1022-8.
[http://dx.doi.org/10.1097/SLA.0000000000001552] [PMID: 26655924]

[56] Kohli R, Bradley D, Setchell KD, Eagon JC, Abumrad N, Klein S. Weight loss induced by Roux-en-Y gastric bypass but not laparoscopic adjustable gastric banding increases circulating bile acids. J Clin Endocrinol Metab 2013; 98(4): E708-12.
[http://dx.doi.org/10.1210/jc.2012-3736] [PMID: 23457410]

[57] Amori RE, Lau J, Pittas AG. Efficacy and safety of incretin therapy in type 2 diabetes: systematic review and meta-analysis. JAMA 2007; 298(2): 194-206.
[http://dx.doi.org/10.1001/jama.298.2.194] [PMID: 17622601]

[58] Heine RJ, Van Gaal LF, Johns D, Mihm MJ, Widel MH, Brodows RG. GWAA Study Group. Exenatide *versus* insulin glargine in patients with suboptimally controlled type 2 diabetes: a randomized trial. Ann Intern Med 2005; 143(8): 559-69.
[http://dx.doi.org/10.7326/0003-4819-143-8-200510180-00006] [PMID: 16230722]

[59] Ding X, Saxena NK, Lin S, Gupta NA, Anania FA, Anania FA. Exendin-4, a glucagon-like protein-1 (GLP-1) receptor agonist, reverses hepatic steatosis in ob/ob mice. Hepatology 2006; 43(1): 173-81.
[http://dx.doi.org/10.1002/hep.21006] [PMID: 16374859]

[60] Petit JM, Cercueil JP, Loffroy R, *et al.* Effect of Liraglutide Therapy on Liver Fat Content in Patients With Inadequately Controlled Type 2 Diabetes: The Lira-NAFLD Study. J Clin Endocrinol Metab 2017; 102(2): 407-15.
[PMID: 27732328]

[61] Valdecantos MP, Pardo V, Ruiz L, *et al.* A novel glucagon-like peptide 1/glucagon receptor dual agonist improves steatohepatitis and liver regeneration in mice. Hepatology 2017; 65(3): 950-68.
[http://dx.doi.org/10.1002/hep.28962] [PMID: 27880981]

[62] Pi-Sunyer X, Astrup A, Fujioka K, *et al.* SCALE Obesity and Prediabetes NN8022-1839 Study Group. A Randomized, Controlled Trial of 3.0 mg of Liraglutide in Weight Management. N Engl J Med 2015; 373(1): 11-22.

[http://dx.doi.org/10.1056/NEJMoa1411892] [PMID: 26132939]

[63] Htike ZZ, Zaccardi F, Papamargaritis D, Webb DR, Khunti K, Davies MJ. Efficacy and Safety of Glucagon-like peptide-1 receptor agonists in type 2 diabetes Systematic review and mixed-treatment comparison analysis. Diabetes Obes Metab 2016.
[PMID: 27981757]

[64] Pfeffer MA, Claggett B, Diaz R, et al. ELIXA Investigators. Lixisenatide in Patients with Type 2 Diabetes and Acute Coronary Syndrome. N Engl J Med 2015; 373(23): 2247-57.
[http://dx.doi.org/10.1056/NEJMoa1509225] [PMID: 26630143]

[65] Marso SP, Bain SC, Consoli A, et al. SUSTAIN-6 Investigators. Semaglutide and Cardiovascular Outcomes in Patients with Type 2 Diabetes. N Engl J Med 2016; 375(19): 1834-44.
[http://dx.doi.org/10.1056/NEJMoa1607141] [PMID: 27633186]

[66] Marso SP, Daniels GH, Brown-Frandsen K, et al. LEADER Steering Committee; LEADER Trial Investigators. Liraglutide and Cardiovascular Outcomes in Type 2 Diabetes. N Engl J Med 2016; 375 (4): 311-22.
[http://dx.doi.org/10.1056/NEJMoa1603827] [PMID: 27295427]

[67] Monami M, Dicembrini I, Nardini C, Fiordelli I, Mannucci E. Glucagon-like peptide-1 receptor agonists and pancreatitis: a meta-analysis of randomized clinical trials. Diabetes Res Clin Pract 2014; 103(2): 269-75.
[http://dx.doi.org/10.1016/j.diabres.2014.01.010] [PMID: 24485345]

[68] Li L, Shen J, Bala MM, et al. Incretin treatment and risk of pancreatitis in patients with type 2 diabetes mellitus: systematic review and meta-analysis of randomised and non-randomised studies. BMJ 2014; 348: g2366.
[http://dx.doi.org/10.1136/bmj.g2366] [PMID: 24736555]

[69] Drucker DJ, Nauck MA. The incretin system: glucagon-like peptide-1 receptor agonists and dipeptidyl peptidase-4 inhibitors in type 2 diabetes. Lancet 2006; 368(9548): 1696-705.
[http://dx.doi.org/10.1016/S0140-6736(06)69705-5] [PMID: 17098089]

[70] Raz I, Hanefeld M, Xu L, Caria C, Williams-Herman D, Khatami H. Sitagliptin Study 023 Group. Efficacy and safety of the dipeptidyl peptidase-4 inhibitor sitagliptin as monotherapy in patients with type 2 diabetes mellitus. Diabetologia 2006; 49(11): 2564-71.
[http://dx.doi.org/10.1007/s00125-006-0416-z] [PMID: 17001471]

[71] Goldstein BJ, Feinglos MN, Lunceford JK, Johnson J, Williams-Herman DE, Group SS. Sitagliptin 036 Study Group. Effect of initial combination therapy with sitagliptin, a dipeptidyl peptidase-4 inhibitor, and metformin on glycemic control in patients with type 2 diabetes. Diabetes Care 2007; 30 (8): 1979-87.
[http://dx.doi.org/10.2337/dc07-0627] [PMID: 17485570]

[72] Nauck MA, Meininger G, Sheng D, Terranella L, Stein PP, Group SS. Sitagliptin Study 024 Group. Efficacy and safety of the dipeptidyl peptidase-4 inhibitor, sitagliptin, compared with the sulfonylurea, glipizide, in patients with type 2 diabetes inadequately controlled on metformin alone: a randomized, double-blind, non-inferiority trial. Diabetes Obes Metab 2007; 9(2): 194-205.
[http://dx.doi.org/10.1111/j.1463-1326.2006.00704.x] [PMID: 17300595]

[73] Rosenstock J, Rendell MS, Gross JL, Fleck PR, Wilson CA, Mekki Q. Alogliptin added to insulin therapy in patients with type 2 diabetes reduces HbA(1C) without causing weight gain or increased hypoglycaemia. Diabetes Obes Metab 2009; 11(12): 1145-52.
[http://dx.doi.org/10.1111/j.1463-1326.2009.01124.x] [PMID: 19758359]

[74] Arjona Ferreira JC, Corry D, Mogensen CE, et al. Efficacy and safety of sitagliptin in patients with type 2 diabetes and ESRD receiving dialysis: a 54-week randomized trial. Am J Kidney Dis 2013; 61 (4): 579-87.
[http://dx.doi.org/10.1053/j.ajkd.2012.11.043] [PMID: 23352379]

[75] Arjona Ferreira JC, Marre M, Barzilai N, et al. Efficacy and safety of sitagliptin *versus* glipizide in

patients with type 2 diabetes and moderate-to-severe chronic renal insufficiency. Diabetes Care 2013; 36(5): 1067-73.
[http://dx.doi.org/10.2337/dc12-1365] [PMID: 23248197]

[76] Li L, Li S, Deng K, et al. Dipeptidyl peptidase-4 inhibitors and risk of heart failure in type 2 diabetes: systematic review and meta-analysis of randomised and observational studies. BMJ 2016; 352: i610.
[http://dx.doi.org/10.1136/bmj.i610] [PMID: 26888822]

[77] Monami M, Dicembrini I, Mannucci E. Dipeptidyl peptidase-4 inhibitors and pancreatitis risk: a meta-analysis of randomized clinical trials. Diabetes Obes Metab 2014; 16(1): 48-56.
[http://dx.doi.org/10.1111/dom.12176] [PMID: 23837679]

[78] Fiorucci S, Mencarelli A, Palladino G, Cipriani S. Bile-acid-activated receptors: targeting TGR5 and farnesoid-X-receptor in lipid and glucose disorders. Trends Pharmacol Sci 2009; 30(11): 570-80.
[http://dx.doi.org/10.1016/j.tips.2009.08.001] [PMID: 19758712]

[79] Sato H, Genet C, Strehle A, et al. Anti-hyperglycemic activity of a TGR5 agonist isolated from Olea europaea. Biochem Biophys Res Commun 2007; 362(4): 793-8.
[http://dx.doi.org/10.1016/j.bbrc.2007.06.130] [PMID: 17825251]

[80] Sato H, Macchiarulo A, Thomas C, et al. Novel potent and selective bile acid derivatives as TGR5 agonists: biological screening, structure-activity relationships, and molecular modeling studies. J Med Chem 2008; 51(6): 1831-41.
[http://dx.doi.org/10.1021/jm7015864] [PMID: 18307294]

[81] Tiwari A, Maiti P. TGR5: an emerging bile acid G-protein-coupled receptor target for the potential treatment of metabolic disorders. Drug Discov Today 2009; 14(9-10): 523-30.
[http://dx.doi.org/10.1016/j.drudis.2009.02.005] [PMID: 19429513]

[82] Pellicciari R, Gioiello A, Macchiarulo A, et al. Discovery of 6alpha-ethyl-23(S)-methylcholic acid (S-EMCA, INT-777) as a potent and selective agonist for the TGR5 receptor, a novel target for diabesity. J Med Chem 2009; 52(24): 7958-61.
[http://dx.doi.org/10.1021/jm901390p] [PMID: 20014870]

[83] Pellicciari R, Gioiello A, Sabbatini P, et al. Avicholic Acid: A Lead Compound from Birds on the Route to Potent TGR5 Modulators. ACS Med Chem Lett 2012; 3(4): 273-7.
[http://dx.doi.org/10.1021/ml200256d] [PMID: 24900463]

[84] Briere DA, Ruan X, Cheng CC, et al. Novel Small Molecule Agonist of TGR5 Possesses Anti-Diabetic Effects but Causes Gallbladder Filling in Mice. PLoS One 2015; 10(8): e0136873.
[http://dx.doi.org/10.1371/journal.pone.0136873] [PMID: 26312995]

[85] Zheng C, Zhou W, Wang T, et al. A Novel TGR5 Activator WB403 Promotes GLP-1 Secretion and Preserves Pancreatic β-Cells in Type 2 Diabetic Mice. PLoS One 2015; 10(7): e0134051.
[http://dx.doi.org/10.1371/journal.pone.0134051] [PMID: 26208278]

[86] National Institutes of Health (US). Clinicaltrialsgov Bethesda (MD): US Department of Health and Human Services, National Institutes of Health 2017. Available at: https://clinicaltrials.gov/ct2/results?term=bile+acids&pg=1

[87] Knop FK. The Impact of Gall Bladder Emptying and Bile Acids on the Human GLP-1-secretion. Bethesda (MD): Clinicaltrialsgov, US Department of Health and Human Services, National Institutes of Health 2015. Available at: https://clinicaltrials.gov/ct2/show/NCT01656057?term=bile+acids&rank=14

[88] Brondon A. Effect of Bile Acid Sequestration on Postprandial GLP-1 Secretion, Glucose Homeostasis and Gut Microbiota. Clinicaltrialsgov, US Department of Health and Human Services, National Institutes of Health 2015. Available at: https://clinicaltrials.gov/ct2/show/NCT02061124?term=bile+acids&rank=33

[89] Pontificia Universidad Catolica de Chile. Effect of Bariatric Surgery on Bile Acid Homeostasis. Clinicaltrialsgov, US Department of Health and Human Services, National Institutes of Health 2015.

Available at: https://clinicaltrials.gov/ct2/show/NCT02366624?term=bile+acids&rank=17

[90] National Institute of Diabetes and Digestive and Kidney Diseases (US). Thyroid Hormones Homeostasis and Energy Metabolism Changes During Stimulation of Endogenously Secreted Bile Acids (BAs). Clinicaltrialsgov, US Department of Health and Human Services, National Institutes of Health 2016. Available at: https://clinicaltrials.gov/ct2/show/NCT00706381?term=bile+acids&rank=35

[91] Yonsei University (KR). 2016.Seum Bile Acid Profile in Type 2 Diabetes and Association Between Bile Cid Profile and Adipokine or Oxidative S. Clinicaltrialsgov, US Department of Health and Human Services, National Institutes of Health Available at: https://clinicaltrials.gov/ct2/show/NCT02650830?term=bile+acids&rank=30

[92] Hvidovre University Hospital (DK). Effect of Bile Acids and Bile Acid Sequstrants on GLP-1 Secretion After Roux-en-Y Gastric Bypass. Clinicaltrialsgov, US Department of Health and Human Services, National Institutes of Health 2016. Available at: https://clinicaltrials.gov/ct2/show/NCT02952963?term=bile+acids&rank=27

[93] Assistance Publique - Hopitaux de Paris (FR). Metabolic Syndrome, Bile Acids, Hepatocellular Carcinoma and Cholangiocarcinoma (METSLIVER). Clinicaltrialsgov, US Department of Health and Human Services, National Institutes of Health 2016. Available at: https://clinicaltrials.gov/ct2/show/NCT02730611?term=bile+acids&rank=13

[94] Ma SY, Ning MM, Zou QA, *et al.* OL3, a novel low-absorbed TGR5 agonist with reduced side effects, lowered blood glucose *via* dual actions on TGR5 activation and DPP-4 inhibition. Acta Pharmacol Sin 2016; 37(10): 1359-69.
[http://dx.doi.org/10.1038/aps.2016.27] [PMID: 27264313]

[95] van Nierop FS, Scheltema MJ, Eggink HM, *et al.* Clinical relevance of the bile acid receptor TGR5 in metabolism. Lancet Diabetes Endocrinol 2017; 5(3): 224-33.
[http://dx.doi.org/10.1016/S2213-8587(16)30155-3] [PMID: 27639537]

[96] Hodge RJ, Lin J, Vasist Johnson LS, Gould EP, Bowers GD, Nunez DJ. Safety, Pharmacokinetics, and Pharmacodynamic Effects of a Selective TGR5 Agonist, SB-756050, in Type 2 Diabetes. Clin Pharmacol Drug Dev 2013; 2(3): 213-22.
[http://dx.doi.org/10.1002/cpdd.34] [PMID: 27121782]

CHAPTER 3

Sodium–Glucose Co-Transporters Inhibitors for Type 2 Diabetes Mellitus: The 'New Kids on the Block' in the Era of Evidence-based Medicine

Cheow Peng Ooi[1,*], Norlaila Mustafa[2], Nor Azmi Kamaruddin[2] and Munn Sann Lye[3]

[1] *Endocrine Unit, Department of Medicine, Faculty of Medicine and Health Sciences, Universiti Putra Malaysia, Malaysia*

[2] *Endocrine Unit, Department of Medicine, Faculty of Medicine, National University of Malaysia, Malaysia*

[3] *Department of Community Health, Faculty of Medicine and Health Sciences, Universiti Putra Malaysia, Malaysia*

Abstract: Type 2 diabetes mellitus (T2DM) is a global crisis. Asia has a young, economically productive population at high risk of the disease. Poor blood glucose control and its associated risk factors resulting in disabling complications will have a catastrophic impact on patients with T2DM, their families, society, the healthcare system and the economy. Inhibiting sodium–glucose co-transporters (SGLT1/SGLT2) in the gastrointestinal tract and kidneys is the latest novel therapeutic pathway in managing the disease. In addition to controlling blood glucose, SGLT inhibitors may also reduce weight and lower blood pressure. However, these drugs are so new that long-term safety data is unavailable. Currently, six SGLT2 inhibitors are available for clinical use; they are continuously monitored for long-term adverse effects by drug regulatory authorities. Although there is some data suggesting benefits favouring Asians, most of the existing evidence from randomised controlled trials (RCTs) level are not applicable to patients in Asia with T2DM. High-quality RCTs reflecting real-world practice in this region are required for evidence-based medicine (EBM) to improve clinical care and justify the significant investments their development have required. Along with the production of high-quality EBM, emerging economies in Asia have the potential to play important roles in developing and facilitating evidence from RCTs for successful clinical practice utilisation. 'First, do not harm' is the fundamental tenet of clinical practice. Educating consumers of EBM about the importance of critical thinking, primary data accuracy, consistency and high-quality EBM for safe clinical practice are essential.

---

[*] **Corresponding author Cheow Peng Ooi:** Endocrine Unit, Faculty of Medicine and Health Sciences, Universiti Putra Malaysia, 43 400 Serdang, Selangor DE, Malaysia; Tel: 603 8947 2557; Fax: 603 8947 2759; E-mail: cpooi2007@gmail.com

Keywords: Anti-hyperglycemic agents, Anti-diabetic agents, Cardiovascular outcomes, Clinical trials, Complications, Dapagliflozin, Diabetes mellitus, Efficacy, Empagliflozin, Euglycemic diabetic ketoacidosis, Evidence-based medicine, Glucose homeostasis, Glucosuria, Glycaemic control, Glycosylated haemoglobin, Hyperglycaemia, Ipragliflozin, Luseogliflozin, Perineal hygiene, Renal function, Risk factors, Sodium–glucose co-transporters inhibitors, SGLT inhibitors, Therapeutics, Tofogliflozin, Type 2, Weight reduction.

INTRODUCTION

Epidemiology

Type 2 diabetes mellitus (T2DM), a chronic disabling disease, is a global crisis. A tsunami of T2DM swept 415 million adults in 2015 and is projected to increase by 55% to 642 million by 2040 [1]. The majority (80%) live in low- and middle-income countries, with more than 55% in Asia. China and India have the most population with diabetes in the world [1]. Southeast Asian and Middle Eastern countries have also observed an alarming rise in the prevalence of T2DM. The high prevalence of diabetes (415 million) is mirrored by undiagnosed T2DM (40–50%) and impaired glucose tolerance [2].

An estimated 5 million people between the ages of 20 and 79 died prematurely from diabetes in 2015 [1]. Cardiovascular (CV) events, such as myocardial infarction and stroke, accounted for approximately 70% of mortality in people with diabetes in both genders, especially those in middle age [3]. Large-scale clinical trials have demonstrated that early intensive therapy reduces the risk of microvascular complications, while individualised, targeted, multifactorial interventions control the CV risk factors [4 - 8]. However, the control of the disease and its associated risk factors is suboptimal [9 - 13].

Globally, an estimated 12% of health expenditures were on diabetes [1]. T2DM is disproportionately prevalent in socially disadvantaged, lower socio-economic groups [1]. Socio-economic development, urbanisation and associated lifestyle changes are associated with increased prevalence of obesity and an ageing population [14 - 23]. Complications and disabilities are common given the progressive nature of the disease and the associated multiple pathophysiological abnormalities. Taken together, the socio-economic impact of this tsunami is devastating, particularly for middle- and low-income countries in Asia [24 - 30].

Pathophysiology of T2DM

T2DM is a complex metabolic disorder of multiple pathophysiologies. Although the alpha (α-cells) and beta cells (β-cells) are central to glucose homoeostasis, complex multiple pathophysiologic defects lead to β-cell failure and insulin resistance. DeFronzo first described the 'ominous octet' of complex pathophysiologic defects. The organs involved are the muscle, liver, β-cells, adipocytes, gastrointestinal tract, α-cells, kidney and brain [31]. Insulin resistance plays a significant role in the pathogenesis of the disease, leading to early β-cell failure, accelerated lipolysis, incretin deficiency or resistance, and hyperglucagonemia [31]. In addition, increased glucose reabsorption and neurotransmitter dysregulation also contribute to the development of the disease.

Glycaemic Control

Early and intensive glycaemic control has been shown to reduce microvascular complications [4 - 8]. Glycaemic targets of glycosylated haemoglobin (A1c) in clinical trials and practice guidelines vary between 6.5 and 7%, while optimal glycaemic control remained elusive [32 - 39]. Given the multiple pathophysiologies of T2DM, multiple treatment modalities are required [31]. Multiple drugs may be needed to achieve glycaemic goals. Adherence to recommended treatments, side effects of medications and hypoglycaemia risk (real or perceived) must be emphasised; the cost of therapy and delay in starting treatment also must be considered.

A pharmacological armamentarium of 11 categories of medications with multiple subtypes for controlling the blood glucose in patients with T2DM has evolved [40]. Also, the myriad of potential permutations of various combinations of these agents allow the clinician to design an optimum treatment regime. Over the past 90 years, morbidity and mortality have been drastically reduced [40]. Nevertheless, there are never-ending struggles with imperfect glycaemic control, as well as questions of efficacy and potentially detrimental effects of these pharmacological agents [41 - 46]. However, these clinical shortcomings spur the continual search for new pharmacological agents to add to the armamentarium for the treatment of T2DM.

NEW TREATMENT OPTION

Currently available pharmacologic agents address the known pathophysiological defects in insulin secretion, reduced peripheral insulin action and incretin system dysfunction of T2DM. These agents operate by decreasing the endogenous glucose load by acting on hepatic glucose production and peripheral glucose uptake. Further reduction of the glucose load has mainly been limited to restricting caloric intake through dietary guidance and increasing energy utilisation through regular exercise and physical activities.

There is an influx of new information from pre-clinical and clinical trials of SGLT inhibitors. Based on this rapid accumulation of information, this new class of agents has been approved in recent years by regulatory authorities to be included as a reasonable choice for second-line or third-line therapy in updated statements and guidelines [47 - 50]. Independent of insulin action and tissue sensitisation, SGLT2 inhibition reduces blood glucose by increasing urinary glucose excretion, whereas selective SGLT1 inhibition reduces postprandial glucose excursions [51]. These drugs also provide the additional benefits of reduced body weight (BW) and lower blood pressure (BP) [51].

Role of SGLTs in Glucose Homoeostasis

Plasma glucose is auto-regulated within the narrow normal range of 3.9–7.8 mmol/L (70-140 mg/dL) by an intricate regulatory and counter-regulatory neurohormonal system [51]. This system involves intestinal absorption, glucose production, renal reabsorption, utilisation and excretion by some tissues in the body. Glucose is the essential, primary source of fuel for producing energy and adequate brain function. The kidneys handle glucose by glomerular filtration and proximal tubule reabsorption of glucose; they also contribute to endogenous glucose production. In a healthy individual without T2DM, approximately 85% of endogenous glucose production is derived from the liver [51]. Basal glucose in the liver is produced equally by glycogenolysis (conversion of glycogen to glucose) and gluconeogenesis (glucose formation) [51]. The kidneys contribute the remaining ~15% of glucose [52]. Renal gluconeogenesis, primarily localised in the proximal tubule, is particularly important for glucose homoeostasis during fasting and is upregulated in diabetes [53]. In T2DM, the kidney contributes equally with sources from the liver to enhance gluconeogenesis [52].

In the kidney, a tubular transport system mediates glucose recovery. Glucose is reabsorbed in combination with sodium (sodium-linked glucose transporters 1 and 2 [SGLT1 and SGLT2]). SGLT1 is expressed predominantly in the gastrointestinal tract. Although both transporters can reabsorb glucose, there are significant differences in affinities and transport capacity. SGLT2 has greater transportability and reabsorbs glucose in combination with sodium in the ratio 1:2 [54]. In contrast, SGLT1 has a higher affinity for glucose and reabsorbs glucose in combination with sodium in the ratio 1:1 [55]. These different transport properties reabsorb all energy, leading to glucose-free urine. SGLT2, localised mainly in the first two segments of the proximal tubular system (S1 and S2 segments), has a high transport capacity, reabsorbing about 90% of glucose from the primary urine [52, 53, 56 - 58]. Because of high affinity SGLT2 for glucose, 10% of initially filtered glucose is recovered in the third section of the proximal tubule (S3 segment) [52, 53, 56, 57]. Both transporters actively remove sodium by the activity of Na+/K+-ATPase in the basolateral membrane. Glucose transporters (GLUT2 and GLUT1) facilitate glucose transport across the basolateral membrane in the early and more distal regions of the proximal tubule.

In T2DM, glucose excretion in urine is increased. SGLT2 expression is often increased, leading to an increase in the renal threshold for glucose absorption in the urine [58]. Consequently, filtration and reabsorption of glucose increase two- to three-fold in hyperglycaemic conditions. The increased capacity of the kidney to reabsorb glucose in patients with T2DM is due to the increased transport maximum for glucose [59, 60]. In turn, glucosuria occurs at higher-than-normal plasma glucose levels, which also means lower urine glucose excretion at a given plasma glucose concentration [56]. Inhibition of the SGLT transport system reduces glucose reabsorption, resulting in increased urinary glucose excretion. Consequently, plasma glucose levels are lowered, with the loss of calories and antihyperglycaemic effects beneficial for patients with T2DM.

Drug Development Status of SGLT Inhibitors

Rapid flow of new information from SGLT2 drugs has increased knowledge about new ways to individualise treatment of diabetes. These novel modes of action appear capable of lowering the glucose of glycosylated haemoglobin (A1C) an additional 0.5 to 1.0% in combination with other classes of antidiabetic agents. Apart from the negligible side effect of hypoglycaemia, the loss of glucose in the urine enhances weight loss and increases sodium clearance, which may complement the control of oedema and hypertension. All these desirable qualities are invaluable additions to clinical resources.

To date, six SGLT2 inhibitors have been developed and tested from pre-clinical models through to post-marketing surveillance (Table 1). Three members of the class (dapagliflozin, canagliflozin and empagliflozin) are approved in the United States and Europe, while the other three (ipragliflozin, luseogliflozin and tofogliflozin) are approved in Japan. None of these new drugs has long-term safety data beyond a median of 3.1 years. The significant adverse effects common to this class of drugs are remediable urinary tract and genital mycotic infections; rare adverse effects include euglycaemic ketoacidosis and lower limb amputation [61].

Dapagliflozin, canagliflozin and empagliflozin are in the intensive pharmacovigilance program of the United States Food and Drug Administration (FDA) and European Medicines Agency (EMA). Besides alleviating the concern for risk of CV events of new antidiabetic pharmacologic agents per FDA guidelines, there is also post-marketing pharmacosurveillance for long-term safety [62]. Further trials are also in progress to demonstrate their effect on paediatric patients [63].

Table 1. Overview of the SGLT inhibitors approved by regulatory authorities.

Drug	SGLT selectivity	Developmental status	Daily dose	CV outcome trial
Canagliflozin	SGLT2	Approved by FDA, EMA	100, 300 mg	Ongoing NCT01032629
Dapagliflozin	SGLT2	Approved by EMA, FDA, PMDA	5, 10 mg	Ongoing NCT01730534
Empagliflozin	SGLT2	Approved by FDA, EMA	10, 25 mg	EMPA-REG [64]
Ipragliflozin	SGLT2	Approved by PMDA, discontinued drug development in US and Europe	25, 50 mg	No
Luseogliflozin (TS-071)	SGLT2	Approved by PMDA	2.5, 5 mg	No
Tofogliflozin	SGLT2	Approved by PMDA	20 mg	No

Dapagliflozin, ipragliflozin, luseogliflozin and tofogliflozin have been approved by the Pharmaceutical and Devices Agency (PMDA) of Japan. Compared to the stringent regulations of FDA and EMA, PMDA provides a wider array of drug development strategies for regulatory approval [65].

Many other SGLT inhibitors are still in the early stages of the drug development process. Table **2** provides an overview of some of the SGLT inhibitors in various stages of the development pipeline before being submitted for evaluation by drug regulatory authorities.

Efficacy of SGLT2 Inhibitors

A survey of the six SGLT2 inhibitors approved by the FDA, EMA and PMDA shows that all the drugs potentially lower glucose levels either alone or synergistically when combined with other antidiabetic agents, including insulin [62, 65]. These drugs may also provide the valuable clinical benefits of reducing weight and lowering BP; they also carry a low risk of hypoglycaemia (Tables **3** and **4**). Available RCTs also confirm a good tolerability profile.

Compared with a placebo, monotherapy for all six SGLT2 inhibitors has been associated with mean reductions in fasting plasma glucose (FPG) ranging from 1.3 to 2.4 mmol/L (24 to 43 mg/dL) and A1c from 0.4 to 1.1% [66 - 71]. When SGLT2 inhibitors are combined with metformin, sulfonylureas, DPP-4 inhibitors, thiazolidinedione and insulin, similar reductions have been observed compared to placebo [66, 68, 70 - 74]. Trials comparing SGLT2 inhibitors with other antidiabetic agents are limited. Canagliflozin, dapagliflozin and empagliflozin appear to be similar in efficacy to sulfonylureas [75 - 78], DPP-4 inhibitors [79 - 82] and metformin [83, 84]. There has not been a clinical trial to assess the efficacy of luseogliflozin, ipragliflozin and tofogliflozin compared to other antidiabetic agents.

Table 2. Overview of the SGLT inhibitors pipeline.

Drug	SGLT selectivity	Developmental status	Daily dose	Clinical trials
Bexagliflozin (EGT0001442; EGT0001474)	SGLT2	Phase III	5, 10, 20 mg	ongoing
Ertugliflozin (PF-04971729)	SGLT2	Phase III	1, 5, 10, 25, 100 mg	ongoing
Sotagliflozin (LX4211)	SGLT1/SGLT2	Phase III	75, 200, 400, 800 mg	ongoing
Henagliflozin (SHR3824)	SGLT2	Phase I	5, 10, 20 mg	ongoing
BI44847	SGLT2	Phase I	Not determined	Ongoing
GSK 1614235	SGLT1	Phase I	Not determined	Ongoing
ISIS-SGLT2Rx	SGLT2	Phase I	Not determined	Ongoing
KGA-2727	SGLT1	Phase I	Not determined	Ongoing

Table 3. Efficacy of SGLT2 inhibitors approved by US FDA and EMA.

Efficacy	Canagliflozin (oral)	Dapagliflozin (oral)	Empagliflozin (oral)
Dosing	100 to 300 mg daily	10 mg daily	10 to 25 mg daily
Glycaemic control – A1c (%) reduction			
1. Monotherapy (compared with placebo)	(26 weeks) 100mg:LSMC -0.91% (95% CI, -1.1 to -0.7; p < 0.001) [85] 300mg: LSMC -1.16% (95% CI, -1.3 to -1.0; p < 0.001) [85]	(24 weeks) WMD –0.5% (95% CI, -0.6% to -0.5; p<0.00001) [86]	(24 weeks) 10 mg MD –0.74% (95% CI, –0.88 to –0.59, p<0001) [87] 25 mg MD –0.85% (–0.99 to –0.71, p<0.0001) [87]

(Table 3) cont.....

Efficacy	Canagliflozin (oral)	Dapagliflozin (oral)	Empagliflozin (oral)
Dosing	100 to 300 mg daily	10 mg daily	10 to 25 mg daily
2. Add-on or combination therapy	Add-on to metformin 100mg: MD -0.59% (95% CI, -0.67 to -0.51; p < 0.00001) [88] 300mg: MD -0.74% (95% CI, -0.82 to -0.66; p < 0.00001) [88] SUs, pioglitazone, incretin-mimetics, insulin: reduction in A1c [89 - 94]	Add-on to metformin (24 weeks) MD –0·84% (95% CI –0·98 to –0·70; p < 0.0001) [95] SUs, pioglitazone, DPP4is, and insulin-reduction in A1c [72]	Add-on to metformin (24 weeks) 10 mg MD –0.57% (95% CI, –0.70 to –0.43, p<0.001) [96] 25 mg MD –0.64% (95% CI, –0.77 to –0.50%, p<0.001) [96] SUs, pioglitazone, sitagliptin, linagliptin and insulin-reduction in A1c [68]
3. Comparative therapy	With glimepiride (52 weeks) 100mg:LSMC -0.01% (95% CI, -0.11 to -0.09; p > 0.05) [75] 300mg:LSMC -0.12% (95% CI, -0.22 to -0.02; p< 0.05) [75] With sitagliptin (52 weeks) 100mg:LSMC 0.0% (95% CI, -0.12 to -0.12; p > 0.05) [97] 300mg: LSMC -0.15% (95% CI, -0.27 to -0.3; p < 0.05) [97]	With glipizide (52 weeks) MD 0.00% (95% CI, 20.11 to 0.11; p > 0.05) [78]	With glimepiride (104 weeks) MD -0.11% (95% CI [0.19 to -0.02] p=0.0153) [98] With sitagliptin (24 weeks) 10 mg: WMD 0·00% (95%CI –0·15 to 0·14; p=0·9697) [87] 20 mg: WMD –0·12% (95%CI –0·26 to 0·03; p=0·1060) [87]
Weight loss (kg)			
1. Monotherapy (compared with placebo)	(26 weeks) 100mg: LSMC –1.9 kg (95%CI, −2.9 to −1.6, p<0.001) [85] 300mg: LSMC-3.3 kg (95%CI, −4.0 to −2.6, p<0.001) [85]	(24 weeks) 10 mg: WMD -1.63 kg (95% CI, -1.83 to -1.43; p<0.00001) [86]	(24 weeks) 10 mg MD –1·93 kg (95%CI –2·41 to –1·45; p<0·0001) [87] 25 mg MD –2·15 kg (95%CI –2·63 to –1·67; p<0·0001) [87]
2. Add-on therapy	Add-on to metformin 100mg: MD -2.09 kg (95% CI, -2.43 to -2.75; p < 0.00001) [88] 300mg: MD -2.66 kg (95% CI, -3.18 to -2.14; p < 0.00001) [88] SUs, pioglitazone, incretin-mimetics, insulin: reduction in HbA1c [75, 89 - 94]	Add-on to metformin (24 weeks) 10 mg: MD –2·9 kg (95%CI, –3·3 to –2·4; p < 0.05) [95] SUs, pioglitazone, DPP4is, and insulin-reduction in weight [72]	Add-on to metformin (24 weeks) 10 mg MD –1.63kg (95% CI, –2.11 to –1.15, p<0.001) [96] 25 mg MD –2.01% (95% CI, –2.49 to –1.53%, p<0.001) [96] SUs, pioglitazone, sitagliptin, linagliptin and insulin-reduction in weight [68]

(Table 3) cont.....

Efficacy	Canagliflozin (oral)	Dapagliflozin (oral)	Empagliflozin (oral)
Dosing	100 to 300 mg daily	10 mg daily	10 to 25 mg daily
3. Comparative therapy	With glimepiride (52 weeks) 100mg: LSMC −5.2 kg (95%CI, −5.7 to −4.7, p<0.0001) [75] 300mg: LSMC −5.7 kg (95%CI, −6.2 to −5.1, p<0.0001) [75] With sitagliptin (52 weeks) 100mg:LSMC 2.4 kg (95% CI, -3.0 to -1.8; p < 0.001) [97] 300mg: LSMC -2.9 kg(95% CI, -3.4 to -2.3; p < 0.001) [97]	With glipizide (52 weeks) WMD -24.65 kg (95% CI, -25.14 to -24.17, p<0.0001) [78]	Glimepiride WMD −4·61 kg (95% CI −5·04 to −4·18, p< 0.0001) [98] With sitagliptin (24 weeks) 10 mg: WMD −2·45 kg (95%CI −2·93 to −1·96; p<0·0001) [87] 20 mg: WMD −2·67 kg (95%CI −3·15 to −2·18; p<0·0001) [87]
Blood pressure			
1. Monotherapy (compared with placebo)	(26 weeks) Systolic BP 100mg: LSMC −3.7 mmHg (95%CI, −5.9 to −1.6, p<0.001) [85] 300mg: LSMC-5.4 mmHg (95%CI, −7.6 to −3.3, p<0.001) [85]	(24 weeks) Systolic BP 10mg: WMD -3.57 mmHg (95% CI, -4.38 to -2.77; p<0.00001) [86]	(24 weeks) Systolic BP 10 mg: MD −2·7 mmHg (95%CI −5·4 to −0·1; p<0·001) [87] 25 mg MD −3·8 kg (95%CI −6·5 to −1·2; p<0·0002) [87]
2. Add-on therapy	Add-on to metformin 100mg: MD -0.6 kg (95% CI, -0.8 to -0.5; p < 0.00001) [97, 99, 100] 300mg: MD -5.1 mmHg (95% CI, -6.1 to -4.1; p < 0.00001) [97, 99, 100] SUs, pioglitazone, incretin-mimetics, insulin: reduction in systolic BP [75, 89 - 94]	Add-on to metformin (24 weeks) Systolic BP: MD -4.90 mmHg (95%CI, -5.22 to -4.58; p< 0.00001) [95] SUs, pioglitazone, DPP4is, and insulin-reduction in systolic BP [72]	Add-on to metformin (24 weeks) Systolic BP 10 mg MD −4.1 mmHg (95% CI, −6.2 to −2.1, p<0.001) [96] 25 mg MD −4.8 mmHg (95% CI, −6.9 to −2.7, p<0.001) [96] SUs, pioglitazone, sitagliptin, linagliption and insulin-reduction in systolic BP [68]

(Table 3) cont.....

Efficacy	Canagliflozin (oral)	Dapagliflozin (oral)	Empagliflozin (oral)
Dosing	100 to 300 mg daily	10 mg daily	10 to 25 mg daily
3. Comparative therapy	With glimepiride (52 weeks) Systolic BP 100mg: LSMC −3.5 mmHg (95%CI, −4.9 to −2.1, p<0.001) [75] 300mg: LSMC-4.8 mmHg (95%CI, −6.2 to −3.4, p<0.001) [75] With sitagliptin (52 weeks) Systolic BP 100mg:LSMC -2.9 mmHg (95% CI, -4.9 to -1.5; p > 0.001) [97] 300mg: LSMC -4.0 mmHg (95% CI, -5.6 to -2.4; p < 0.001) [97]	With glipizide (52 weeks) Systolic BP −5.0 mmHg (95% CI of difference −6.7, −3.4) [78]	With glimepiride SBP: WMD −6·1 mmHg (95% CI [−7·6 to −4·7 mmHg] p<0·0001) [98] With sitagliptin (24 weeks) Systolic BP 10 mg: WMD −3·4 mmHg (95%CI −5·7 to −1·2; p=0·0031) [87] 20 mg: WMD −4·2 mmHg (95%CI −6·5 to −2·0; p=0·0003) [87]
Renal impairment: discontinue/do not initiate	US: eGFR ≤ 45 mL/min/1.73 m². Europe: eGFR ≤60 mL/min/1.73m² [104]	eGFR ≤60 mL/min/1.73m² [105]	Empagliflozin 10 mg daily - eGFR ≤ 45 mL/min/1.73 m² [106]
NB Dapagliflozin is also approved by PMDA CI: confidence interval DBP: diastolic blood pressure eGFR: estimated glomerular filtration rate HbA1c: glycosylated haemoglobin SBP: systolic blood pressure LSMC: least square mean change WMD: weighted mean difference			

Table 4. Efficacy of SGLT2 inhibitors approved by PMDA.

Efficacy	Ipragliflozin	Luseogliflozin	Tofogliflozin
Glycaemic control			
1. Monotherapy (compared with placebo)	A1c: MD -0.5% to --% [68]	A1c: WMD - 0.5 to -0.75% [69]	A1c -0.4 to 0.8% [70]
2. Add-on therapy	Metformin, sulfonylurea and pioglitazone reduction in A1c and FPG [72 - 74]	Metformin, sulfonylurea, pioglitazone, α-glucosidase inhibitor and pioglitazone- reduction in A1c and FPG [69]	Combinations of oral hypoglycaemic agents - reduction in A1c (−0.77 to −0.87) [70]

(Table 4) cont.....

Efficacy	Ipragliflozin	Luseogliflozin	Tofogliflozin
Weight loss			
3. Monotherapy (compared with placebo)	WMD -1.1 to -1.7 kg [68]	WMD -1.8 kg [69]	WMD –0.77 to –3.44 kg [70]
4. Add-on therapy	Metformin, sulfonylurea and pioglitazone reduction in weight [72 - 74]	Metformin, sulfonylurea, pioglitazone, α-glucosidase inhibitor and pioglitazone - reduction in weight [69]	Combinations of oral hypoglycaemic agents - reduction in weight (WMD -1.6 to -2.8 kg) [70]
Blood pressure			
5. Monotherapy (compared with placebo)	No statistical significant reduction in BP [68]	Decrease BP	SBP: WMD -3.3 to -9.4 mmHg DBP: WMD -1.0 to -5.6 mmHg [70]
6. Add-on therapy	Sulfonylureas and pioglitazone reduction in BP [72 - 74]. No statistical significant reduction in BP with metformin	Metformin, sulfonylurea, pioglitazone, α-glucosidase inhibitor and pioglitazone - reduction in BP [69]	Combinations of oral hypoglycaemic agents - reduction in bp SBP: WMD -3.1 to -5.2 mmHg DBP: WMD -2.1 to -2.2 mmHg [70]
Renal impairment	Insufficient data	Decrease efficacy in moderate renal impairment [69]	Insufficient data
BP: blood pressure; CI: confidence interval; DBP: diastolic blood pressure; eGFR: estimated glomerular filtration rate; A1c: glycosylated haemoglobin; SBP: systolic blood pressure; WMD: weighted mean difference			

Patients with poor glycaemic control at the baseline benefitted the most, showing reductions in both FPG and A1C. Only specific SGLT2 inhibitors (canagliflozin, dapagliflozin, empagliflozin and luseogliflozin) have evidence from clinical trials demonstrating attenuated effects in individuals with advanced renal disease [70, 101 - 103]. As such, these agents are not used in patients with estimated glomerular filtration rate (GFR) < 45 ml/min/1.73 m^2.

In addition to blood glucose control, SGLT2 inhibition compared with placebo or metformin was associated with an approximate 2–3 kg reduction in BW, that apparently was sustained beyond 24–52 weeks [85, 105, 106]. A modest reduction of 3–4 mmHg in systolic BP and a 1–2 mmHg reduction in diastolic BP due to mild osmotic diuresis were also observed [107, 108].

Safety of SGLT2 Inhibitors

Data from clinical trials suggest a well-defined safety profile of adverse events, primarily consistent with their mechanism of action (Tables **5** and **6**). These well-recognised adverse effects include infections and volume-associated adverse events. Nevertheless, serious safety issues such as bone fractures, pyelonephritis, urosepsis, lower limb amputation and euglycaemic ketoacidosis have been reported. However, these were uncommon in RCTs. Please refer to Table **5** and **6**.

Table 5. Adverse effects of SGLT2 inhibitors drugs approved by US FDA and EMA.

Adverse effects	Canagliflozin	Dapagliflozin	Empagliflozin
Hypoglycaemia	Low intrinsic hypoglycaemic predisposition due to the mechanism of action. Higher risk when add-on to insulin or sulphonylureas.		
Major cardiovascular (CV) events	Insufficient information. Ongoing trials CANVAS and CREDENCE [109]	Insufficient information. Ongoing trial DECLARE-TIMI 58 [109]	Significant CV morbidity and mortality benefit [64]
CV risk	Ongoing investigation [110] Meta-analysis: no increase CV risk [109]	No suggestion of increased risk [111]. Ongoing investigation [112]	Reduced MACE by 14 %; reduced cv mortality by 38% [64]
Diabetes ketoacidosis	Rare [113 - 115]		
Urinary tract/genital infections	Urinary tract infections and genital mycotic infections, especially in women. Higher risk than placebo or comparators [66, 115, 116]		
Volume depletions	Infrequent but increased risk of volume depletions compared to placebo or comparators [115 - 117]		
Renal effects	Early transient reduction in eGFR [115, 116, 118]		
Lipids	Conflicting results [66, 115, 116]		
Haematocrit	Dose-related increase response [115 - 117]		
Malignancy risk	No suggestion	Ongoing post-marketing surveillance [118]	No suggestion [115]
Bone effects	Decreases in BMD at the total hip [119]; higher fracture incidence within 12 weeks, primarily in distal extremities [120]; ongoing trials to provide additional information.	Higher incidence in dapagliflozin group with moderate renal impairment. No imbalance between dapagliflozin and controlled group in pooled data [116]	Insufficient data [115]
CANVAS: Canagliflozin Cardiovascular Assessment Study CREDENCE: Evaluation of the Effects of Canagliflozin on Renal and Cardiovascular Outcomes in Participants With Diabetic Nephropathy DECLARE-TIMI 58: Multicenter Trial to Evaluate the Effect of Dapagliflozin on the Incidence of Cardiovascular Events			

Table 6. Adverse effects of SGLT2 inhibitors approved by PMDA.

Adverse effects	Ipragliflozin	Luseogliflozin	Tofogliflozin
Hypoglycaemia	Low incidence. Higher risk when add-on to insulin or sulphonylureas [121].	Higher risk when add-on to sulphonylureas [122]. No study with insulin yet.	Low incidence. Higher risk when add-on to insulin or sulphonylureas [71].
Major CV events	Insufficient information		
Diabetes ketoacidosis	Rare [71, 122, 123].		
Urinary tract/genital infections	Urinary tract infections and genital mycotic infections, especially in women. Higher risk than placebo or comparators [71, 73, 104, 121, 122, 124 - 127].		
Volume depletions	Infrequent. Higher rate in the ipragliflozin group [121]	No appreciable change [128].	Infrequent. Higher risk than comparators [71].
Renal effects	Early transient reduction in eGFR [121, 129]		No reduction in renal function noted [71].
Lipids	Conflicting results [71, 121, 128]		
Haematocrit	Increased in ipraglifozin treatment arm compared with placebo [121]	No appreciable change [128].	No information
Malignancy risk	No reporting [116, 121]		
CV risk	Insufficient long-term data [111, 112, 130].		No cardiovascular events in trials up to 52 weeks [71].
Bone effects	Suggestion of increased fracture risk in older population with T2DM [120]	No suggestion of appreciable change [128]	

Recently, a robust improvement in CV outcomes in patients with T2DM treated with empagliflozin was reported [64]. For two other SGLT2 inhibitors, canagliflozin and dapagliflozin, on-going trials will provide evidence about whether this positive influence on CV outcomes may contribute to a class effect. There is still insufficient information on CV outcomes for ipragliflozin, luseogliflozin and tofogliflozin.

Even though the specific SGLT2 inhibitors currently available are efficacious and relatively safe, there is a limited comparative assessment of their long-term efficacy and safety. Nevertheless, the progressive loss of efficacy in patients with reduced glomerular function needs to be balanced against the possibility of renal protection.

The hard lesson learned from the thiazolidinedione era (well into the life cycles of troglitazone and rosiglitazone) is that favourable metabolic effects do not guarantee improved clinical outcomes. There is no substitute for good quality, adequately powered, clinical trials with meaningful and carefully adjudicated clinical endpoints. Thus, the financial and regulatory pressures that have emerged in recent years have escalated the challenges of bringing new glucose-lowering drugs to market.

Post-regulatory Approval and Long-term Safety Concerns

The continuously expanding population with T2DM, especially in Asia, present exciting opportunities (and challenges) for the pharmaceutical industry. Because drugs for this chronic condition must be taken for a significantly prolonged period, many patients with T2DM can be affected by even rare side effects. Due to the prevalence of obesity and co-morbidities, vulnerabilities such as heart disease, deranged lipid profile and some cancers amplify this risk.

Since 2008 the FDA has required developers of new diabetes medications to conduct clinical trials to test the heart risks, because rosiglitazone was linked to heart failure and stroke eight years after it was approved [131]. These trials are expensive as numerous participants from the required population must be recruited and followed for some years. In addition to the CV concerns, the approved SGLT2 inhibitors in the US and European markets are subject to intense post-marketing safety monitoring for other long-term side effects [61, 132 - 134]. Similarly, the six SGLT2 inhibitors approved in Japan in 2014 (canagliflozin, dapagliflozin, ipragliflozin, tofogliflozin and luseogliflozin) are subject to intense post-marketing pharmacosurveillance [135 - 138].

EVIDENCE-BASED MEDICINE

David Sackett, the father of evidence-based medicine (EBM), asserted that "Evidence-based medicine is the conscientious, explicit and judicious use of current best evidence in making decisions about the care of individual patients. The practice of EBM means integrating individual clinical expertise with the best available external clinical evidence from systematic research. Increased expertise is reflected in many ways, but especially in more effective and efficient diagnosis and in the more thoughtful identification and compassionate use of individual patients' predicaments, rights and preferences in making clinical decisions about their care" [139].

Since its introduction, EBM has developed unabated. The evolution of EBM focusses on the importance of critical appraisal in medical care and the need for greater objectivity in medical decision-making. The aim is to increase the use of high-quality clinical research in clinical decision-making. Medical knowledge expands every day, and researchers are at risk of drowning in an ocean of data of uncertain relevance. The busy doctor does not have time to read and critically evaluate all the literature. A concerted effort to harness the information and critically summarise it into the best current evidence led to the evolution of systematic reviews and meta-analyses. These reviews and meta-analyses are used to make clinical decisions about the care of patients. The Cochrane Collaboration leads the movement in producing systematic summaries of clinical trials for EBM.

The EBM movement has infiltrated the United States and Europe. In almost all developed countries, doctors apply EBM in treatment for every patient with the support of their respective governments, ministries of health and pharmaceutical industries. EBM also intimately bolsters clinical practice guidelines, leading to improved care and patient outcomes. There is a host of EBM spinoffs, including comparative effectiveness research, personalised medicine, precision medicine, shared-decision-making, patient-centred care, and a renewed emphasis on professionalism.

EBM uses a quantitative methodological approach to develop new knowledge. As in any quantitative research, controlling possible variables or conditions (e.g., exercise and diet) allow the variable under study to be elucidated. Therefore, it is a strategy to tease out facts from values and context to identify the 'reality'. Randomised controlled clinical trials from phase I to phase IV undergo a process of determining the effects, efficacy, adverse effects (both short- and long-term) and their interaction with the body. Pre-clinical trials, as well as phase I, II, and III clinical trials have shown that the current SGLT2 inhibitors available for prescription are efficacious and relatively safe (Tables **5** and **6**).

Evidence-based Medicine at the Crossroads

Medical knowledge is continuously evolving in many directions. This development is exhilarating when such knowledge is used beneficially. An example is the case of phlorizin, which was found in the bark of the apple tree and led to the development of SGLT2 inhibitors and the investigation of its use in T2DM management. There is no doubt that EBM has contributed tremendously to higher quality evidence, leading to the improvement of diabetes care in recent years [140]. Patients are better off and healthier than in the past.

At other times, this evolution is frustrating when selection pressure allows substandard evidence to flourish in ecological niches. An example is a research community accepts existing reports as an established and unquestioned foundation. Unfortunately, such disturbing aberrant evolutionary changes distorting the EBM landscape are not uncommon and are of grave concern.

Dilemmas in Evidence-based Medicine

Although the suggested panacea for medical uncertainty is EBM, it has failed to be a solution to many problems in medicine [141]. From the clinician's perspective, it is the 'uncertainty', 'failure', 'noise' or 'interference' that is of interest and great value for understanding. Patients with T2DM will continue to require long-term pharmacological agents for glycaemic and risk-factor control. For long-term use of these agents, long-term follow-up will be required to fully understand the effects of the drugs and the drug regimen on the patients' quality of life. In fact, to deliver care according to the standards set by various medical professional bodies, one needs to look beyond EBM as a single source or type of knowledge upon which to base clinical practice decisions. Seamless integration of the ethical, intellectual, socio-political, technical and practical clinical practice components is essential [142]. The evidence required to make decisions in the care of these participants differs depending on the questions being considered. Therefore, it is important to understand the dilemmas of EBM about how to integrate knowledge from other sources (especially the ethos of clinical practice) to contextualise the knowledge produced via EBM for use in patients with T2DM and their families.

Restricted View of Evidence

Systematic review and meta-analysis have a restricted view of the evidence. The evaluation of average effects and determination of intervention effects are meant for groups or populations, not the individual. Thus, only information about the population's common characteristics rather than insight into an individual's complexities and important critical differences is collected. When a doctor uses the clinical practice guideline to manage T2DM, there is a host of evidence-based interventions to consider. The guideline cannot lead the doctor to understand the complex interactions of psychosocial factors such as ethics, health literacy and social support.

Limitations of Clinical Trials

Although RCTs are the 'gold standard' for evaluating the effectiveness of interventions, constraints when translating the evidence to the 'real' world of clinical medicine are increasingly recognised. This is evident in the published RCTs report on SGLT2 inhibitors. These issues are discussed below.

Not Representative of Patients with T2DM

The targeted group of patients with T2DM is not represented in the clinical trials. RCTs on dapagliflozin, canagliflozin and empagliflozin have clearly defined inclusion and exclusion criteria. Only adults above 18 years, with A1c ≥ 7.0% and ≤ 10%, BMI ≤ 45 kg/m2 and having a specific set of other clinical and biochemical criteria were recruited as participants. These extensive criteria mean only a minority of those with T2DM are eligible to be recruited. The RCTs of ipragliflozin, luseogliflozin and tofogliflozin were conducted predominantly in Japan. Therefore, the participants recruited were not representative of the T2DM population in the other Asian countries where most of the population with T2DM resides.

Side-effects

The findings of the Empagliflozin, Cardiovascular Outcomes, and Mortality in Type 2 Diabetes (EMPA-REG OUTCOME) clinical trial may provide reassurance of CV safety up to a median of 3.1 years [64]. However, it is questionable whether these findings apply to all Asians. The geographical representation of participants with T2DM was disproportionate [64]. Only about 30% of the participants in the EMPA-REG OUTCOME clinical trial were Asians, whereas more than 60% of the population with T2DM live in Asia, with its diverse ethnicities and cultures [2].

Body fat content and distribution between East Asians, South Asians, and Caucasians is different; thus, cardiometabolic risks also differ [143]. In addition, the mean age of the recruited participants was over 60 years, whereas younger Asians with diabetes are high CV risks. These younger Asians with T2DM have a shorter duration of the disease with no clinical evidence of CV complications. Even among the high-risk CV patients who participated in EMPA-REG OUTCOME, it is unclear whether there is a distinct sub-group with unique CV abnormalities that render them particularly susceptible to the beneficial effects of empagliflozin, especially when Asians appear to have the optimal benefits. The sub-group analysis did not have adequate statistical power to tease out these issues of responses to interventions among diverse individuals, while simultaneously inflating the type 1 error [144]. This sub-group analysis may, however, generate hypotheses for future research. The greatest metabolic benefit from empagliflozin is the reduction of CV risk and not CV mortality [145]. Therefore, the hypothesis that haemodynamics effects (specifically reduced BP and decreased extracellular volume) are responsible for the reduction in CV mortality and heart failure requiring hospitalisation needs to be tested.

In a pooled analysis of the efficacy of canagliflozin in the Hispano/Latino population, there were significantly greater dose-related reductions in A1c, BW and systolic BP observed with both canagliflozin doses compared with a placebo [146]. Therefore, in the presence of health disparities among T2DM patients of different ethnic groups in Asia, the use of SGLT2 inhibitors in these heterogeneous populations may not translate to similar efficacy and safety generated by the clinical trials data.

Despite a large body of evidence demonstrating their safety and efficacy in non-fasting populations, only one trial reported data for dapagliflozin compared to sulphonylureas in a fasting Muslim cohort during Ramadan [147]. Twenty-three percent of the world population of 6.8 billion is Muslim [148]. Fasting during Ramadan, a holy month of Islam, is observed by all healthy adult Muslims [148]. Although exempted from fasting during this holy month, many of the 6.6% of adults aged 20-79 with T2DM in Muslim countries still observed this religious tradition, against medical advice [1]. The Epidemiology of Diabetes and Ramadan (EPIDIAR) study captured data from 12,243 patients with diabetes in 13 Muslim countries who fasted during this holy month [149]. Of those, 78.7% of T2DM patients who fasted for at least 15 days during Ramadan had 7.5 times higher risk of hypoglycaemia, leading to hospitalisation during the month of Ramadan compared to the preceding months [149]. Given the low risk of hypoglycaemia, SGLT2 may be an additional option for diabetes management during Ramadan. However, abstaining from drinking during the fasting period and its associated fluid depletion, which may aggravate diabetes ketoacidosis, is a concern [150].

Lack of Data on Long-term Treatment

All the trials reported short to intermediate end-of-treatment responses, ranging from 4 weeks to the median observation time of 3.1 years [64, 147]. Therefore, long-term responses beyond this period are not known. New therapies for T2DM have the potential to alter the natural history of the disease. Landmark trials like the United Kingdom Prospective Diabetes Study (UKPDS) may no longer be relevant for some patients [151]. Moreover, continuing long-term treatment with a drug that achieved the targeted goals is common in clinical practice. With the progression of T2DM over time, treatment needs to be intensified with more antidiabetic agents to achieve the same degree of glycaemic control. No studies to date have compared the benefits and risks of continuing treatment versus withdrawal or analysed the treatment effects with the intensification of antihyperglycaemic therapy over time. Thus, there is a significant gap in the knowledge of the benefits versus risks of long-term treatment.

Co-morbidity

Multi-morbidity is common in patients with T2DM, especially with the progression of age and duration of the disease. It is common for patients in this group to be prescribed several drugs aimed at treating multiple conditions. Since this is a potential confounding effect of treatment, patients (especially in the older age group) with multiple co-morbidities receiving multiple, concurrent medications are nearly always excluded from clinical trials [152, 153]. Thus, the safe application of existing data from RCTs of SGLT2 inhibitors to this group of patients with T2DM is unclear.

Study Design

The variations in the prevalence of diabetes by global region and country, race/ethnicity, and rates of complications are well established [1]. The currently available RCTs on SGLT2 inhibitors are of the most rigorous experimental design to assess the efficacy and safety of the drugs in the short and intermediate term. However, there is controversy surrounding whether using standard or usual care impacts the quality of evidence. 'Standard care' has no standardised definition and is heterogeneous [154]. Publication typically represents medical recognition in a peer-reviewed journal or some form of recognition by a professional medical society. Thus, standard care varies among different population and countries, especially in Asia. The level of evidence for these recognised standards of care also varies. Clinical practice guidelines that represent standards of care are based on the best evidence available. Best evidence at all levels of research may be included: one or more appropriate RCTs, cohort or case-controlled analytical studies, expert opinions of respected authorities based on clinical experience, descriptive studies or reports of expert committees. EMPA-REG OUTCOME is a multicentre international trial. The varying standard of care among the trial participants from different population and countries with different clinical practice guidelines and routine clinical practices is a potential confounder with high or unclear risks of bias.

The clinical trials for SGLT2 inhibitors were designed based on clinical practice in developed countries. Similarly, clinical trial sites must meet the criteria set by regulators from developed countries, like the US FDA. These criteria are also expected in centres from developing countries. For many countries in Asia, this means trials are confined to certain secondary and tertiary centres that are rich in resources, such as integrated diabetes care and trained researchers, for enhancing the quality care for the participants with T2DM. Not only are many participants recruited but expensive resources for complex quality management of the participants with T2DM are also utilised. Therefore, it is not surprising that clinical trials for translating new drugs to the clinical arena for T2DM are among the most expensive [155]. Minimising the confounding variables reveals a drug's true efficacy. However, the irony is that the participants recruited in these resource-rich centres often do not represent the potential patients in the real-world clinical practice of many of these developing countries.

The majority of the RCTs for the SGLT inhibitors were explanatory trials prioritising internal validity; that is, focusing on the effect of the drugs in T2DM patients by minimising clinician and patient biases. In doing so, these drugs are generalisable only to 'ideal' conditions. Moreover, the primary and secondary outcomes were standard: change in A1c, change in weight or BMI, change in lipid profile, change in blood profile and adverse effects.

The FDA required the EMPA-REG OUTCOME trial to comply with pre-market CV safety requirements by ensuring a risk ratio of < 1.3 for major adverse CV events (MACE) [64]. This pragmatic design prioritised external validity by studying interventions in the context of broadly based and routine clinical practice conditions of standard care. Pooled data from the previous RCTs of empagliflozin suggested a CV safety of risk ratio of < 1.8 for MACE [156]. To compete with the pharmacological armamentarium of antidiabetic drugs that already have confirmed CV safety outcome data, benefits needed to extend beyond glycaemic control for the new drug to be competitive. This requirement is especially true in developing countries in Asia with competing health priorities and where most the T2DM population resides. To date, there have been no clinical trials of SGLT2 inhibitors assessing health-related quality of life, functional capacity (including cognition) and socio-economic effects. These outcomes would help patients, health providers and policymakers to evaluate their priorities in improving management of the disease.

Head-to-head trials comparing SGLT2 inhibitors with other antidiabetic agents would be more useful to real-life clinical practice. Currently, these comparative studies have been limited to metformin, sulphonylureas and dipeptidyl peptidase 4 (DPP-4) inhibitors [66, 68, 157].

Statistical Issues

In the EMPA-REG OUTCOME trial, more than 20% of the participants prematurely discontinued the study medication; this jeopardises the validity of the methodology used [64, 158]. The potentially misleading interferences that can arise from a flawed methodology are cause for concern. One mitigating factor, however, is that the authors conducted the following sensitivity analysis: "A sensitivity analysis of death from any cause in which it was assumed that all patients who were lost to follow-up in the empagliflozin group died and all patients who were lost to follow-up in the placebo group were alive showed a significant benefit of empagliflozin versus placebo (hazard ratio, 0.77; 95% CI, 0.65 to 0.93; $P = 0.005$)" [64]. Having said that, it would be in the interest of transparency if the authors revealed the following information: distribution of patients who discontinued therapy by stage of trial treatment; number of patients who discontinued therapy who did not attend follow-up scheduled visits; completeness of data after treatment discontinuation and type of treatment given after the discontinuation of trial medication. The authors were contacted for clarification but have yet to reply. Modified data of the magnitude of more than 20% (even with imputation using the last observation carried forward (LOCF) or restricted maximum likelihood (REML)-based mixed model repeated measures (MMRM) approaches) may threaten the validity of the data [159 - 161]. In addition, there was no adjustment for premature discontinuation of treatment in the sample size calculation in the protocol [160]. Therefore, the inferences made from these data may be compromised.

Misleading Results

Industry challenges that necessitate evidence to demonstrate product value have led to aberrant practices in the development of clinical evidence from research. Further, for a particular topic, not all research findings are published; the remainder are either not reported, withheld, 'buried in plain sight' or published in places not traditionally viewed. Research with unfavourable findings being 'lost en route' is also well-known. Invalid evidence and distortion of research findings are also creating chaos in the EBM scene [162 - 164]. Inappropriate research questions, analysis and inferences are not uncommon.

In the authors' view of medical knowledge, evidence in original research reports accepted and published in peer-reviewed journals or indexed in MEDLINE or a database such as EMBASE is more likely to be identified, evaluated, synthesised and referenced than evidence not published in this manner. However, through natural selection, many clinical research findings on a similar topic were not accepted for publication. Hence, publication bias will have confounding effects, especially for secondary and tertiary research analysis [164, 165].

Combining incompatible studies in the evidence synthesis misleads the view of evidence [166]. Together with an incomplete view of the evidence landscape, these flaws are compounded in evidence synthesis exercises such as meta--,
leaving the bias of medical research strong and consistent [167]. A meta-analysis showing consistency and increased precision when a substantial proportion of studies have the same bias affecting them does not increase the reliability of the evidence.

Issues in Translating RCTs to the Clinical Care Arena

In the Hippocratic oath, doctors pledge *'Primum non nocere'* which means 'First, do no harm' to patients. Consideration of the patients' safety and well-being is paramount. Further, appropriate reporting is central to the proper translation of findings from clinical research into clinical practice. The importance of evidence from medical research contributing to clinical decision-making is accepted on an intellectual level and is largely unquestioned. Given the glucose-lowering efficacy, low risk of hypoglycaemia and other benefits associated with SGLT2 inhibitor therapy, this class of oral glucose-lowering medication may be a valuable addition to treatment options for T2DM. These pharmacological agents may play an increasingly prominent therapeutic role as emerging data are revealed. However, along the way, there are many challenges for all stakeholders in disseminating this information.

The Efficacy-Efficiency Gap

An important role of the healthcare professional is to provide patients with the best care available. Best care requires high-quality evidence that is up-to-date. Therefore, the research evidence needs to be translated to the clinical practice arena efficiently and timely to enable patients to receive the most effective therapies and achieve improved outcomes.

However, acceptance and adoption of a new intervention take time. Understanding the factors that could influence the adoption of new ideas and innovations is an important step in the efficient dissemination of potential innovations [168]. Besides the pharmaceutical effort in promoting a new drug, a systematic review suggested that doctors played a major role [169]. Specifically, doctors' interest in particular therapeutic areas, participation in clinical trials and volume of prescribing in the therapeutic class of the new drug increase the likelihood of early adoption [169]. Younger patients with better health literacy and those with poorer health status are likely to be receptive to new therapy. Unfortunately, financial and regulatory pressures have emerged in recent years that threaten to exert an adverse impact on this pressing, unmet need. As a result, despite the presence of evidence-based recommendations, good clinical evidence may not necessarily be adopted in routine clinical practice [153].

On the one hand, it is a reasonable expectation that the short- and intermediate-term benefits and safety of SGLT2 inhibitors may provide a major opportunity for doctors to improve the quality of care while decreasing the risk of adverse events. On the other hand, for the busy doctor, sieving through the endless literature and critically appraising the vast ocean of information is nearly impossible. To bridge this gap, researchers have developed up-to-date systematic reviews or other syntheses of research to identify key messages for targeted audiences [170].

Secondary Research of Evidence Synthesis

Systematic reviews are attempts to synthesise the results and conclusions of two or more publications on a given topic [134, 139]. The process starts by identifying and formulating a clear clinical question. After selecting and describing the studies, meta-analysis follows. Meta-analysis deals quantitatively with varying study results and uses a statistical approach to combine these findings [171]. The major advantages of meta-analysis are improvements in the precision and accuracy of effect estimates and increases in the statistical power to detect an effect on the accumulation of evidence from the combined findings of selected studies. These findings may facilitate the generalisation of results to a larger population. The critical appraisals in these translation products are fashioned in language and knowledge assimilated by different audiences. Many researchers jumped on the EBM bandwagon to produce systematic reviews and meta-analyses on a wide range of topics. The industry leader is the Cochrane Library, known for independent, quality and unbiased reviews [172]. In the Cochrane Library platform, systematic reviews and meta-analyses are regularly updated as new studies on a particular topic become available.

Even with the existence of systematic reviews and meta-analyses of therapeutic agents, many doctors do not have the time or skills to appraise these secondary research findings, which are of variable quality. Busy physicians who find statistics and clinical epidemiology less relevant may find Cochrane Clinical Answers, Cochrane Learning and Cochrane Journal Club to be simplified versions of the Cochrane systematic reviews and meta-analyses, relevant for easy reading and continuing medical education.

Over the years, the medical community, including doctors and regulatory agencies, has relied on systematic reviews and meta-analyses to set clinical guidelines and make clinical decisions about drugs. Policymakers, as well as grant funders, take advantage of this form of secondary research in their decision-making. A well-performed meta-analysis may revive treatment options once considered ineffective or reveal drawbacks of practices previously considered the gold standard.

However, when the quality of data from the primary research is poor, the product of the analysis is also poor. Flaws in the design, conduct, analysis and reporting of RCTs may cause the effect of an intervention to be underestimated or overestimated. Assessing the quality of the primary studies included in the reviews is the best way to deal with this issue. Many quality assessment tools for quantitative studies have been developed [173]. Cochrane has also developed a risk-of-bias assessment tool to facilitate the evaluation of the quality of primary RCTs for reviews produced by the collaboration platform [174].

Others issues have emerged over the years that have undermined the quality of systematic reviews and meta-analyses. These include selective publication of primary studies on the topic, inadequate descriptions or unclear study methodology, and data dredging through undisclosed use of multiple analytical strategies [175]. In addition, small trials tend to report larger treatment benefits than larger trials [176, 177].

Identifying the small-study effect may be difficult if there is no evidence of heterogeneity from the usual sources. These sources are the lower methodological quality of small trials, publication bias or other reporting biases [178]. Contributing to publication bias, small studies with negative effects are unpublished or less accessible than larger studies [178]. More recently, discrepancies between prespecified outcomes in the protocol and reported outcomes in high-impact medical journals were found [179, 180]. These issues undermine the integrity of published data and increase the risk of exaggerated or positive findings. Such issues of incomplete reporting are common among trials of SGLT2 inhibitors [64, 70, 73, 123, 124, 181 - 183]. In such circumstances, it is difficult for the medical community to ascertain relevant information that may be beneficial in clinical practice.

Despite the dramatic increase in published meta-analyses in clinical trials and their contribution to improving healthcare decisions, not all the findings of a synthesis may be relevant to the clinical questions that may arise. More pertinent to clinical practice is the consideration of existing drugs in the T2DM armamentarium with more safety data over these new arrivals. Comparative research in the form of head-to-head trials would be more appropriate, given the competing treatments available. Clinicians, especially doctors, need to know what works best for their patients. Unfortunately, for the SGLT2 inhibitors, comparative trials with other oral hypoglycaemics are limited to metformin, sulphonylureas and DPP4 inhibitors [66, 68, 157]. Therefore, a gap exists for evidence-based clinical decisions pertaining to switching from other antidiabetics to SGLT2 inhibitors.

When more than two primary studies compare two interventions, conventional pair-wise meta-analysis is more appropriate. However, for the SGLT2 inhibitors at the immediate post-regulatory approval stage, available RCTs of interest do not all compare the same interventions. Instead, each trial compares only a subset of the interventions of interest. In this case, evaluating a network of RCTs where all of the trials have at least one intervention in common with another may be more appropriate [184, 185]. Depending on the clinical question and conceptual model, indirect comparisons of interventions not studied in a head-to-head fashion may allow assessment of plausible effects and generation of hypotheses for relevant and more specific future research to bridge the knowledge translation gap. This recommendation assumes that the resources for research are competitive and limited [186].

The Harm of Poor-quality Evidence

Thousands of articles are published every day, and submissions to most mainstream journals are increasing exponentially [187]. A preliminary search in PubMed found 160 RCTs and 950 secondary and tertiary research publications on sodium-glucose transport inhibitors and SGLT inhibitors. Six of the SGLT2 inhibitors are at the post-authorisation marketing stage. These RCTs focussed predominantly on glycaemic control in the treatment of T2DM, along with secondary weight loss and lowering of BP. However, more important patient-centred diabetes care outcomes addressing enhancement and maintenance of health leading to disability-free years were not addressed.

RCTs offer extremely valuable evidence of the effectiveness of these new interventions. Nevertheless, this does not mean that they are perfect [187]. Significant biases can erode their credibility and relevance. There is growing concern about the credibility of the majority or even the vast majority of published research claims [188]. Problems with the power of sample size, biases affecting the internal validity, biases affecting the total randomised evidence on a particular topic, lack of relevance, poor generalisability and biases related to the interpretation of the results are some of the major threats to the credibility and relevance of the results [187]. Biases threatening internal validity include poor design, poor conduct and inappropriate analysis. Publication bias, as well as selective outcome and analysis reporting biases, may skew the evidence landscape. In addition, outcome switching, leading to more favourable reporting of statistically significant results, is highly prevalent even in credible journals [175]. Under-reporting is a common phenomenon, as is research misconduct [189]. Overestimating the benefits of treatment and underestimating its harmful effects are severe consequences that can put patients at risk and waste precious healthcare resources. Publication bias of unpublished negative research findings can add a misleading skew to the literature and impede clinicians' understanding of a drug [164, 190].

Low-quality primary data will continue to inflate the 'garbage in, garbage out' trend when it is used to develop secondary research (reviews and meta-analyses) and tertiary research (overviews of reviews and meta-analyses). Undermining the integrity of published data increases the risk of exaggerated or even false-positive findings. Systematic reviews that fail to provide complete and up-to-date evidence of treatment may also have a detrimental effect on treatment decisions and future research planning [162, 191]. Given the abundance of data, research on research (i.e., meta-research) empirical estimates of the prevalence of risk factors for false-positive rates is high. Through these subtle statistical signals of benefit and risk in these noisy, messy trial data, the misled physicians, researchers and patients using it to make informed decisions about treatments are putting lives at stake.

In the past, most published clinical trials and systematic reviews were not useful [192]. Such waste across medical research has been estimated to be more than 85% of the billions spent annually [193, 194]. RCTs are an important part of the research and development (R & D) process to produce new interventions. The innovative biopharmaceutical industry accounts for the vast majority of investments in RCTs of potential new medicines at the clinical site level [195]. In addition, further investments in trials of appropriate design will be required to successfully translate and utilise new drugs in the clinical arena. For the biopharmaceutical industry, the stakes are high. These multibillion-dollar investments will be lost should the primary data be flawed and of low quality for clinical utilisation [195].

Potential conflicts of interest (financial, corporate, academic, scientific) may lead to publication bias, lag bias, selective outcome bias, analysis reporting bias and even fraudulent results [162, 196]. These biases may have a major impact on the credibility of the available, visible and randomised evidence [197]. Most of the time, many are unaware these biases. In particular, publication bias is common with unpublished trial data [197].

Clinicians are expected to practice medicine based on the best available evidence. Usually, the source of this evidence is the literature in medical journals. However, when the credibility and quality of literature involve ghost management of research and publication, clinicians cannot trust what they read [198]. Usually undertaken by the pharmaceutical industries, ghost management of research involves controlling or shaping several crucial steps in the research, writing and publication of scientific literature in ways that serve industry interests [198, 199]. If left unchecked, gift, guest and ghost authorship will erode trust in the peer-review system, academic research and healthcare paradigms.

CURRENT EFFORTS IN BRIDGING THE EFFICACY-EFFICIENCY GAP

EBM needs to evolve to remain valid. There are ongoing efforts to address issues of credibility and efficient scientific investigations. Understanding and identifying the major threats to the clinical trial research data, such statistical issues and biases of internal validity, allows rectification. Groups led by John Ioannidis and Ben Goldacre have been actively advocating methods to improve the credibility and relevance of RCTs [175, 179, 192, 200]. Identifying appropriate research questions and improving the credibility as well as reliability of primary research data will not only improve the utilisation in the clinical arena but also reduce waste [201].

Selective reporting of trials distorts the body of evidence available for clinical decision-making. To address this issue, the International Committee of Medical Journal Editors (ICMJE) proposes comprehensive trial registration [202]. If all trials were registered in a public repository at their inception, every trial's existence would be part of the public record. Therefore, the various stakeholders in clinical research can explore the full range of clinical evidence. Even though to date the response is still unsatisfactory on both sides of the Atlantic, without registration, selective reporting would be much worse [203 - 205].

Having the full range of evidence on a particular topic available is important to prevent reporting bias. Contributing to this effort is the AllTrials initiative started by Ben Goldacre from Oxford, advocating and promoting transparency of data from clinical trials [206]. This lead to OpenTrial, an open, collaborative database of all available information from all past and present clinical trials, including full methods and summary results [207]. This comprehensive information allows interested groups, such as researchers and policymakers, to access all available data.

New methods or groups are continuously evolving in EBM to tackle the problems of misleading evidence. Over the years, EBM has encouraged researchers to become evidence sleuths and archaeologists to dig and unearth buried evidence, as well as encouraging clinical trial data transparency [206, 208 - 214]. To overcome the problem of passing forward invalid interpretations of evidence, scrutinising behind the scenes of published reports for insight into evidence of fraud or inappropriate analyses is tremendously important. This scrutiny includes how the data were selected and how they were combined in meta-analyses, not how the statistical results were reported. Besides rigorous critical appraisal for validity and relevance, other relevant evidence-quality issues, such as letters to the editor and discussion lists, must go under the investigators' scrutinising lenses. Furthermore, when the preponderance of data comes from a single source, such as the same institution or overlapping authors, there may be bias or a conflict of interest. Similarly, if a substantial proportion of studies are affected by the same bias, a meta-analysis showing consistency and increased precision does not make the evidence more reliable.

Direct 'head-to-head' data comparing different antidiabetic pharmacologic agents provide more useful evidence for real-life clinical practice. However, when these data are unavailable, the usefulness of traditional pairwise meta-analyses is limited [215]. The methodology in network or multiple treatments meta-analysis (MTAs) may be used to estimate of the comparative effectiveness of multiple treatments. Currently, the GetReal consortium in the Europe Union (EU) has a three-year project to show how robust new methods of real-world evidence collection and synthesis could be adopted earlier in pharmaceutical R & D and the healthcare decision-making process [216]. This EU public-private consortium consists of pharmaceutical companies, academia, health technology assessment (HTA) agencies and regulators, patient organisations and small and medium enterprise (SMEs). One of the goals of this consortium is to develop and improve the methodology of MTAs [215]. The regulators include the National Institute for Health and Care Excellence (NICE), health and social care agencies (HAS), the EMA and Zorginstituut Nederland (ZIN).

To enhance trust and discard the 'villain' role in industry-sponsored research, many pharmaceutical companies have made rapid and important changes to their publication guidelines, policies and procedures [217]. There is compelling evidence that industry practices are improving. However, there is certainly more to do, both in industry and academia [218]. To this end, the third edition of the Good Publication Practice for Communicating Company-Sponsored Medical Research has been updated to further address the related issues with improved clarity on good publication practices [219]. Although the ICMJE Recommendations set relevant research and reporting standards in publications, they are not fool proof [220]. Without these recommendations, commercial bias would likely be a significantly greater problem than it is today.

Developing new pharmaceutical agents and medical devices is necessary for continual improvement in patient care. Developers and manufacturers of pharmaceutical agents assist physicians in the pursuit of their educational goals and objectives through financial support of various medical, research and educational programs. The major goal of the pharmaceutical industry is optimising profit by providing useful goods and services to maximise the sale of their products. Contrastingly, physicians' primary responsibility is to act as protectors of their patients' interests and not enhance the utilisation of the pharmacological agents in the clinical arena [221].

The industry's goals and the physician's duty of care may converge or significantly diverge, creating an opportunity for bias. The latter is of concern as a conflict of interest adversely impacts the delivery of care and erodes the trust of patients [222, 223]. Further, when there is a high prevalence of physician panel members with a conflict of interest in a guideline development group, the quality of the evidence cited raises questions about the quality of the evidence and validity of the recommendations [224, 225]. Over the years, such grave threats to professionalism and questions about conflicts of interest in the realm of patient care have sparked discussions about professional codes of conduct and statements on conflicts of interest in medical practice. These commendable responses from professional societies have resulted in changes to national legislation [226], subsequently leading to recommendations on how to address issues at all levels of the hierarchy of medical practice: the physician community; healthcare providers; and pharmaceutical, medical device and biotechnology companies [227].

For national and international health organisations, health regulatory authorities, policy makers, and pharmaceutical reimbursements, major decisions affecting the health of a population or community require a sound basis of evidence. Health technology assessment (HTA) products are evaluations of clinical effectiveness and/or cost-effectiveness. Such documents may include the ethical, legal and social implications of health technologies on the health of the patient and the healthcare system [228 - 232]. This form of evidence translates primary data for use in decision-making at all levels of care. Synthesis of these data was carried out by a multidisciplinary team consisting of experts in their respective fields. Given its importance, the integrity and credibility of the primary data are of utmost importance.

While it had been challenging for the pharmaceutical industries (with an economic agenda focussing on profits) to collaborate with academia, the concern is unfounded. A clinical trial may have pre-specified protocol and standard operation procedures to guide the trial process. However, issues may arise along the way and necessitate error management from those with particular expertise, which need to be in place promptly. Understanding and respecting each other's organisational culture as well as combining the intellectual and technological assets to answer big clinical research questions accelerates and improves the quality of every collaboration [233]. Also, an institution with the appropriate expertise can also help improve protocol design, including cost containments [234].

Although the approved SGLT2 inhibitors were recommended for diabetes care in many clinical guidelines, there are still many barriers to their utilisation in the actual diabetes service delivery. Facilitating the uptake of evidence in clinical practice may need to consider quality improvement systems, tools and workforce to achieve optimal diabetes care outcomes. As the knowledge of diabetes care increases in complexity, not only new therapies need to be considered but also other aspects of diabetes care such as clinical monitoring, self-monitoring, patient compliance, screening for complications, education, lifestyle management and social support [235]. While evidence from clinical research is necessary, it is insufficient on its own to deliver optimal clinical care. SGLT2 inhibitors may be prescribed as an add-on medication to existing oral antidiabetics and other medications in an overweight individual with T2DM and poor glycaemic control, at least for an intermediate period. However, many developing countries in Asia with competing health priorities do not have the optimal resources to support the care of the patients with T2DM [1].

FUTURE RESEARCH

Fortunately, the forces driving the production and dissemination of poor quality clinical research data vary and are mostly identifiable and modifiable. Moreover, there is evidence that all involved parties have introduced initiatives towards good practice. Nevertheless, continuous challenges will confront researchers in the future.

Asia is a region with large and diverse populations, increasing prosperity and an ageing population. As economies have grown and diversified, T2DM and obesity have also increased. These emerging economies offer vast, new, untapped markets. However, the different health infrastructures pose a challenge. Different levels of public funding, reimbursement and private healthcare structures raise the issue of affordability.

Landmark trials have proven that the natural course of T2DM is modifiable [7]. SGLT2 inhibitors have largely proven their efficacy in glycaemic control with additional weight and BP reduction benefits in the short and intermediate terms under experimental conditions. It is also possible that there are differential effects on different populations that favour Asians. Therefore, more evidence is required to translate these benefits into effective clinical utilisation practices in Asia.

Secondary research using new evolving methodology (e.g., network or multiple treatments meta-analysis, cost-effective analysis) in the context of the Asian population will enable issues for developing clinical questions and strategies for pragmatic translational research in the local context to be identified [236, 237]. The majority of the RCTs for SGLT2 inhibitors are studying treatment effects in idealised clinical trial conditions. Questions remain on the generalisability of the findings in real-life circumstances, particularly in the context of clinical practice in Asia. Pragmatic research in SGLT2 inhibitors should be designed to provide significant amounts of evidence to ensure patients, clinicians and other decision makers can be confident about the magnitude and specifics of benefits and risk of harms. These studies should be judged on clinical impact and their ability to change practice.

Comparative, appropriate, 'head-to-head' studies identified from secondary research, additional patient-oriented outcomes, functional capacity (including cognition), quality of life and socio-economic costs that have relevance in improving clinical practice and increasing the opportunity for clinical utility should be conducted. Nevertheless, conducting a clinical trial is costly, and many clinical trials have been terminated because of futility [238, 239]. A feasibility assessment may be useful for investigators and funders to grasp the 'real' situation [240]. The benefits of valuable clinical research that extend to reducing the burden of the disease justify the cost of performing the research [241].

The clinical utility increases with high-quality research and transparent research findings. Study data, protocols and other processes need to be available for verification and further use of others. Trust erodes when significant biases occur in the design, conduct and reporting of research. The Good Clinical Practice and Good Publication Practice for Communicating Company-Sponsored Medical Research guidelines can assist in optimising quality [219, 242]. Registering clinical trials in national and international clinical trial registries and developing protocols with pre-specified analysis plans as laid out by the Standard Protocol Items: Recommendations for Interventional Trials (SPIRIT) statement will also further this goal [243]. The SPIRIT statement is an international initiative that aims to improve the quality of clinical trial protocols by defining an evidence-based set of items on which to focus in a protocol. This initiative will further enhance research transparency [243].

Good research practice will improve the quality and usability of primary research data. Such data propagated in the EBM chain through meta-analyses, decision analyses and guidelines will be reliable and trustworthy. Secondary research cannot salvage a situation based on useless studies and may add other problems and biases [244 - 246].

The opportunity to improve the EBM process for successful utilisation of medical research in the clinical arena rests not only with researchers but also with institutions, funding mechanisms, the pharmaceutical industry, journals and other stakeholders, including patients and the public. Collaborations with multiple stakeholders to address issues will enable solutions that are more widely acceptable and adopted and thus are more successfully implemented in the clinical arena [216].

There are many models in the vast clinical research and clinical practice workforce. Different stakeholders have different goals and perspectives for publications. Ultimately, at the end of the EBM chain are consumers: physicians, allied health professionals, patients and policymakers. Many are not prolific researchers. Providing training in instilling healthy scepticism and critical thinking skills in understanding research methods and EBM at all level of their professional training will help improve the translation of research to clinical utilisation.

T2DM is a multifaceted disease with complex management which encompasses the biopsychosocial domain. Study design relevant to the population, local culture and clinical practice will yield more valuable information generalisable to the local population. Asia is home to more than 55% (450 million) of all T2DM patients worldwide. Over the next 15 years, the population with T2DM is projected to increase by 55% to 642 million. Many emerging economies in Asia invest heavily in empowering human capital and developing the education and health sectors. Appropriate skills in R & D are the key to increasing the translation and clinical utilisation of new drugs like SGLT2 inhibitors to improve the health of the population with T2DM. Many of the academic institutions in the region are developing or have developed clinical research workforces with appropriate expertise and strong relationships with local governments and stakeholders. Therefore, with strategic planning and involving local researchers in their R & D, pharmaceutical companies will find not only value for their investments but also an upward trajectory for their profits in the long-term.

CONCLUSION

The multiple pathophysiological mechanisms of T2DM enhance the niche for SGLT2 inhibitors in the disease treatment arsenal. However, evidence of efficacy from RCTs exists only for the short and intermediate term. Most patients with T2DM reside in Asia. Many of the RCTs are difficult to generalise for use in this population. Asian populations have different cultures, physical characteristics and CV risks; this diversity necessitates further pragmatic evaluation of SGLT2 inhibitors for them to be optimally translated and utilised by the local population in the clinical arena. Engaging local researchers from emerging economies in Asia at all levels of the research process will further enhance EBM's ability to effectively translate and utilise SGLT2 inhibitors.

Costly clinical research is wasted if not planned appropriately. Robust primary clinical research data are the basis of translation to high-quality EBM through systematic reviews, meta-analyses and HTAs. Most quality issues originate in the clinical trial process. Accurate and consistent primary data and data integrity are important for conversion to safe clinical practice. Many of the problems affecting research quality have been identified. Stakeholders, including pharmaceutical companies, have shown their commitment to rectifying the current situation and achieving a healthy environment in which EBM can progress and improve.

Health professionals, patients and policymakers are usually the end consumers of EBM in therapeutics. Doctors prescribe the medicine and lead the utilisation chain. Their commitment to providing safe, effective care, the core principle of EBM and the issues raised in this chapter demonstrate that stakeholders need to be more educated about the importance of critical thinking, data integrity and high-quality EBM for safe clinical practice.

CONSENT FOR PUBLICATION

Not applicable.

CONFLICT OF INTEREST

The authors (editor) declare no conflict of interest, financial or otherwise.

ACKNOWLEDGEMENTS

Declared none.

REFERENCES

[1] International Diabetes Federation. IDF Diabetes Atlas. Belgium: International Diabetes Federation 2015. http://www.diabetesatlas.org/

[2] Nanditha A, Ma RC, Ramachandran A, *et al.* Diabetes in Asia and the Pacific: Implications for the global epidemic. Diabetes Care 2016; 39(3): 472-85.
[http://dx.doi.org/10.2337/dc15-1536] [PMID: 26908931]

[3] Laakso M. Benefits of strict glucose and blood pressure control in type 2 diabetes: lessons from the UK Prospective Diabetes Study. Circulation 1999; 99(4): 461-2.
[http://dx.doi.org/10.1161/01.CIR.99.4.461] [PMID: 9927388]

[4] Gerstein HC, Miller ME, Byington RP, *et al.* Action to Control Cardiovascular Risk in Diabetes Study Group. Effects of intensive glucose lowering in type 2 diabetes. N Engl J Med 2008; 358(24): 2545--.
[http://dx.doi.org/10.1056/NEJMoa0802743] [PMID: 18539917]

[5] Duckworth W, Abraira C, Moritz T, *et al.* VADT Investigators. Glucose control and vascular complications in veterans with type 2 diabetes. N Engl J Med 2009; 360(2): 129-39.
[http://dx.doi.org/10.1056/NEJMoa0808431] [PMID: 19092145]

[6] Patel A, MacMahon S, Chalmers J, *et al.* ADVANCE Collaborative Group. Intensive blood glucose control and vascular outcomes in patients with type 2 diabetes. N Engl J Med 2008; 358(24): 2560-72.
[http://dx.doi.org/10.1056/NEJMoa0802987] [PMID: 18539916]

[7] UK Prospective Diabetes Study Group. Tight blood pressure control and risk of macrovascular and microvascular complications in type 2 diabetes: UKPDS 38. BMJ 1998; 317(7160): 703-13.
[http://dx.doi.org/10.1136/bmj.317.7160.703] [PMID: 9732337]

[8] Dailey G. Early and intensive therapy for management of hyperglycemia and cardiovascular risk factors in patients with type 2 diabetes. Clin Ther 2011; 33(6): 665-78.
[http://dx.doi.org/10.1016/j.clinthera.2011.04.025] [PMID: 21704233]

[9] Chan JC, Malik V, Jia W, *et al.* Diabetes in Asia: epidemiology, risk factors, and pathophysiology. JAMA 2009; 301(20): 2129-40.
[http://dx.doi.org/10.1001/jama.2009.726] [PMID: 19470990]

[10] Cheung BM, Ong KL, Cherny SS, Sham PC, Tso AW, Lam KS. Diabetes prevalence and therapeutic target achievement in the United States, 1999 to 2006. Am J Med 2009; 122(5): 443-53.
[http://dx.doi.org/10.1016/j.amjmed.2008.09.047] [PMID: 19375554]

[11] International Diabetes Federation. IDF Diabetes Atlas. Belgium: International Diabetes Federation 2011. http://www.diabetesatlas.org/resources/previous-editions.html

[12] Kolding Kristensen J, Lauritzen T. Inadequate treatment of dyslipidemia in people with type 2 diabetes: quality assessment of diabetes care in a Danish County. Scand J Prim Health Care 2006; 24(3): 181-5.
[http://dx.doi.org/10.1080/02813430600819710] [PMID: 16923628]

[13] Yang ZJ, Liu J, Ge JP, Chen L, Zhao ZG, Yang WY. Prevalence of cardiovascular disease risk factor in the Chinese population: the 2007-2008 China National Diabetes and Metabolic Disorders Study. Eur Heart J 2011. [Epub 02/7/2011]-[Epub 02/07/].

[14] Al-Daghri NM, Alkharfy KM, Al-Attas OS, *et al*. Association between type 2 diabetes mellitus-related SNP variants and obesity traits in a Saudi population. Mol Biol Rep 2014; 41(3): 1731-40.
[http://dx.doi.org/10.1007/s11033-014-3022-z] [PMID: 24435973]

[15] Ma RC, Lin X, Jia W. Causes of type 2 diabetes in China. Lancet Diabetes Endocrinol 2014; 2(12): 980-91.
[http://dx.doi.org/10.1016/S2213-8587(14)70145-7] [PMID: 25218727]

[16] Praveen PA, Kumar SR, Tandon N. Type 2 diabetes in youth in South Asia. Curr Diab Rep 2015; 15(2): 571.
[http://dx.doi.org/10.1007/s11892-014-0571-4] [PMID: 25620404]

[17] Zhang J, Chaaban J. The economic cost of physical inactivity in China. Prev Med 2013; 56(1): 75-8.
[http://dx.doi.org/10.1016/j.ypmed.2012.11.010] [PMID: 23200874]

[18] Chen Y, Copeland WK, Vedanthan R, *et al*. Association between body mass index and cardiovascular disease mortality in east Asians and south Asians: pooled analysis of prospective data from the Asia Cohort Consortium. BMJ 2013; 347: f5446.
[http://dx.doi.org/10.1136/bmj.f5446] [PMID: 24473060]

[19] Cho YS, Lee JY, Park KS, Nho CW. Genetics of type 2 diabetes in East Asian populations. Curr Diab Rep 2012; 12(6): 686-96.
[http://dx.doi.org/10.1007/s11892-012-0326-z] [PMID: 22993124]

[20] Kato N. Ethnic diversity in type 2 diabetes genetics between East Asians and Europeans. J Diabetes Investig 2012; 3(4): 349-51.
[http://dx.doi.org/10.1111/j.2040-1124.2012.00222.x] [PMID: 24843588]

[21] Ma RC, Chan JC. Type 2 diabetes in East Asians: similarities and differences with populations in Europe and the United States. Ann N Y Acad Sci 2013; 1281: 64-91.
[http://dx.doi.org/10.1111/nyas.12098] [PMID: 23551121]

[22] Misra A. Ethnic-specific criteria for classification of body-mass index: A perspective for Asian Indians and American Diabetes Association position statement. Diabetes Technol Ther 2015; 17(9): 667-71.
[http://dx.doi.org/10.1089/dia.2015.0007] [PMID: 25902357]

[23] Pan A, Malik VS, Hu FB. Exporting diabetes mellitus to Asia: the impact of Western-style fast food. Circulation 2012; 126(2): 163-5.
[http://dx.doi.org/10.1161/CIRCULATIONAHA.112.115923] [PMID: 22753305]

[24] Agardh E, Allebeck P, Hallqvist J, Moradi T, Sidorchuk A. Type 2 diabetes incidence and socio-economic position: a systematic review and meta-analysis. Int J Epi 2011. [Epb 22/02/2011]-[Epb 22/02/].
[http://dx.doi.org/10.1093/ije/dyr029]

[25] Bruno G, Landi A. Epidemiology and costs of diabetes. Transplant Proc 2011; 43(1): 327-9.
[http://dx.doi.org/10.1016/j.transproceed.2010.09.098] [PMID: 21335215]

[26] Dall TM, Zhang Y, Chen YJ, Quick WW, Yang WG, Fogli J. The economic burden of diabetes. Health Aff (Millwood) 2010; 29(2): 297-303.
[http://dx.doi.org/10.1377/hlthaff.2009.0155] [PMID: 20075080]

[27] Pan C, Shang S, Kirch W, Thoenes M. Burden of diabetes in the adult Chinese population: A systematic literature review and future projections. Int J Gen Med 2010; 3: 173-9.
[PMID: 20689690]

[28] Abegunde DO, Mathers CD, Adam T, Ortegon M, Strong K. The burden and costs of chronic diseases in low-income and middle-income countries. Lancet 2007; 370(9603): 1929-38.
[http://dx.doi.org/10.1016/S0140-6736(07)61696-1] [PMID: 18063029]

[29] Tharkar S, Devarajan A, Kumpatla S, Viswanathan V. The socioeconomics of diabetes from a developing country: a population based cost of illness study. Diabetes Res Clin Pract 2010; 89(3): 334-40.
[http://dx.doi.org/10.1016/j.diabres.2010.05.009] [PMID: 20538363]

[30] Zhang P, Zhang X, Brown J, *et al.* Global healthcare expenditure on diabetes for 2010 and 2030. Diabetes Res Clin Pract 2010; 87(3): 293-301.
[http://dx.doi.org/10.1016/j.diabres.2010.01.026] [PMID: 20171754]

[31] Defronzo RA. Banting Lecture. From the triumvirate to the ominous octet: a new paradigm for the treatment of type 2 diabetes mellitus. Diabetes 2009; 58(4): 773-95.
[http://dx.doi.org/10.2337/db09-9028] [PMID: 19336687]

[32] Mathur S, Zammitt NN, Frier BM. Optimal glycaemic control in elderly people with type 2 diabetes: what does the evidence say? Drug Saf 2015; 38(1): 17-32.
[http://dx.doi.org/10.1007/s40264-014-0247-7] [PMID: 25481812]

[33] Cavalot F. Daily glycaemic variability: benefits of optimal control. Diabetes Obes Metab 2013; 15 (Suppl. 2): 1-2.
[http://dx.doi.org/10.1111/dom.12139] [PMID: 24034512]

[34] Frier BM. Hypoglycaemia in diabetes mellitus: epidemiology and clinical implications. Nat Rev Endocrinol 2014; 10(12): 711-22.
[http://dx.doi.org/10.1038/nrendo.2014.170] [PMID: 25287289]

[35] Kauppila T, Laine MK, Honkasalo M, Raina M, Eriksson JG. Contacting dropouts from type 2 diabetes care in public primary health care: description of the patient population. Scand J Prim Health Care 2016; 34(3): 267-73.
[http://dx.doi.org/10.1080/02813432.2016.1207144] [PMID: 27404014]

[36] Osborn CY, Mayberry LS, Kim JM. Medication adherence may be more important than other behaviours for optimizing glycaemic control among low-income adults. J Clin Pharm Ther 2016; 41(3): 256-9.
[http://dx.doi.org/10.1111/jcpt.12360] [PMID: 26939721]

[37] Ceriello A. The glucose triad and its role in comprehensive glycaemic control: current status, future management. Int J Clin Pract 2010; 64(12): 1705-11.
[http://dx.doi.org/10.1111/j.1742-1241.2010.02517.x] [PMID: 20860758]

[38] Leiter LA, Fitchett D. Optimal care of cardiovascular disease and type 2 diabetes patients: shared responsibilities between the cardiologist and diabetologist. Atheroscler Suppl 2006; 7(1): 37-42.
[http://dx.doi.org/10.1016/j.atherosclerosissup.2006.01.006] [PMID: 16516560]

[39] Lindskog M, Kärvestedt L, Fürst CJ. Glycaemic control in end-of-life care. Curr Opin Support Palliat Care 2014; 8(4): 378-82.
[http://dx.doi.org/10.1097/SPC.0000000000000095] [PMID: 25259543]

[40] White JR Jr. A brief history of the development of diabetes medications. Diabetes Spectr 2014; 27(2): 82-6.
[http://dx.doi.org/10.2337/diaspect.27.2.82] [PMID: 26246763]

[41] Kobayashi M, Yamazaki K, Hirao K, *et al.* Japan Diabetes Clinical Data Management Study Group. The status of diabetes control and antidiabetic drug therapy in Japan--a cross-sectional survey of 17,000 patients with diabetes mellitus (JDDM 1). Diabetes Res Clin Pract 2006; 73(2): 198-204.
[http://dx.doi.org/10.1016/j.diabres.2006.01.013] [PMID: 16621117]

[42] Mafauzy M, Hussein Z, Chan SP. The status of diabetes control in Malaysia: results of DiabCare 2008. Med J Malaysia 2011; 66(3): 175-81.
[PMID: 22111435]

[43] Mohamed M. Diabcare-Asia 2003 Study Group. An audit on diabetes management in Asian patients treated by specialists: the Diabcare-Asia 1998 and 2003 studies. Curr Med Res Opin 2008; 24(2): 507-14.
[http://dx.doi.org/10.1185/030079908X261131] [PMID: 18184454]

[44] Mohan V, Shah SN, Joshi SR, *et al.* DiabCare India 2011 Study Group. Current status of management, control, complications and psychosocial aspects of patients with diabetes in India: Results from the DiabCare India 2011 Study. Indian J Endocrinol Metab 2014; 18(3): 370-8.
[http://dx.doi.org/10.4103/2230-8210.129715] [PMID: 24944934]

[45] Omar MS, Khudada K, Safarini S, Mehanna S, Nafach J. DiabCare survey of diabetes management and complications in the Gulf countries. Indian J Endocrinol Metab 2016; 20(2): 219-27.
[http://dx.doi.org/10.4103/2230-8210.176347] [PMID: 27042419]

[46] Pan C, Yang W, Jia W, Weng J, Tian H. Management of Chinese patients with type 2 diabetes, 1998-2006: the Diabcare-China surveys. Curr Med Res Opin 2009; 25(1): 39-45.
[http://dx.doi.org/10.1185/03007990802586079] [PMID: 19210137]

[47] American Diabetes Association. 7. Approaches to Glycemic Treatment. Diabetes Care 2016; 39 (Suppl. 1): S52-9.
[http://dx.doi.org/10.2337/dc16-S010] [PMID: 26696682]

[48] Clinical Practice Guidelines Development Group. Clinical Practice Guidelines: Management of type 2 diabetes mellitus Malaysia: Malaysian Endocrine and Metabolic Society (MEMS). Malaysia: Ministry of Health 2015.

[49] Canadian Diabetes Association Clinical Practice Guidelines Expert Committee. Pharmacologic management of type 2 diabetes: 2016 interim update. Can J Diabetes 2016; 40(3): 193-5.
[http://dx.doi.org/10.1016/j.jcjd.2016.02.006] [PMID: 27032548]

[50] Gunton JE, Cheung NW, Davis TM, Zoungas S, Colagiuri S. Australian Diabetes Society. A new blood glucose management algorithm for type 2 diabetes: a position statement of the Australian Diabetes Society. Med J Aust 2014; 201(11): 650-3.
[http://dx.doi.org/10.5694/mja14.01187] [PMID: 25495309]

[51] Ferrannini E, Solini A. SGLT2 inhibition in diabetes mellitus: rationale and clinical prospects. Nat Rev Endocrinol 2012; 8(8): 495-502.
[http://dx.doi.org/10.1038/nrendo.2011.243] [PMID: 22310849]

[52] Gerich JE. Role of the kidney in normal glucose homeostasis and in the hyperglycaemia of diabetes mellitus: therapeutic implications. Diabet Med 2010; 27(2): 136-42.
[http://dx.doi.org/10.1111/j.1464-5491.2009.02894.x] [PMID: 20546255]

[53] Mather A, Pollock C. Glucose handling by the kidney. Kidney Int Suppl 2011; (120): S1-6.
[http://dx.doi.org/10.1038/ki.2010.509] [PMID: 21358696]

[54] Wright EM, Loo DD, Hirayama BA. Biology of human sodium glucose transporters. Physiol Rev 2011; 91(2): 733-94.
[http://dx.doi.org/10.1152/physrev.00055.2009] [PMID: 21527736]

[55] Dyer J, Daly K, Salmon KS, *et al.* Intestinal glucose sensing and regulation of intestinal glucose absorption. Biochem Soc Trans 2007; 35(Pt 5): 1191-4.
[http://dx.doi.org/10.1042/BST0351191] [PMID: 17956309]

[56] Jabbour SA, Goldstein BJ. Sodium glucose co-transporter 2 inhibitors: blocking renal tubular reabsorption of glucose to improve glycaemic control in patients with diabetes. Int J Clin Pract 2008; 62(8): 1279-84.
[http://dx.doi.org/10.1111/j.1742-1241.2008.01829.x] [PMID: 18705823]

[57] DeFronzo RA, Davidson JA, Del Prato S. The role of the kidneys in glucose homeostasis: a new path towards normalizing glycaemia. Diabetes Obes Metab 2012; 14(1): 5-14.
[http://dx.doi.org/10.1111/j.1463-1326.2011.01511.x] [PMID: 21955459]

[58] Rahmoune H, Thompson PW, Ward JM, Smith CD, Hong G, Brown J. Glucose transporters in human renal proximal tubular cells isolated from the urine of patients with non-insulin-dependent diabetes. Diabetes 2005; 54(12): 3427-34.
[http://dx.doi.org/10.2337/diabetes.54.12.3427] [PMID: 16306358]

[59] Lajara R. The potential role of sodium glucose co-transporter 2 inhibitors in combination therapy for type 2 diabetes mellitus. Expert Opin Pharmacother 2014; 15(17): 2565-85.
[http://dx.doi.org/10.1517/14656566.2014.968551] [PMID: 25316597]

[60] Abdul-Ghani MA, Norton L, Defronzo RA. Role of sodium-glucose cotransporter 2 (SGLT 2) inhibitors in the treatment of type 2 diabetes. Endocr Rev 2011; 32(4): 515-31.
[http://dx.doi.org/10.1210/er.2010-0029] [PMID: 21606218]

[61] European Medicine Agency. SGLT2 inhibitors: information on potential risk of toe amputation to be included in prescribing information 2017 2017. http://www.ema.europa.eu/ema/index.jsp?curl=pages/medicines/human/referrals/SGLT2_inhibitors_(previously_Canagliflozin)/human_referral_prac_000059.jsp&mid=WC0b01ac05805c516f

[62] U S Food and Drug Administration. Dapagliflozin. Label and approval history 2016 2016. http://www.accessdata.fda.gov/scripts/cder/drugsatfda/index.cfm?fuseaction=Search.Label_ApprovalHistory#apphist

[63] U S Food and Drug Administration. Canagliflozin. Label and approval history 2016 2016. http://www.accessdata.fda.gov/scripts/cder/drugsatfda/index.cfm?fuseaction=Search.DrugDetails

[64] Zinman B, Wanner C, Lachin JM, et al. EMPA-REG OUTCOME Investigators. Empagliflozin, cardiovascular outcomes, and mortality in type 2 diabetes. N Engl J Med 2015; 373(22): 2117-28.
[http://dx.doi.org/10.1056/NEJMoa1504720] [PMID: 26378978]

[65] Asano K, Tanaka A, Sato T, Uyama Y. Regulatory challenges in the review of data from global clinical trials: the PMDA perspective. Clin Pharmacol Ther 2013; 94(2): 195-8.
[http://dx.doi.org/10.1038/clpt.2013.106] [PMID: 23872835]

[66] Yang XP, Lai D, Zhong XY, Shen HP, Huang YL. Efficacy and safety of canagliflozin in subjects with type 2 diabetes: systematic review and meta-analysis. Eur J Clin Pharmacol 2014; 70(10): 1149-58.
[http://dx.doi.org/10.1007/s00228-014-1730-x] [PMID: 25124541]

[67] Zhang M, Zhang L, Wu B, Song H, An Z, Li S. Dapagliflozin treatment for type 2 diabetes: a systematic review and meta-analysis of randomized controlled trials. Diabetes Metab Res Rev 2014; 30(3): 204-21.
[http://dx.doi.org/10.1002/dmrr.2479] [PMID: 24115369]

[68] Liakos A, Karagiannis T, Athanasiadou E, et al. Efficacy and safety of empagliflozin for type 2 diabetes: a systematic review and meta-analysis. Diabetes Obes Metab 2014; 16(10): 984-93.
[http://dx.doi.org/10.1111/dom.12307] [PMID: 24766495]

[69] Fonseca VA, Ferrannini E, Wilding JP, et al. Active- and placebo-controlled dose-finding study to assess the efficacy, safety, and tolerability of multiple doses of ipragliflozin in patients with type 2 diabetes mellitus. J Diabetes Complications 2013; 27(3): 268-73.
[http://dx.doi.org/10.1016/j.jdiacomp.2012.11.005] [PMID: 23276620]

[70] Seino Y, Sasaki T, Fukatsu A, Sakai S, Samukawa Y. Efficacy and safety of luseogliflozin

monotherapy in Japanese patients with type 2 diabetes mellitus: a 12-week, randomized, placebo-controlled, phase II study. Curr Med Res Opin 2014; 30(7): 1219-30.
[http://dx.doi.org/10.1185/03007995.2014.901943] [PMID: 24597840]

[71] Rosenwasser RF, Rosenwasser JN, Sutton D, Choksi R, Epstein B. Tofogliflozin: a highly selective SGLT2 inhibitor for the treatment of type 2 diabetes. Drugs Today (Barc) 2014; 50(11): 739-45.
[http://dx.doi.org/10.1358/dot.2014.50.11.2232267] [PMID: 25525634]

[72] Yacoub T. Dapagliflozin combination therapy in type 2 diabetes mellitus. Postgrad Med 2016; 128(1): 124-36.
[http://dx.doi.org/10.1080/00325481.2016.1111119] [PMID: 26571022]

[73] Kashiwagi A, Kazuta K, Goto K, Yoshida S, Ueyama E, Utsuno A. Ipragliflozin in combination with metformin for the treatment of Japanese patients with type 2 diabetes: ILLUMINATE, a randomized, double-blind, placebo-controlled study. Diabetes Obes Metab 2015; 17(3): 304-8.
[http://dx.doi.org/10.1111/dom.12331] [PMID: 24919820]

[74] Kashiwagi A, Akiyama N, Shiga T, *et al.* Efficacy and safety of ipragliflozin as an add-on to a sulfonylurea in Japanese patients with inadequately controlled type 2 diabetes: results of the randomized, placebo-controlled, double-blind, phase III EMIT study. Diabetology International 2015; 6(2): 125-38.
[http://dx.doi.org/10.1007/s13340-014-0184-9]

[75] Cefalu WT, Leiter LA, Yoon KH, *et al.* Efficacy and safety of canagliflozin versus glimepiride in patients with type 2 diabetes inadequately controlled with metformin (CANTATA-SU): 52 week results from a randomised, double-blind, phase 3 non-inferiority trial. Lancet 2013; 382(9896): 941--.
[http://dx.doi.org/10.1016/S0140-6736(13)60683-2] [PMID: 23850055]

[76] Leiter LA, Yoon KH, Arias P, *et al.* Canagliflozin provides durable glycemic improvements and body weight reduction over 104 weeks versus glimepiride in patients with type 2 diabetes on metformin: a randomized, double-blind, phase 3 study. Diabetes Care 2015; 38(3): 355-64.
[http://dx.doi.org/10.2337/dc13-2762] [PMID: 25205142]

[77] Nauck MA, Del Prato S, Durán-García S, *et al.* Durability of glycaemic efficacy over 2 years with dapagliflozin versus glipizide as add-on therapies in patients whose type 2 diabetes mellitus is inadequately controlled with metformin. Diabetes Obes Metab 2014; 16(11): 1111-20.
[http://dx.doi.org/10.1111/dom.12327] [PMID: 24919526]

[78] Nauck MA, Del Prato S, Meier JJ, *et al.* Dapagliflozin versus glipizide as add-on therapy in patients with type 2 diabetes who have inadequate glycemic control with metformin: a randomized, 52-week, double-blind, active-controlled noninferiority trial. Diabetes Care 2011; 34(9): 2015-22.
[http://dx.doi.org/10.2337/dc11-0606] [PMID: 21816980]

[79] Schernthaner G, Gross JL, Rosenstock J, *et al.* Canagliflozin compared with sitagliptin for patients with type 2 diabetes who do not have adequate glycemic control with metformin plus sulfonylurea: a 52-week randomized trial. Diabetes Care 2013; 36(9): 2508-15. [Erratum appears in Diabetes Care. 2013 Dec;36(12):4172].
[http://dx.doi.org/10.2337/dc12-2491] [PMID: 23564919]

[80] Mathieu C, Ranetti AE, Li D, *et al.* Randomized, double-blind, phase 3 trial of triple therapy with dapagliflozin add-on to saxagliptin plus metformin in type 2 diabetes. Diabetes Care 2015; 38(11): 2009-17.
[http://dx.doi.org/10.2337/dc15-0779] [PMID: 26246458]

[81] Lewin A, DeFronzo RA, Patel S, *et al.* Initial combination of empagliflozin and linagliptin in subjects with type 2 diabetes. Diabetes Care 2015; 38(3): 394-402. [Erratum appears in Diabetes Care. 2015 Jun;38(6):1173; PMID: 25998299].
[http://dx.doi.org/10.2337/dc14-2365] [PMID: 25633662]

[82] Rosenstock J, Seman LJ, Jelaska A, *et al.* Efficacy and safety of empagliflozin, a sodium glucose cotransporter 2 (SGLT2) inhibitor, as add-on to metformin in type 2 diabetes with mild

hyperglycaemia. Diabetes Obes Metab 2013; 15(12): 1154-60.
[http://dx.doi.org/10.1111/dom.12185] [PMID: 23906374]

[83] List JF, Woo V, Morales E, Tang W, Fiedorek FT. Sodium-glucose cotransport inhibition with dapagliflozin in type 2 diabetes. Diabetes Care 2009; 32(4): 650-7.
[http://dx.doi.org/10.2337/dc08-1863] [PMID: 19114612]

[84] Araki E, Tanizawa Y, Tanaka Y, et al. Long-term treatment with empagliflozin as add-on to oral antidiabetes therapy in Japanese patients with type 2 diabetes mellitus. Diabetes Obes Metab 2015; 17 (7): 665-74.
[http://dx.doi.org/10.1111/dom.12464] [PMID: 25772548]

[85] Stenlöf K, Cefalu WT, Kim KA, et al. Efficacy and safety of canagliflozin monotherapy in subjects with type 2 diabetes mellitus inadequately controlled with diet and exercise. Diabetes Obes Metab 2013; 15(4): 372-82.
[http://dx.doi.org/10.1111/dom.12054] [PMID: 23279307]

[86] Ferrannini E, Ramos SJ, Salsali A, Tang W, List JF. Dapagliflozin monotherapy in type 2 diabetic patients with inadequate glycemic control by diet and exercise: a randomized, double-blind, placebo-controlled, phase 3 trial. Diabetes Care 2010; 33(10): 2217-24.
[http://dx.doi.org/10.2337/dc10-0612] [PMID: 20566676]

[87] Roden M, Weng J, Eilbracht J, et al. EMPA-REG MONO trial investigators. Empagliflozin monotherapy with sitagliptin as an active comparator in patients with type 2 diabetes: a randomised, double-blind, placebo-controlled, phase 3 trial. Lancet Diabetes Endocrinol 2013; 1(3): 208-19.
[http://dx.doi.org/10.1016/S2213-8587(13)70084-6] [PMID: 24622369]

[88] Yang T, Lu M, Ma L, Zhou Y, Cui Y. Efficacy and tolerability of canagliflozin as add-on to metformin in the treatment of type 2 diabetes mellitus: a meta-analysis. Eur J Clin Pharmacol 2015; 71(11): 1325-32.
[http://dx.doi.org/10.1007/s00228-015-1923-y] [PMID: 26282731]

[89] Fulcher G, Matthews DR, Perkovic V, et al. Efficacy and safety of canagliflozin used in conjunction with sulfonylurea in patients with type 2 diabetes mellitus: A randomized, controlled trial. Diabetes Ther 2015; 6(3): 289-302.
[http://dx.doi.org/10.1007/s13300-015-0117-z] [PMID: 26081793]

[90] Fulcher G, Matthews DR, Perkovic V, et al. CANVAS trial collaborative group. Efficacy and safety of canagliflozin when used in conjunction with incretin-mimetic therapy in patients with type 2 diabetes. Diabetes Obes Metab 2016; 18(1): 82-91.
[http://dx.doi.org/10.1111/dom.12589] [PMID: 26450639]

[91] Neal B, Perkovic V, de Zeeuw D, et al. CANVAS Trial Collaborative Group. Efficacy and safety of canagliflozin, an inhibitor of sodium-glucose cotransporter 2, when used in conjunction with insulin therapy in patients with type 2 diabetes. Diabetes Care 2015; 38(3): 403-11.
[http://dx.doi.org/10.2337/dc14-1237] [PMID: 25468945]

[92] Polidori D, Sha S, Mudaliar S, et al. Canagliflozin lowers postprandial glucose and insulin by delaying intestinal glucose absorption in addition to increasing urinary glucose excretion: results of a randomized, placebo-controlled study. Diabetes Care 2013; 36(8): 2154-61.
[http://dx.doi.org/10.2337/dc12-2391] [PMID: 23412078]

[93] Rodbard HW, Seufert J, Aggarwal N, et al. Efficacy and safety of titrated canagliflozin in patients with type 2 diabetes mellitus inadequately controlled on metformin and sitagliptin. Diabetes Obes Metab 2016; 18(8): 812-9.
[http://dx.doi.org/10.1111/dom.12684] [PMID: 27160639]

[94] Wilding JP, Charpentier G, Hollander P, et al. Efficacy and safety of canagliflozin in patients with type 2 diabetes mellitus inadequately controlled with metformin and sulphonylurea: a randomised trial. Int J Clin Pract 2013; 67(12): 1267-82.
[http://dx.doi.org/10.1111/ijcp.12322] [PMID: 24118688]

[95] Bailey CJ, Gross JL, Pieters A, Bastien A, List JF. Effect of dapagliflozin in patients with type 2 diabetes who have inadequate glycaemic control with metformin: a randomised, double-blind, placebo-controlled trial. Lancet 2010; 375(9733): 2223-33.
[http://dx.doi.org/10.1016/S0140-6736(10)60407-2] [PMID: 20609968]

[96] Häring HU, Merker L, Seewaldt-Becker E, *et al.* EMPA-REG MET Trial Investigators. Empagliflozin as add-on to metformin in patients with type 2 diabetes: a 24-week, randomized, double-blind, placebo-controlled trial. Diabetes Care 2014; 37(6): 1650-9.
[http://dx.doi.org/10.2337/dc13-2105] [PMID: 24722494]

[97] Lavalle-González FJ, Januszewicz A, Davidson J, *et al.* Efficacy and safety of canagliflozin compared with placebo and sitagliptin in patients with type 2 diabetes on background metformin monotherapy: a randomised trial. Diabetologia 2013; 56(12): 2582-92.
[http://dx.doi.org/10.1007/s00125-013-3039-1] [PMID: 24026211]

[98] Ridderstråle M, Andersen KR, Zeller C, Kim G, Woerle HJ, Broedl UC. EMPA-REG H2H-SU trial investigators. Comparison of empagliflozin and glimepiride as add-on to metformin in patients with type 2 diabetes: a 104-week randomised, active-controlled, double-blind, phase 3 trial. Lancet Diabetes Endocrinol 2014; 2(9): 691-700. [Erratum appears in Lancet Diabetes Endocrinol. 2015 Sept;3(9):e7].
[http://dx.doi.org/10.1016/S2213-8587(14)70120-2] [PMID: 24948511]

[99] Qiu R, Capuano G, Meininger G. Efficacy and safety of twice-daily treatment with canagliflozin, a sodium glucose co-transporter 2 inhibitor, added on to metformin monotherapy in patients with type 2 diabetes mellitus. J Clin Transl Endocrinol 2014; 1(2): 54-60.
[http://dx.doi.org/10.1016/j.jcte.2014.04.001]

[100] Rosenstock J, Aggarwal N, Polidori D, *et al.* Canagliflozin DIA 2001 Study Group. Dose-ranging effects of canagliflozin, a sodium-glucose cotransporter 2 inhibitor, as add-on to metformin in subjects with type 2 diabetes. Diabetes Care 2012; 35(6): 1232-8.
[http://dx.doi.org/10.2337/dc11-1926] [PMID: 22492586]

[101] Rosenthal N, Meininger G, Ways K, *et al.* Canagliflozin: a sodium glucose co-transporter 2 inhibitor for the treatment of type 2 diabetes mellitus. Ann N Y Acad Sci 2015; 1358: 28-43.
[http://dx.doi.org/10.1111/nyas.12852] [PMID: 26305874]

[102] Hinnen D. Glucuretic effects and renal safety of dapagliflozin in patients with type 2 diabetes. Ther Adv Endocrinol Metab 2015; 6(3): 92-102.
[http://dx.doi.org/10.1177/2042018815575273] [PMID: 26137213]

[103] Zhong X, Lai D, Ye Y, Yang X, Yu B, Huang Y. Efficacy and safety of empagliflozin as add-on to metformin for type 2 diabetes: a systematic review and meta-analysis. Eur J Clin Pharmacol 2016; 72 (6): 655-63.
[http://dx.doi.org/10.1007/s00228-016-2010-8] [PMID: 26832915]

[104] Kashiwagi A, Shiga T, Akiyama N, *et al.* Efficacy and safety of ipragliflozin as an add-on to pioglitazone in Japanese patients with inadequately controlled type 2 diabetes: a randomized, double-blind, placebo-controlled study (the SPOTLIGHT study). Diabetology International 2015; 6(2): 104-16.
[http://dx.doi.org/10.1007/s13340-014-0182-y]

[105] Kadowaki T, Haneda M, Inagaki N, *et al.* Efficacy and safety of empagliflozin monotherapy for 52 weeks in Japanese patients with type 2 diabetes: a randomized, double-blind, parallel-group study. Adv Ther 2015; 32(4): 306-18.
[http://dx.doi.org/10.1007/s12325-015-0198-0] [PMID: 25845768]

[106] Bolinder J, Ljunggren Ö, Johansson L, *et al.* Dapagliflozin maintains glycaemic control while reducing weight and body fat mass over 2 years in patients with type 2 diabetes mellitus inadequately controlled on metformin. Diabetes Obes Metab 2014; 16(2): 159-69.
[http://dx.doi.org/10.1111/dom.12189] [PMID: 23906445]

[107] Chilton R, Tikkanen I, Cannon CP, et al. 4B.02: The sodium glucose cotransporter 2 inhibitor empagliflozin reduces blood pressure and markers of arterial stiffness and vascular resistance in type 2 diabetes. J Hypertens 2015; 33 (Suppl. 1): e53.
[PMID: 26102848]

[108] Chilton R, Tikkanen I, Cannon CP, et al. Effects of empagliflozin on blood pressure and markers of arterial stiffness and vascular resistance in patients with type 2 diabetes. Diabetes Obes Metab 2015; 17(12): 1180-93.
[http://dx.doi.org/10.1111/dom.12572] [PMID: 26343814]

[109] Ghosh RK, Bandyopadhyay D, Hajra A, Biswas M, Gupta A. Cardiovascular outcomes of sodium-glucose cotransporter 2 inhibitors: A comprehensive review of clinical and preclinical studies. Int J Cardiol 2016; 212: 29-36.
[http://dx.doi.org/10.1016/j.ijcard.2016.02.134] [PMID: 27017118]

[110] Neal B, Perkovic V, de Zeeuw D, et al. Rationale, design, and baseline characteristics of the Canagliflozin Cardiovascular Assessment Study (CANVAS)--a randomized placebo-controlled trial. Am Heart J 2013; 166(2): 217-223.e11.
[http://dx.doi.org/10.1016/j.ahj.2013.05.007] [PMID: 23895803]

[111] Sonesson C, Johansson PA, Johnsson E, Gause-Nilsson I. Cardiovascular effects of dapagliflozin in patients with type 2 diabetes and different risk categories: a meta-analysis. Cardiovasc Diabetol 2016; 15: 37.
[http://dx.doi.org/10.1186/s12933-016-0356-y] [PMID: 26895767]

[112] NCT01730534. Multicenter trial to evaluate the effect of dapagliflozin on the incidence of cardiovascular events (DECLARE-TIMI58): U.S. National Institutes of Health; 2012. Available from: https://clinicaltrials.gov/ct2/show/NCT01730534

[113] Erondu N, Desai M, Ways K, Meininger G. Diabetic ketoacidosis and related events in the canagliflozin type 2 diabetes clinical program. Diabetes Care 2015; 38(9): 1680-6.
[http://dx.doi.org/10.2337/dc15-1251] [PMID: 26203064]

[114] Scheen AJ. SGLT2 inhibition: efficacy and safety in type 2 diabetes treatment. Expert Opin Drug Saf 2015; 14(12): 1879-904.
[http://dx.doi.org/10.1517/14740338.2015.1100167] [PMID: 26513131]

[115] Kohler S, Salsali A, Hantel S, et al. Safety and tolerability of empagliflozin in patients with type 2 diabetes. Clin Ther 2016; 38(6): 1299-313.
[http://dx.doi.org/10.1016/j.clinthera.2016.03.031] [PMID: 27085585]

[116] Ptaszynska A, Johnsson KM, Parikh SJ, de Bruin TW, Apanovitch AM, List JF. Safety profile of dapagliflozin for type 2 diabetes: pooled analysis of clinical studies for overall safety and rare events. Drug Saf 2014; 37(10): 815-29.
[http://dx.doi.org/10.1007/s40264-014-0213-4] [PMID: 25096959]

[117] Weir MR, Januszewicz A, Gilbert RE, et al. Effect of canagliflozin on blood pressure and adverse events related to osmotic diuresis and reduced intravascular volume in patients with type 2 diabetes mellitus. J Clin Hypertens (Greenwich) 2014; 16(12): 875-82.
[http://dx.doi.org/10.1111/jch.12425] [PMID: 25329038]

[118] Perkovic V, Jardine M, Vijapurkar U, Meininger G. Renal effects of canagliflozin in type 2 diabetes mellitus. Curr Med Res Opin 2015; 31(12): 2219-31.
[http://dx.doi.org/10.1185/03007995.2015.1092128] [PMID: 26494163]

[119] Alba M, Xie J, Fung A, Desai M. The effects of canagliflozin, a sodium glucose co-transporter 2 inhibitor, on mineral metabolism and bone in patients with type 2 diabetes mellitus. Curr Med Res Opin 2016; 32(8): 1375-85.
[http://dx.doi.org/10.1080/03007995.2016.1174841] [PMID: 27046479]

[120] Watts NB, Bilezikian JP, Usiskin K, et al. Effects of canagliflozin on fracture risk in patients with type

2 diabetes mellitus. J Clin Endocrinol Metab 2016; 101(1): 157-66.
[http://dx.doi.org/10.1210/jc.2015-3167] [PMID: 26580237]

[121] Hedrington MS, Davis SN. Ipragliflozin, a sodium-glucose cotransporter 2 inhibitor, in the treatment of type 2 diabetes. Expert Opin Drug Metab Toxicol 2015; 11(4): 613-23.
[http://dx.doi.org/10.1517/17425255.2015.1009893] [PMID: 25643044]

[122] Seino Y. Luseogliflozin for the treatment of type 2 diabetes. Expert Opin Pharmacother 2014; 15(18): 2741-9.
[http://dx.doi.org/10.1517/14656566.2014.978290] [PMID: 25359155]

[123] Ishihara H, Yamaguchi S, Nakao I, Okitsu A, Asahina S. Efficacy and safety of ipragliflozin as add-on therapy to insulin in Japanese patients with type 2 diabetes mellitus (IOLITE): a multi-centre, randomized, placebo-controlled, double-blind study. Diabetes Obes Metab 2016; 18(12): 1207-16.
[http://dx.doi.org/10.1111/dom.12745] [PMID: 27436788]

[124] Kashiwagi A, Kawano H, Kazuta K, Utsuno A, Yoshida S. Long-term safety, tolerability and efficacy of ipragliflozin in Japanese patients with type 2 diabetes mellitus: -IGNITE study-. Japanese Pharmacology and Therapeutics 2015; 43(1): 85-100. [Japanese].

[125] Kashiwagi A, Kazuta K, Yoshida S, Nagase I. Randomized, placebo-controlled, double-blind glycemic control trial of novel sodium-dependent glucose cotransporter 2 inhibitor ipragliflozin in Japanese patients with type 2 diabetes mellitus. J Diabetes Investig 2014; 5(4): 382-91.
[http://dx.doi.org/10.1111/jdi.12156] [PMID: 25411597]

[126] Kashiwagi A, Takahashi H, Ishikawa H, et al. A randomized, double-blind, placebo-controlled study on long-term efficacy and safety of ipragliflozin treatment in patients with type 2 diabetes mellitus and renal impairment: results of the long-term ASP1941 safety evaluation in patients with type 2 diabetes with renal impairment (LANTERN) study. Diabetes Obes Metab 2015; 17(2): 152-60.
[http://dx.doi.org/10.1111/dom.12403] [PMID: 25347938]

[127] Kashiwagi A, Yoshida S, Nakamura I, et al. Efficacy and safety of ipragliflozin in Japanese patients with type 2 diabetes stratified by body mass index: A subgroup analysis of five randomized clinical trials. J Diabetes Investig 2016; 7(4): 544-54.
[http://dx.doi.org/10.1111/jdi.12471] [PMID: 27181576]

[128] Haneda M, Seino Y, Inagaki N, et al. Influence of renal function on the 52-week efficacy and safety of the sodium glucose cotransporter 2 inhibitor luseogliflozin in Japanese patients with type 2 diabetes mellitus. Clin Ther 2016; 38(1): 88e66-88e120.

[129] Jinnouchi H, Nozaki K, Watase H, Omiya H, Sakai S, Samukawa Y. Impact of Reduced Renal Function on the Glucose-Lowering Effects of Luseogliflozin, a Selective SGLT2 Inhibitor, Assessed by Continuous Glucose Monitoring in Japanese Patients with Type 2 Diabetes Mellitus. Adv Ther 2016; 33(3): 460-79.
[http://dx.doi.org/10.1007/s12325-016-0291-z] [PMID: 26846284]

[130] Yokote K, Terauchi Y, Nakamura I, Sugamori H. Real-world evidence for the safety of ipragliflozin in elderly Japanese patients with type 2 diabetes mellitus (STELLA-ELDER): final results of a post-marketing surveillance study. Expert Opin Pharmacother 2016; 17(15): 1995-2003.
[http://dx.doi.org/10.1080/14656566.2016.1219341] [PMID: 27477242]

[131] Food and Drug Administration. Guidance for Industry: Diabetes mellitus — evaluating cardiovascular risk in new antidiabetic therapies to treat type 2 diabetes. US Department of Health and Human Services, editor USA: Food and Drug Administration, Center for Drug Evaluation and Research (CDER). 2008.

[132] Burki TK. FDA rejects novel diabetes drug over safety fears. Lancet 2012; 379(9815): 507.
[http://dx.doi.org/10.1016/S0140-6736(12)60216-5] [PMID: 22334883]

[133] U S Food and Drug Administration. Drugs@FDA. Canagliflozin USA: Food and Drug Administration 2016. http://www.accessdata.fda.gov/scripts/cder/drugsatfda/index.cfm?fuseaction=Search.Drug Details

[134] Paulsell D, Thomas J, Monahan S, Seftor NS. A trusted source of information: How systematic reviews can support user decisions about adopting evidence-based programs. Eval Rev 2016; 0193841X16665963.
[PMID: 27590676]

[135] Yale JF, Bakris G, Cariou B, *et al.* DIA3004 Study Group. Efficacy and safety of canagliflozin over 52 weeks in patients with type 2 diabetes mellitus and chronic kidney disease. Diabetes Obes Metab 2014; 16(10): 1016-27.
[http://dx.doi.org/10.1111/dom.12348] [PMID: 24965700]

[136] Pharmaceutical and Devices Agency of Japan. Ipragliflozin. Report on the deliberation results Japan: Pharmaceuticals and Medical Devices Agency 2013. http://www.pmda.go.jp/files/000206796.pdf

[137] Pharmaceutical and Devices Agency of Japan. Revision of precautions Canagliflozin hydrate, dapagliflozin propylene glycolate hydrate, and empagliflozin Japan Pharmaceuticals and Medical Devices Agency 2015. http://www.pmda.go.jp/files/000207400.pdf

[138] Pharmaceutical and Devices Agency of Japan. Revision of precautions Ipragliflozin l-proline, luseogliflozin hydrate, and tofogliflozin hydrate Japan: Pharmaceuticals and Medical Devices Agency 2015. http://www.pmda.go.jp/files/000207399.pdf

[139] Sackett DL, Rosenberg WM, Gray JA, Haynes RB, Richardson WS. Evidence based medicine: what it is and what it isn't. BMJ 1996; 312(7023): 71-2.
[http://dx.doi.org/10.1136/bmj.312.7023.71] [PMID: 8555924]

[140] Grant RW, Kirkman MS. Trends in the evidence level for the American Diabetes Association's "Standards of Medical Care in Diabetes" from 2005 to 2014. Diabetes Care 2015; 38(1): 6-8.
[http://dx.doi.org/10.2337/dc14-2142] [PMID: 25538309]

[141] Fernandez A, Sturmberg J, Lukersmith S, *et al.* Evidence-based medicine: is it a bridge too far? Health Res Policy Syst 2015; 13(1): 66.
[http://dx.doi.org/10.1186/s12961-015-0057-0] [PMID: 26546273]

[142] Kulkarni AV. The challenges of evidence-based medicine: a philosophical perspective. Med Health Care Philos 2005; 8(2): 255-60.
[http://dx.doi.org/10.1007/s11019-004-7345-8] [PMID: 16215804]

[143] Haldar S, Chia SC, Henry CJ. Body composition in Asians and Caucasians: Comparative analyses and influences on cardiometabolic outcomes. Adv Food Nutr Res 2015; 75: 97-154.
[http://dx.doi.org/10.1016/bs.afnr.2015.07.001] [PMID: 26319906]

[144] Wang R, Lagakos SW, Ware JH, Hunter DJ, Drazen JM. Statistics in medicine--reporting of subgroup analyses in clinical trials. N Engl J Med 2007; 357(21): 2189-94.
[http://dx.doi.org/10.1056/NEJMsr077003] [PMID: 18032770]

[145] Abdul-Ghani M, Del Prato S, Chilton R, DeFronzo RA. SGLT2 inhibitors and cardiovascular risk: Lessons learned from the EMPA-REG OUTCOME Study. Diabetes Care 2016; 39(5): 717-25.
[http://dx.doi.org/10.2337/dc16-0041] [PMID: 27208375]

[146] Davidson JA, Aguilar R, Lavalle González FJ, *et al.* Efficacy and safety of canagliflozin in type 2 diabetes patients of different ethnicity. Ethn Dis 2016; 26(2): 221-8.
[http://dx.doi.org/10.18865/ed.26.2.221] [PMID: 27103773]

[147] Wan Seman WJ, Kori N, Rajoo S, *et al.* Switching from sulphonylurea to a sodium-glucose cotransporter2 inhibitor in the fasting month of Ramadan is associated with a reduction in hypoglycaemia. Diabetes Obes Metab 2016; 18(6): 628-32.
[http://dx.doi.org/10.1111/dom.12649] [PMID: 26889911]

[148] Niazi AK, Kalra S. Patient centered care in Islam: distinguishing between religious and sociocultural factors. J Diabetes Metab Disord 2013; 12(1): 30.
[http://dx.doi.org/10.1186/2251-6581-12-30] [PMID: 23800351]

[149] Salti I, Bénard E, Detournay B, *et al.* EPIDIAR study group. A population-based study of diabetes and its characteristics during the fasting month of Ramadan in 13 countries: results of the epidemiology of diabetes and Ramadan 1422/2001 (EPIDIAR) study. Diabetes Care 2004; 27(10): 2306-11.
[http://dx.doi.org/10.2337/diacare.27.10.2306] [PMID: 15451892]

[150] Rashid O, Farooq S, Kiran Z, Islam N. Euglycaemic diabetic ketoacidosis in a patient with type 2 diabetes started on empagliflozin. BMJ Case Rep 2016; 2016
[http://dx.doi.org/10.1136/bcr-2016-215340]

[151] Rodríguez-Gutiérrez R, Montori VM. Glycemic control for patients with type 2 diabetes mellitus: Our evolving faith in the face of evidence. Circ Cardiovasc Qual Outcomes 2016; 9(5): 504-12.
[http://dx.doi.org/10.1161/CIRCOUTCOMES.116.002901] [PMID: 27553599]

[152] Jadad AR, To MJ, Emara M, Jones J. Consideration of multiple chronic diseases in randomized controlled trials. JAMA 2011; 306(24): 2670-2.
[http://dx.doi.org/10.1001/jama.2011.1886] [PMID: 22203536]

[153] Van Spall HG, Toren A, Kiss A, Fowler RA. Eligibility criteria of randomized controlled trials published in high-impact general medical journals: a systematic sampling review. JAMA 2007; 297(11): 1233-40.
[http://dx.doi.org/10.1001/jama.297.11.1233] [PMID: 17374817]

[154] Kim SY, Miller FG. Varieties of standard-of-care treatment randomized trials: ethical implications. JAMA 2015; 313(9): 895-6.
[http://dx.doi.org/10.1001/jama.2014.18528] [PMID: 25591061]

[155] Kocher R, Roberts B. The calculus of cures. N Engl J Med 2014; 370(16): 1473-5.
[http://dx.doi.org/10.1056/NEJMp1400868] [PMID: 24571723]

[156] Wu JH, Foote C, Blomster J, *et al.* Effects of sodium-glucose cotransporter-2 inhibitors on cardiovascular events, death, and major safety outcomes in adults with type 2 diabetes: a systematic review and meta-analysis. Lancet Diabetes Endocrinol 2016; 4(5): 411-9.
[http://dx.doi.org/10.1016/S2213-8587(16)00052-8] [PMID: 27009625]

[157] Rosenstock J, Hansen L, Zee P, *et al.* Dual add-on therapy in type 2 diabetes poorly controlled with metformin monotherapy: a randomized double-blind trial of saxagliptin plus dapagliflozin addition versus single addition of saxagliptin or dapagliflozin to metformin. Diabetes Care 2015; 38(3): 376-83.
[http://dx.doi.org/10.2337/dc14-1142] [PMID: 25352655]

[158] Abdul-Ghani MA, Norton L, DeFronzo RA. Renal sodium-glucose cotransporter inhibition in the management of type 2 diabetes mellitus. Am J Physiol Renal Physiol 2015; 309(11): F889-900.
[PMID: 26354881]

[159] Kristman V, Manno M, Côté P. Loss to follow-up in cohort studies: how much is too much? Eur J Epidemiol 2004; 19(8): 751-60.
[http://dx.doi.org/10.1023/B:EJEP.0000036568.02655.f8] [PMID: 15469032]

[160] O'Neill RT, Temple R. The prevention and treatment of missing data in clinical trials: an FDA perspective on the importance of dealing with it. Clin Pharmacol Ther 2012; 91(3): 550-4.
[http://dx.doi.org/10.1038/clpt.2011.340] [PMID: 22318615]

[161] Altman DG. Missing outcomes in randomized trials: addressing the dilemma. Open Med 2009; 3(2): e51-3.
[PMID: 19946393]

[162] Every-Palmer S, Howick J. How evidence-based medicine is failing due to biased trials and selective publication. J Eval Clin Pract 2014; 20(6): 908-14.
[http://dx.doi.org/10.1111/jep.12147] [PMID: 24819404]

[163] Wyer P, da Silva SA. 'One mission accomplished, more important ones remain': commentary on Every-Palmer, S., Howick, J. (2014) How evidence-based medicine is failing due to biased trials and selective publication. Journal of Evaluation in Clinical Practice, 20 (6), 908-914. J Eval Clin Pract

2015; 21(3): 518-28.
[http://dx.doi.org/10.1111/jep.12330] [PMID: 25720797]

[164] Lundh A, Sismondo S, Lexchin J, Busuioc OA, Bero L. Industry sponsorship and research outcome. Cochrane Database Syst Rev 2012; 12: MR000033.
[PMID: 23235689]

[165] Turner EH, Matthews AM, Linardatos E, Tell RA, Rosenthal R. Selective publication of antidepressant trials and its influence on apparent efficacy. N Engl J Med 2008; 358(3): 252-60.
[http://dx.doi.org/10.1056/NEJMsa065779] [PMID: 18199864]

[166] Dijkers M. Four types of evidence abuse KT Update 2016; 4(3) http://ktdrr.org/products/update/v4n3/index.html

[167] McGauran N, Wieseler B, Kreis J, Schüler Y-B, Kölsch H, Kaiser T. Reporting bias in medical research - a narrative review. Trials 2010; 11(1): 37.
[http://dx.doi.org/10.1186/1745-6215-11-37] [PMID: 20388211]

[168] Mittman BS, Øvretveit J, Plsek P, Salem-Schatz S. How to think about evidence when deciding whether to adopt an innovation: Agency for Health Research and Quality Health Care. Innovations Exchange 2013.

[169] Lublóy Á. Factors affecting the uptake of new medicines: a systematic literature review. BMC Health Serv Res 2014; 14: 469.
[http://dx.doi.org/10.1186/1472-6963-14-469] [PMID: 25331607]

[170] Grimshaw JM, Eccles MP, Lavis JN, Hill SJ, Squires JE. Knowledge translation of research findings. Implement Sci 2012; 7(1): 50.
[http://dx.doi.org/10.1186/1748-5908-7-50] [PMID: 22651257]

[171] O'Rourke K. An historical perspective on meta-analysis: dealing quantitatively with varying study results. J R Soc Med 2007; 100(12): 579-82.
[http://dx.doi.org/10.1177/0141076807100012020] [PMID: 18065712]

[172] Wiley. Cochrane Library. 2017. http://as.wiley.com/WileyCDA/Brand/id-6.html?category=For+Working

[173] Olivo SA, Macedo LG, Gadotti IC, Fuentes J, Stanton T, Magee DJ. Scales to assess the quality of randomized controlled trials: a systematic review. Phys Ther 2008; 88(2): 156-75.
[http://dx.doi.org/10.2522/ptj.20070147] [PMID: 18073267]

[174] Higgins JP, Altman DG, Gøtzsche PC, et al. Cochrane Bias Methods Group; Cochrane Statistical Methods Group. The Cochrane Collaboration's tool for assessing risk of bias in randomised trials. BMJ 2011; 343: d5928.
[http://dx.doi.org/10.1136/bmj.d5928] [PMID: 22008217]

[175] Goldacre B. Make journals report clinical trials properly. Nature 2016; 530(7588): 7.
[http://dx.doi.org/10.1038/530007a] [PMID: 26842021]

[176] Sterne JA, Gavaghan D, Egger M. Publication and related bias in meta-analysis: power of statistical tests and prevalence in the literature. J Clin Epidemiol 2000; 53(11): 1119-29.
[http://dx.doi.org/10.1016/S0895-4356(00)00242-0] [PMID: 11106885]

[177] Sterne JA, Egger M, Smith GD. Systematic reviews in health care: Investigating and dealing with publication and other biases in meta-analysis. BMJ 2001; 323(7304): 101-5.
[http://dx.doi.org/10.1136/bmj.323.7304.101] [PMID: 11451790]

[178] Higgins J, Green S. Cochrane Handbook for Systematic Reviews of Interventions: The Cochrane Collaboration; 2011. Available from: http://handbook.cochrane.org/

[179] Drysdale H, Slade E, Goldacre B, Heneghan C. COMPare trials team. Outcomes in the trial registry should match those in the protocol. Lancet 2016; 388(10042): 340-1.
[http://dx.doi.org/10.1016/S0140-6736(16)31128-X] [PMID: 27477160]

[180] Slade E, Drysdale H, Goldacre B. COMPare Team. Discrepancies between prespecified and reported outcomes. Ann Intern Med 2016; 164(5): 374.
[http://dx.doi.org/10.7326/L15-0614] [PMID: 26720309]

[181] Seino Y, Inagaki N, Haneda M, *et al.* Efficacy and safety of luseogliflozin added to various oral antidiabetic drugs in Japanese patients with type 2 diabetes mellitus. J Diabetes Investig 2015; 6(4): 443-53.
[http://dx.doi.org/10.1111/jdi.12316] [PMID: 26221523]

[182] Kaku K, Watada H, Iwamoto Y, *et al.* Tofogliflozin 003 Study Group. Efficacy and safety of monotherapy with the novel sodium/glucose cotransporter-2 inhibitor tofogliflozin in Japanese patients with type 2 diabetes mellitus: a combined Phase 2 and 3 randomized, placebo-controlled, double-blind, parallel-group comparative study. Cardiovasc Diabetol 2014; 13: 65.
[http://dx.doi.org/10.1186/1475-2840-13-65] [PMID: 24678906]

[183] Tanizawa Y, Kaku K, Araki E, *et al.* Tofogliflozin 004 and 005 Study group. Long-term safety and efficacy of tofogliflozin, a selective inhibitor of sodium-glucose cotransporter 2, as monotherapy or in combination with other oral antidiabetic agents in Japanese patients with type 2 diabetes mellitus: multicenter, open-label, randomized controlled trials. Expert Opin Pharmacother 2014; 15(6): 749-66.
[http://dx.doi.org/10.1517/14656566.2014.887680] [PMID: 24512053]

[184] Wygant G, Barnett AH, Orme ME, Fenici P, Townsend R, Roudaut M. Systematic review and network meta-analysis to compare dapagliflozin with other diabetes medications in combination with metformin for adults with type 2 diabetes. Diabetologia 2014; S346.

[185] Zaccardi F, Webb DR, Htike ZZ, Youssef D, Khunti K, Davies MJ. Efficacy and safety of sodium-glucose co-transporter-2 inhibitors in type 2 diabetes mellitus: systematic review and network meta-analysis. Diabetes Obes Metab 2016; 18(8): 783-94.
[http://dx.doi.org/10.1111/dom.12670] [PMID: 27059700]

[186] Chalmers I, Bracken MB, Djulbegovic B, *et al.* Research: increasing value, reducing waste. 1. How to increase value and reduce waste when research priorities are set. Lancet 2014; 383.

[187] Ioannidis JP. Some main problems eroding the credibility and relevance of randomized trials. Bull NYU Hosp Jt Dis 2008; 66(2): 135-9.
[PMID: 18537784]

[188] Ioannidis JP. Why most published research findings are false. PLoS Med 2005; 2(8): e124.
[http://dx.doi.org/10.1371/journal.pmed.0020124] [PMID: 16060722]

[189] Chalmers I, Glasziou P, Godlee F. All trials must be registered and the results published. BMJ 2013; 346: f105.
[http://dx.doi.org/10.1136/bmj.f105] [PMID: 23303893]

[190] Page MJ, McKenzie JE, Kirkham J, *et al.* Bias due to selective inclusion and reporting of outcomes and analyses in systematic reviews of randomised trials of healthcare interventions. Cochrane Database Syst Rev 2014; (10): MR000035.
[PMID: 25271098]

[191] Créquit P, Trinquart L, Yavchitz A, Ravaud P. Wasted research when systematic reviews fail to provide a complete and up-to-date evidence synthesis: the example of lung cancer. BMC Med 2016; 14: 8.
[http://dx.doi.org/10.1186/s12916-016-0555-0] [PMID: 26792360]

[192] Ioannidis JP. How to make more published research true. PLoS Med 2014; 11(10): e1001747.
[http://dx.doi.org/10.1371/journal.pmed.1001747] [PMID: 25334033]

[193] Al-Shahi Salman R, Beller E, Kagan J, *et al.* Increasing value and reducing waste in biomedical research regulation and management. Lancet 2014; 383(9912): 176-85.
[http://dx.doi.org/10.1016/S0140-6736(13)62297-7] [PMID: 24411646]

[194] Moher D, Glasziou P, Chalmers I, *et al.* Increasing value and reducing waste in biomedical research:

who's listening? Lancet 2016; 387(10027): 1573-86.
[http://dx.doi.org/10.1016/S0140-6736(15)00307-4] [PMID: 26423180]

[195] Battelle Technology Partnership Practice. Biopharmaceutical industry-sponsored clinical trials: Impact on state economies. Pharmaceutical Research and Manufacturers of America (PhRMA) 2015.

[196] Ioannidis JP. Effect of the statistical significance of results on the time to completion and publication of randomized efficacy trials. JAMA 1998; 279(4): 281-6.
[http://dx.doi.org/10.1001/jama.279.4.281] [PMID: 9450711]

[197] Ioannidis JP, Trikalinos TA. An exploratory test for an excess of significant findings. Clin Trials 2007; 4(3): 245-53.
[http://dx.doi.org/10.1177/1740774507079441] [PMID: 17715249]

[198] Schofferman J, Wetzel FT, Bono C. Ghost and guest authors: you can't always trust who you read. Pain Med 2015; 16(3): 416-20.
[http://dx.doi.org/10.1111/pme.12579] [PMID: 25338945]

[199] Sismondo S, Doucet M. Publication ethics and the ghost management of medical publication. Bioethics 2010; 24(6): 273-83.
[http://dx.doi.org/10.1111/j.1467-8519.2008.01702.x] [PMID: 19222451]

[200] Ioannidis JP. Why most clinical research is not useful. PLoS Med 2016; 13(6): e1002049.
[http://dx.doi.org/10.1371/journal.pmed.1002049] [PMID: 27328301]

[201] Chan A-W, Song F, Vickers A, et al. Increasing value and reducing waste: addressing inaccessible research. Lancet 2014; 383(9913): 257-66.
[http://dx.doi.org/10.1016/S0140-6736(13)62296-5] [PMID: 24411650]

[202] De Angelis C, Drazen JM, Frizelle FA, et al. International Committee of Medical Journal Editors. Clinical trial registration: a statement from the International Committee of Medical Journal Editors. N Engl J Med 2004; 351(12): 1250-1.
[http://dx.doi.org/10.1056/NEJMe048225] [PMID: 15356289]

[203] Chen R, Desai NR, Ross JS, et al. Publication and reporting of clinical trial results: cross sectional analysis across academic medical centers. BMJ 2016; 352: i637.
[http://dx.doi.org/10.1136/bmj.i637] [PMID: 26888209]

[204] Brænd AM, Straand J, Jakobsen RB, Klovning A. Publication and non-publication of drug trial results: a 10-year cohort of trials in Norwegian general practice. BMJ Open 2016; 6(4): e010535.
[http://dx.doi.org/10.1136/bmjopen-2015-010535] [PMID: 27067893]

[205] Bourgeois FT, Murthy S, Mandl KD. Outcome reporting among drug trials registered in ClinicalTrials.gov. Ann Intern Med 2010; 153(3): 158-66.
[http://dx.doi.org/10.7326/0003-4819-153-3-201008030-00006] [PMID: 20679560]

[206] Brown T. It's time for alltrials registered and reported. Cochrane Database Syst Rev 2013; (4): ED000057.
[PMID: 23728702]

[207] Goldacre B, Gray J. OpenTrials: towards a collaborative open database of all available information on all clinical trials. Trials 2016; 17: 164.
[http://dx.doi.org/10.1186/s13063-016-1290-8] [PMID: 27056367]

[208] Young T, Hopewell S. Methods for obtaining unpublished data. Cochrane Database Syst Rev 2011; 9(11): Mr000027.
[http://dx.doi.org/10.1002/14651858.MR000027.pub2]

[209] Boutron I, Dechartres A, Baron G, Li J, Ravaud P. Sharing of data from industry-funded registered clinical trials. JAMA 2016; 315(24): 2729-30.
[http://dx.doi.org/10.1001/jama.2016.6310] [PMID: 27367768]

[210] Chan AW, Laupacis A, Moher D. Registering results from clinical trials. JAMA 2010; 303(21): 2138-

9.
[http://dx.doi.org/10.1001/jama.2010.702] [PMID: 20516411]

[211] Cressey D. Drug-company data vaults to be opened. Nature 2013; 495(7442): 419-20.
[http://dx.doi.org/10.1038/495419a] [PMID: 23538802]

[212] Egger GF, Herold R, Rodriguez A, Manent N, Sweeney F, Saint Raymond A. European Union Clinical Trials Register: on the way to more transparency of clinical trial data. Expert Rev Clin Pharmacol 2013; 6(5): 457-9.
[http://dx.doi.org/10.1586/17512433.2013.827404] [PMID: 23971872]

[213] Herder M. Unlocking Health Canada's cache of trade secrets: mandatory disclosure of clinical trial results. CMAJ 2012; 184(2): 194-9.
[http://dx.doi.org/10.1503/cmaj.110721] [PMID: 21876028]

[214] Strahlman E, Rockhold F, Freeman A. Public disclosure of clinical research. Lancet 2009; 373(9672): 1319-20.
[http://dx.doi.org/10.1016/S0140-6736(09)60613-9] [PMID: 19321201]

[215] Efthimiou O, Debray TP, van Valkenhoef G, *et al.* GetReal Methods Review Group. GetReal in network meta-analysis: a review of the methodology. Res Synth Methods 2016; 7(3): 236-63.
[http://dx.doi.org/10.1002/jrsm.1195] [PMID: 26754852]

[216] Egger M, Moons KG, Fletcher C. GetReal Workpackage 4. GetReal: from efficacy in clinical trials to relative effectiveness in the real world. Res Synth Methods 2016; 7(3): 278-81.
[http://dx.doi.org/10.1002/jrsm.1207] [PMID: 27390256]

[217] Matheson A. The disposable author: How pharmaceutical marketing is embraced within medicine's scholarly literature. Hastings Cent Rep 2016; 46(4): 31-7.
[http://dx.doi.org/10.1002/hast.576] [PMID: 27417868]

[218] Woolley KL, Gertel A, Hamilton CW, Jacobs A, Snyder GP. Global Alliance of Publication Professionals. Time to finger point or fix? An invitation to join ongoing efforts to promote ethical authorship and other good publication practices. Ann Pharmacother 2013; 47(7-8): 1084-7.
[http://dx.doi.org/10.1345/aph.1S178] [PMID: 23800751]

[219] Battisti WP, Wager E, Baltzer L, *et al.* International Society for Medical Publication Professionals. Good publication practice for communicating company-sponsored medical research: GPP3. Ann Intern Med 2015; 163(6): 461-4.
[http://dx.doi.org/10.7326/M15-0288] [PMID: 26259067]

[220] Matheson A. The ICMJE Recommendations and pharmaceutical marketing--strengths, weaknesses and the unsolved problem of attribution in publication ethics. BMC Med Ethics 2016; 17: 20.
[http://dx.doi.org/10.1186/s12910-016-0103-7] [PMID: 27044283]

[221] Haines JD. The Hippocratic Oath: still relevant after 2,500 years. J Okla State Med Assoc 2003; 96(5): 233-4.
[PMID: 12833723]

[222] Lee D, Begley CE. Physician report of industry gifts and quality of care. Health Care Manage Rev 2016; 41(3): 275-83.
[http://dx.doi.org/10.1097/HMR.0000000000000042] [PMID: 25427138]

[223] Grande D, Shea JA, Armstrong K. Pharmaceutical industry gifts to physicians: patient beliefs and trust in physicians and the health care system. J Gen Intern Med 2012; 27(3): 274-9.
[http://dx.doi.org/10.1007/s11606-011-1760-3] [PMID: 21671130]

[224] Cosgrove L, Bursztajn HJ, Erlich DR, Wheeler EE, Shaughnessy AF. Conflicts of interest and the quality of recommendations in clinical guidelines. J Eval Clin Pract 2013; 19(4): 674-81.
[http://dx.doi.org/10.1111/jep.12016] [PMID: 23731207]

[225] Neuman J, Korenstein D, Ross JS, Keyhani S. Prevalence of financial conflicts of interest among panel members producing clinical practice guidelines in Canada and United States: cross sectional

study. BMJ 2011; 343: d5621.
[http://dx.doi.org/10.1136/bmj.d5621] [PMID: 21990257]

[226] Katayama AC. The Sunshine Act: it's for real now. WMJ 2013; 112(2): 96-7.
[PMID: 23758019]

[227] Lo B, Field M. Conflict of interest in medical research, education, and practice Washington (DC): National Academies Press (US): Institute of Medicine. US: Committee on Conflict of Interest in Medical Research, Education, and Practice 2009.

[228] Augustovski F, Alcaraz A, Caporale J, García Martí S, Pichon Riviere A. Institutionalizing health technology assessment for priority setting and health policy in Latin America: from regional endeavors to national experiences. Expert Rev Pharmacoecon Outcomes Res 2015; 15(1): 9-12.
[http://dx.doi.org/10.1586/14737167.2014.963560] [PMID: 25420744]

[229] O'Reilly D, Campbell K, Vanstone M, *et al.* Evidence-based decision-making 3: Health technology assessment. Methods Mol Biol 2015; 1281: 417-41.
[http://dx.doi.org/10.1007/978-1-4939-2428-8_25] [PMID: 25694325]

[230] Trigg A, Howells R. Patient-reported outcomes within health technology assessment decision making: Current status and implications for future policy. Value Health 2015; 18(7): A739.
[http://dx.doi.org/10.1016/j.jval.2015.09.2837] [PMID: 26534140]

[231] Wang T. Benchmarking the impact of HTA on new medicines development and coverage decision making. Value Health 2014; 17(7): A798.
[http://dx.doi.org/10.1016/j.jval.2014.08.478] [PMID: 27202997]

[232] Velasco-Garrido M, Busse R. Health technology assessment An introduction to objectives, role of evidence, and structure in Europe Brussels, Belgium: European Observatory on Health Systems and Policies World Health Organisation 2005. http://www.euro.who.int/__data/assets/pdf_file/0018/90432/E87866.pdf?ua=1

[233] Ehrismann D, Patel D. University - industry collaborations: models, drivers and cultures. Swiss Med Wkly 2015; 145: w14086.
[PMID: 25658854]

[234] Getz K. Improving protocol design feasibility to drive drug development economics and performance. Int J Environ Res Public Health 2014; 11(5): 5069-80.
[http://dx.doi.org/10.3390/ijerph110505069] [PMID: 24823665]

[235] Busetto L, Luijkx KG, Elissen AM, Vrijhoef HJ. Intervention types and outcomes of integrated care for diabetes mellitus type 2: a systematic review. J Eval Clin Pract 2016; 22(3): 299-310.
[http://dx.doi.org/10.1111/jep.12478] [PMID: 26640132]

[236] Gandjour A. Avoiding research waste through cost-effectiveness analysis: the example of medication adherence-enhancing interventions. Expert Rev Pharmacoecon Outcomes Res 2015; 15(1): 43-6.
[http://dx.doi.org/10.1586/14737167.2014.927313] [PMID: 24910291]

[237] Krummenauer F, Landwehr I. Incremental cost effectiveness evaluation in clinical research. Eur J Med Res 2005; 10(1): 18-22.
[PMID: 15737949]

[238] Chapman SJ, Shelton B, Mahmood H, Fitzgerald JE, Harrison EM, Bhangu A. Discontinuation and non-publication of surgical randomised controlled trials: observational study. BMJ 2014; 349: g6870.
[http://dx.doi.org/10.1136/bmj.g6870] [PMID: 25491195]

[239] Kasenda B, von Elm E, You J, *et al.* Prevalence, characteristics, and publication of discontinued randomized trials. JAMA 2014; 311(10): 1045-51.
[http://dx.doi.org/10.1001/jama.2014.1361] [PMID: 24618966]

[240] Abu-Arafeh A, Andrews PJ. Conducting feasibility studies in clinical trials are an investment to ensure a good study. Resuscitation 2016; 104: A1-2.
[http://dx.doi.org/10.1016/j.resuscitation.2016.04.015] [PMID: 27155545]

[241] Detsky AS. Are clinical trials a cost-effective investment? JAMA 1989; 262(13): 1795-800.
[http://dx.doi.org/10.1001/jama.1989.03430130071037] [PMID: 2506366]

[242] Mentz RJ, Hernandez AF, Berdan LG, *et al.* Good Clinical Practice guidance and pragmatic clinical trials: Balancing the best of both worlds. Circulation 2016; 133(9): 872-80.
[http://dx.doi.org/10.1161/CIRCULATIONAHA.115.019902] [PMID: 26927005]

[243] Chan AW, Tetzlaff JM, Altman DG, *et al.* SPIRIT 2013 statement: defining standard protocol items for clinical trials. Ann Intern Med 2013; 158(3): 200-7.
[http://dx.doi.org/10.7326/0003-4819-158-3-201302050-00583] [PMID: 23295957]

[244] Lenzer J, Hoffman JR, Furberg CD, Ioannidis JP. Guideline Panel Review Working Group. Ensuring the integrity of clinical practice guidelines: a tool for protecting patients. BMJ 2013; 347: f5535.
[http://dx.doi.org/10.1136/bmj.f5535] [PMID: 24046286]

[245] Bell CM, Urbach DR, Ray JG, *et al.* Bias in published cost effectiveness studies: systematic review. BMJ 2006; 332(7543): 699-703.
[http://dx.doi.org/10.1136/bmj.38737.607558.80] [PMID: 16495332]

[246] Jørgensen AW, Hilden J, Gøtzsche PC. Cochrane reviews compared with industry supported meta-analyses and other meta-analyses of the same drugs: systematic review. BMJ 2006; 333(7572): 782.
[http://dx.doi.org/10.1136/bmj.38973.444699.0B] [PMID: 17028106]

CHAPTER 4

The Effects of Traditional Chinese Medicine on Inflammatory Cytokines in Diabetic Nephropathy: The Progress in the Past Decades

Xiang Tu[1], YuanPing Deng[2], Ming Chen[3], James B. Jordan[1] and Sen Zhong[1,*]

[1] *National Traditional Chinese Medicine Clinical Research Base for Diabetes Mellitus/Teaching Hospital of Chengdu University of Traditional Chinese Medicine, Chengdu 610072, Sichuan Province, China*

[2] *Department of Internal Medicine, Traditional Chinese Medicine Hospital of Fushun County, Fushun 643200, Sichuan Province, China*

[3] *Department of Nephrology, Teaching Hospital of Chengdu University of Traditional Chinese Medicine, Chengdu 610072, Sichuan Province, China*

Abstract: Current research has revealed that inflammation as the cardinal pathophysiological mechanism in the development and progression of diabetic nephropathy (DN). Diverse inflammatory cytokines interact with each other and together with other mechanisms; result in the progression of DN. It is now widely accepted that many inflammatory cytokines, such as transforming growth factor (TGF)-β_1, tumor necrosis factor (TNF)-α, monocyte chemoattractant protein (MCP)-1, interleukin (IL)-1, IL-18, *etc.* are involved in the pathogenesis and development of DN. Published articles reported that a large number of single Traditional Chinese Medicine (TCM) and their extracts could confer benefits to DN by regulating inflammatory cytokines; such as Astragalus, Danshen Root, Szechuan Lovage Rhizome, Kudzuvine Root, Common Threewingnut Root, *etc*. TCM formulae also could regulate inflammatory cytokines in DN; such as Buyang Huanwu Decoction, Fufang Danshen Diwan, Bushen Tongluo Formular, *etc*. This is usually associated with improved kidney histomorphology, proteinuria and renal function parameters in DN. Although some researches concerning TCM for inflammatory cytokines in DN are in-depth, most researches, if not all, are still very superficial. The signaling pathways and underlying mechanisms of TCM affecting inflammatory cytokines remain unclear. Although there are many shortcomings in this field, it is still fortunate that the experimental reno-benefits of TCM can be translated for human into clinical treatment. This indicates new, effective and promising therapeutic drugs.

---

* **Correspondence author Sen Zhong:** Teaching Hospital of Chengdu University of Traditional Chinese Medicine, Chengdu 610072, Sichuan Province, China; Tel: 8618980086606; Fax: 8602887732407; E-mail: zhongsen6606@163.com

Atta-ur-Rahman (Ed.)
All rights reserved-© 2019 Bentham Science Publishers

Keywords: Diabetic nephropathy, Diabetes Review, Inflammatory Cytokines, TCM Formulae, Traditional Chinese Medicine.

INTRODUCTION

As the population with diabetes mellitus (DM) is rapidly increasing worldwide [1, 2], the complications of DM have been widely studied. The incidence of diabetic nephropathy (DN), the major micro-vascular complication of DM, usually grows with the progression of DM. Clinically, DN is characterized by abnormal glomerular filtration rate (GFR) and albuminuria. Albuminuria is measured by the value of urine albumin to creatinine ratio (ACR); with ACR< 300 mg/g referred to as micro-albuminuria and ACR≥ 300 mg/g referred to as macro-albuminuria.

Traditionally, DN is classified as 5 stages, according to the Mogensen Classification [3]. Stage 1 is characterized by early hyperfunction and hypertrophy; Stage 2 is characterized by morphologic lesions without signs of clinical disease; Stage 3 is incipient DN; Stage 4 is overt DN; Stage 5 is end-stage renal failure with uremia due to DN. The glomerular classification of DN pathology has evolved for decades and has finally reached consensus [4]. The American Renal Pathology Society has developed a practical uniform classification scheme [4]. Biopsies diagnosed as diabetic nephropathy are classified as follows: Class I - Glomerular basement membrane thickening; Class II - Mesangial expansion, mild (IIa) or severe (IIb); Class III - Nodular sclerosis (Kimmelstiel-Wilson lesions); and Class IV- Advanced diabetic glomerulosclerosis.

The mechanisms of the pathogenesis of DN are complex and until recently, unclear. The primary pathophysiological mechanism of the development of DN is traditionally considered as the interaction between hemodynamic and metabolic factors [5]. The metabolic disorders in patients with DN usually involve a high level of blood sugar and lipid metabolic disorder. It is now widely accepted that the pathogenesis of DN is multifactorial. Numerous articles have disclosed the involvement of pathogenesis and development of DN such as: genetic factors; oxidative stress; advanced glycosylated end products (AGEs); sorbitol and protein kinase C pathways; and the immune system, *etc.* [6].

Current research has revealed "inflammation has become a cardinal pathophysiological mechanism in the development and progression of DN." [7]. It is believed that inflammatory cytokines play important roles in this complex scenario [8, 9]. Researches on inflammatory cytokines in DN not only help in understanding the pathogenesis and development of DN but also help to develop new therapeutic strategies.

MAJOR INFLAMMATORY CYTOKINES INVOLVED IN DN

Transforming Growth Factor (TGF)-β_1

TGF-β_1 is a member of the TGF-β superfamily and widely accepted that TGF-β/Smad signaling is a major pathway leading to kidney disease. It plays a diverse role in renal fibrosis and inflammation [10]. The high expression of TGF-β_1 results in the excretion of extracellular matrix (ECM) and the fibrosis of renal interstitium and glomerulosclerosis results from the accumulation of ECM regulated by the TGF-β1/Smad pathway.

TGF-β signals through Smad2/3 to mediate renal fibrosis [11] and TGF-β_1 activates Smad3 to regulate microRNAs that mediate renal fibrosis [12]. Targeting downstream TGF-β/Smad signaling by overexpressing Smad7- or Smad3-dependent microRNA related to fibrosis - may provide a therapeutic strategy in the treatment of DN [12, 13].

Current literature has disclosed the crosstalk of TGF-β_1 signaling is very complex. A number of molecules and signaling, such as Janus kinase (JAK)/signal transduction, and activator of transcription (STAT) signaling, are involved in the crosstalk signaling [14, 15]. One *in vitro* study demonstrates that one of the suppressors of cytokine signaling (SOCS) proteins (SOCS-1), inhibits high glucose-induced overexpression of TGF-β_1 and synthesis of fibronectin in human mesangial cells (MCs) - which may be *via* the JAK/STAT pathway [16].

Tumor Necrosis Factor (TNF)-α

TNF-α "amplifies the inflammatory network of cytokines, leading to worsening of the progression of diabetic nephropathy" [17]. TNF-α may cause direct cytotoxicity to renal cells, produce alterations of intraglomerular blood flow and reduction of glomerular filtration, and directly induce the formation of reactive oxygen species (ROS) by renal cells, *etc*. An article published in 2016 reported that "higher circulating soluble tumor necrosis factor receptors 1 and 2 (sTNFR1 & sTNFR2) are associated with diabetic kidney disease; and predicts incident cardiovascular disease and mortality independently of microalbuminuria and kidney function" [18]. Awad *et al.* [19] generated macrophage-specific TNF-α-deficient mice [CD11b(Cre)/TNF-α (Flox/Flox)] to assess the direct role of macrophage-derived TNF-α in DN. Their results demonstrated that compared to diabetic TNF-α(Flox/Flox) control mice, after 12 weeks of streptozotocin (STZ)-induced diabetes, albuminuria; the increase in plasma creatinine and blood urea nitrogen (BUN), histopathologic changes, and kidney macrophage recruitment - were significantly reduced by conditional ablation of TNF-α in macrophages [19].

Generally speaking, many researchers believed that TNF-α may be a therapeutic target due to its key role in the pathogenesis and development of DN [20, 21].

Monocyte Chemoattractant Protein-1 (MCP-1)

MCP-1 significantly affects the macrophage infiltration. A number of articles have reported that high glucose enhances mesangial cells (MC) proliferation and MCP-1 expression through the ROS/ nuclear factor-κB (NF-κB) signaling pathways [22, 23]. An article published by Chung *et al.* [24], reported there was a robust increase (900%) in the secretion of MCP-1 in podocytes treated with TNF-α. The authors concluded "both TNF-α and MCP-1 levels were increased in the urine of diabetic db/db mice, correlating with the severity of diabetic albuminuria". Shoukry *et al.* [25], included seventy-five type 2 diabetic patients with normoalbuminuria (n=25), microalbuminuria (n=25), macroalbuminuria (n=25), and 25 healthy controls in their clinical study. Their results suggest that urinary MCP-1 and urinary vitamin D-binding protein (uVDBP) may be potential "diagnostic biomarkers for the early detection of diabetic nephropathy". Another article published in 2015 assayed MCP-1 and morphometric variables in kidney biopsies in type 1 diabetes mellitus (T1DM) [26]. The authors noticed that elevated urinary MCP-1 concentration was associated with early interstitial changes in women with T1DM changes - but there was no association between morphometric variables and another potential biomarker of renal inflammation - hepcidin.

Interleukin (IL)-1

A meta-analysis included 34 studies where 11 genetic variants including IL-1, IL-8, and IL-10 were researched showing significant positive association with DN [27]. Stefanidis *et al.* [28], examined if variants of the IL1B are implicated in the development of DN and their results provided evidence that IL1B C511T variant may be associated with development of DN. The synthesis of ICAM-1 specifically acts on the development of abnormalities in intraglomerular hemodynamics related to prostaglandin synthesis by MCs [9].

IL-6

There was a direct correlation between urinary levels and renal expression of IL-6 with urinary albumin excretion(UAE) and the urinary IL-6 level seems to be a good indicator of DN [9, 29]. The intracellular cytokines-associated signaling pathways include: JAK1/2/3; tyrosine kinase (Tyk) -2; and STAT3 [9]. Rakitianskaia *et al.* [30], confirmed that IL-6 has significant influence on the development of morphological (glomerular & tubulointerstitial) changes in the biopsy of diabetic kidney tissues.

IL-18

IL-18, a member of the IL-1 family of inflammatory cytokines, has been proven to be involved in the development and progression of DN [31]. It is a potent inflammatory cytokine that induces IFN-γ, leading to production of other inflammatory cytokines (including IL-1 & TNF-α), up-regulation of ICAM-1, as well as apoptosis of endothelial cells. Therapies which down-regulate IL-18 expression could confer reno-protective effects to DN, such as alprostadil treatment [32], 5'-(N-ethylcarboxamido)-adenosine (NECA) [33].

Targeting inflammatory cytokines has attracted intense attention and numerous studies have investigated the effects of various inflammatory cytokines with the signaling pathways in DN [34, 35]. Taking into account the pathogenic and therapeutic value of inflammatory cytokines in DN, some promising therapeutic strategies based on anti-inflammatory effects may be translated into clinical treatment for DN in the future.

THE EFFECTS OF TRADITIONAL CHINESE MEDICINE (TCM) ON INFLAMMATORY CYTOKINES IN DN

Single TCM and Their Extracts

A large number of articles have reported the laboratory or clinical effects of TCM on DN [36, 37] and interestingly, many of them focused on the effects of TCM on inflammatory cytokines. Let us take a closer look so that we can have a better understanding in this field.

Astragalus [Plant name: *Astragalus membranaceus* (Fisch.) Bunge]

Astragalus is one of the most widely used TCM for the management of DN and a large number of articles supported its use for DN. Kim *et al.* [38], stated that "in the early phase of diabetic nephropathy, administration of Astragalus membranaceus can be a therapeutic option". A meta-analysis included 25 studies involving 1804 patients and the results demonstrated that, compared with the control group - astragalus confers more positive benefits to DN patients including: renal protective effect [(*BUN, serum creatinine(SCr), creatinine clearance rate (CCr) and urine protein*)]; and systemic state improvement *(serum albumin level)* [39]. Another systematic review examined the effects of Astragalus on DN in animal models. The results of meta-analysis demonstrated that, compared with controls, Astragalus could confer benefits to diabetic rats, including "reducing fasting blood glucose and albuminuria levels, reversing the glomerular hyperfiltration state, and ameliorating the pathological changes of early DN" [40]. The combination therapy with Astragalus and angiotensin-converting enzyme inhibitors (ACEIs), *i.e.*, captopril, could significantly improve proteinuria and renal function parameters in a randomized controlled trial (RCT) [41].

In streptozotocin (STZ)-induced DM rats, administration of Astragalus resulted in improved histological kidney changes, serum creatinine (SCr), and 24h urinary albumin excretion (UAE). The results of immunohistochemistry assay showed that compared with DM rats, expressions of NF-κB and MCP-1 were significantly decreased in the DM rats receiving Astragalus [42]. Another animal experiment reported that, in STZ-induced DM rats, Astragalus polysaccharides (APS) not only decreased the levels of blood sugar but also reduced the expression of TGF-$β_1$ which was also measured by immunohistochemistry assay [43]. Sun *et al.* [44], used immunohistochemistry and western blotting method to measure the expression of cyclooxygenase (COX) -2 and NF-κB at protein level and reverse transcription-polymerase chain reaction (RT-PCR) to measure their expression at the genetic level in STZ-induced DM rats. Their results showed that compared with controls, administration of Astragalus resulted in improved SCr, BUN and 24h UAE. The expression levels of COX-2 and NF-κB at both protein level and genetic level were significantly ameliorated by Astragalus. Yin *et al.* [45], assessed the reno-protective effects of Astragalus Saponin I (AS I) in STZ-induced early stage DN rats. Compared with vehicle-treated DN rats, the renal hypertrophy, the oxidative stress intensity, and the blood glucose level of DN rats were ameliorated by AS I. The rats given AS I had reduced microalbuminuria levels, AGEs - either in serum or in kidney cortex, and the aldose reductase (AR) activity. Although the authors focused on the effects of AS I on oxidative stress, AGEs, AR activity, and renal fibrosis, it is notable that one of the major inflammatory cytokines, *i.e.*, TGF-$β_1$, was obviously regulated by AS I.

Deng *et al.* [46], randomly assigned 36 elderly patients (> 65 years old) with early DN to an APS group (receiving APS injection, 250mg ivgtt, for three weeks) or a placebo group (receiving saline). By the end of the trial, the 24h UAE of the APS group was significantly improved. The average levels of TNF-α decreased from 2.86 to 2.59 U/ml and the levels of IL-6 decreased from 2.61 to 1.35 U/ml of the APS group. The authors concluded that APS is beneficial to elderly patients with early stage DN and one of the underlying mechanisms may be declining inflammation. The pharmacological effects of Astragalus may be multi-faceted. A recent article published by Wang *et al.* [47] disclosed that Astragaloside-IV (AS-IV) reduced proteinuria and attenuated diabetes in STZ-induced DN rats, which is associated with decreased endoplasmic reticulum (ER) stress. Yin *et al.* [48], proved that AS I possess antioxidative effects.

According to the Chinese Pharmacopoeia, AS-IV ($C_{41}H_{68}O_{14}$, Fig. 1) shall not be less than 0.040% and Calycosin7-O-β-D-Glucopyranosid ($C_{22}H_{22}O_{10}$, Fig. 2) shall not be less than 0.020% in anhydrous astragalus assayed by HPLC [49].

Fig. (1). Astragaloside-IV.

Fig. (2). Calycosin7-O-β-D-Glucopyranosid.

Danshen Root (Plant name: Salvia miltiorrhiza Bge.)

J Ethnopharmacol published two articles concerning Danshen Root for DN. In 2011, Lee *et al*. [50], reported in STZ-induced DM rats, Danshen Root could significantly improve 24h UAE, reduce the serum and kidney levels of TGF-$β_1$ and the kidney levels of collagen IV, monocytes/macrophages and the receptor for AGE. The results of another article published in 2014 were largely consistent with Lee *et al*. After receiving the intraperitoneal administration of Danshen Injection, STZ-induced DM rats had significantly improved 24h UAE ((48.21 ± 8.04)%), SCr((39.4 ± 3.7)%), and BUN((43.37 ± 6.74)%), abnormalities of hypertrophy, matrix expansion, and fibrosis in glomerulus. The level of TGF-$β_1$ expression was significantly reduced and the activity of superoxide dismutase (SOD) and glutathione peroxidase (GSH-Px) was significantly increased by Danshen Injection. Albeit anti-inflammatory effect is not the focus of these researches, the significant regulation of TGF-$β_1$ by Danshen Root suggested its potential effects on inflammatory cytokines [51].

Tanshinone IIA, a compound extracted from Danshen Root, has very similar pharmacological effects to Danshen Root, reported by Kim et al. [52]. Chen et al. [53], examined the genetic and protein levels of TGF-$β_1$ and NF-κB p65 in STZ-induced DM rats. In addition to improved 24h UAE, SCr, BUN, renal histomorphology, the expression levels of TGF-$β_1$ and NF-κB p65 were significantly reduced by Tanshinone IIA. The results suggested Tanshinone IIA may act as an inhibitor of NF-κB activation, and subsequently, effectively regulate the release of some systemic inflammatory mediators.

When translated into clinical treatment, Danshen Root appeared to be promising for DN. Salvianolate, is a compound processed from Danshen Root. Lu et al. [54], carried out a RCT to evaluate the effects of Salvianolate on inflammatory cytokines and renal vascular endothelial function in early stage DN patients. Their results demonstrated that the levels of serum IL-6, TNF-α, ICAM-1 and endothelin (ET)-1 were significantly reduced by Salvianolate. The authors concluded that Salvianolate could improve endothelial function in these patients by inhibiting inflammation of renal vascular endothelium.

According to the Chinese Pharmacopoeia, the total content of Tanshinone II_A ($C_{19}H_{18}O_3$, Fig. 3), Crptotanshinone ($C_{19}H_{20}O_3$, Fig. 4) and Tanshinone I($C_{18}H_{12}O_3$, Fig. 5) shall not be less than 0.25% and Salvianolic acid B ($C_{36}H_{30}O_{16}$, Fig. 6) shall not be less than 3.0% in anhydrous Danshen Root assayed by HPLC [49].

Fig. (3). Tanshinone II_A.

Fig. (4). Crptotanshinone.

Fig. (5). Tanshinone.I.

Fig. (6). Salvianolic acid B.

Szechuan Lovage Rhizome (Plant name: *Ligusticum chuanxiong* Hort.)

Szechuan Lovage Rhizome is a TCM frequently used in the treatment of DN in the TCM arena and researchers have extracted some compounds from this herb, such as ligustrazine. A meta-analysis showed that ligustrazine could confer benefits to patients with DN, including improving renal function and reducing proteinuria [55]. Su *et al.* [56], reported ligustrazine treatment could result in more significant reduction of TGF-β_1, creactive protein (CRP), IL-6 levels; but the reduction in TNF-α was not significant. The authors concluded that ligustrazine possesses reno-protective effects by exerting anti-inflammation activity.

Interestingly, there is a combination of Danshen Root and ligustrazine in China, in the form of injection. Lan (2013) carried out an RCT where 94 patients with early stage DN were randomly assigned to receive the combination plus Benazepril or Benazepril alone. Compared with Benazepril alone, the combination of Danshen Root + ligustrazine + Benazepril resulted in more significant reduction in the levels of IL-6, IL-18, TNF-α, as well as proteinuria [57].

According to the Chinese Pharmacopoeia, Ferulic acid ($C_{10}H_{10}O_4$, Fig. 7) shall not be less than 0.10% in anhydrous Szechuan Lovage Rhizome assayed by HPLC [49].

Fig. (7). Ferulic acid.

Kudzuvine Root (Plant name: *Pueraria thomsonii* Benth.)

Puerarin is a flavonoid extracted from Kudzuvine Root. A meta-analysis included ten RCTs involving 669 participants and evaluated the potential benefits and harms comparing puerarin plus angiotensin converting enzyme inhibitor (ACEI) and ACEI alone in the treatment of patients with stage III DN. The results of meta-analysis demonstrated that compared with ACEI alone, the combination therapy with puerarin and ACEI resulted in a more significant reduction in urinary albumin excretion rate (UAER), but no significant reduction in 24h UAE, SCr, BUN [58]. Pan *et al.* [59] examined the effects of puerarin on the expression of ICAM-1 and TNF-α in early stage DN rats, at doses of 0.25, 0.5 and 1.0 mg/(kg·d). The results showed that puerarin significantly improved proteinuria, renal function parameters, as well as histopathological damages. Puerarin at each dosage also significantly eliminated elevations of ICAM-1 and TNF-α levels in model rats. In STZ-induced DM rats, administration of puerarin could result in significant reduction in TNF-α and NF-κB65 expression assayed by immunohistochemistry in the kidney tissues [60].

Sanchi [Plant name: *Panax notoginseng* (Burkill) F.H.Chen]

Du *et al.*, reported Panax notoginseng saponins (PNS), which is extracted from Sanchi, could reduce both blood glucose and lipid levels. The authors silenced the Silent information regulator 1 (SIRT1) in rat MCs by RNA interference. Their experimental results showed that PNS could up-regulate the expression of SIRT1 ($P<0.01$) and in turn suppress the transcription of TGF-$β_1$ ($P<0.05$) and MCP-1 ($P<0.05$). As a result, the authors concluded that the underlying mechanisms of reno-protective effects of PNS are up-regulating SIRT1 and subsequently, inhibiting inflammation by decreasing the induction of inflammatory cytokines.

Common Threewingnut Root (Plant Name: *Tripterygium wilfordii* Hook. F.)

Tripterygium Glycosides is a compound extracted from Common Threewingnut Root and in China, it is generally accepted that this herb possesses anti-inflammatory and immunosuppressive properties [61]. In STZ-induced DM rats, compared with vehicle controls, Tripterygium Glycosides was able to significantly reduce the levels of blood sugar, Hemoglobin A1c (HbA1c), SCr and BUN. It also reduced the levels of IL-6 by 28.51%, TNF-α by 30.72% and chemokine (C-C motif) ligand 5 (CCL5) by 41.22% in kidney homogenate and serum high sensitivity CRP (hs-CRP) level by 31.31% [62]. This clearly indicated that Tripterygium Glycosides exerts reno-protective effects by inhibiting the expression of inflammatory cytokines in serum and the kidneys. The authors published another article in Chinese, which reported Tripterygium Glycosides could significantly reduce the serum IL-1β, IL-17 and IFN-γ levels [63]. Two teams of researchers investigated the effects of Tripterygium Glycosides on p38 mitogenactivated protein kinase (MAPK) signaling. Their results were consistent. Tripterygium Glycosides could down-regulate the expression of key molecules in p38MAPK signaling pathway including p38 MAPK, phosphorylated p38 (p-p38MAPK) and TGF-$β_1$, as well as the expression of TNF-α, IL-1β [64, 65].

Triptolide, also a compound extracted from Common Threewingnut Root, is also considered as a novel immunosuppressive and anti-inflammatory agent [66]. It can dramatically improve albuminuria and renal lesion accompanied with dyslipidaemia and obesity in db/db diabetic mice and seems to be a promising drug for the management of DN [67]. Triptolide also exerts reno-protective effects by reducing macrophage infiltration and down-regulating MCP-1 and osteopontin (OPN) [68].

The results of a prospective, randomized, controlled clinical trial showed that the anti-proteinuric efficacy of *Tripterygium wilfordii* extract was superior to valsartan and the authors suggested that *Tripterygium wilfordii* extract "represents a novel, potentially effective, and safe drug" in the treatment of proteinuric patients with DN [69].

Published literature reported the many single TCM could confer experimental or clinical reno-protective effects, such as Heartleaf Houttuynia Herb (Plant name: *Houttuynia cordata* Thunb.) [70, 71], Red Paeony Root (Plant name: *Paeonia lactiflora* Pall.) [72], Common Macrocarpium Fruit (Plant name: *Cornus officinalis* Sieb. et Zucc.) [73], Barbary Wolfberry Fruit (Plant name:*Lycium barbarum* L.) [74] and the underlying mechanisms may be associated with modulating inflammatory cytokines. There are many reported extracts from TCM which may regulate inflammatory cytokines, such as: Ginsenoside Rg1 [75], tannic acid [76 - 78], Breviscapine (Bre) [79, 80], berberine [81, 82], *etc.* Many TCM were used for the management of DN and the top 20 TCM summarization from two Chinese reviews are listed in frequency and shown in Table **1** [37].

Table 1. The top 20 Chinese medicinals which were summarized in two Chinese review articles (listed in decrease frequency).

Wang SF 2009 [83]	Pinyin name	English name	Plant name
#1	Huangqi	Astragalus	*Astragalus membranaceus* (Fisch.) Bunge
#2	Fuling	Indian Buead	*Poria cocos* (Schw.) Wolf
#3	Shanyao	Common Yam Rhizome / Wingde Yan Rhizome	*Dioscorea opposita* Thunb.
#4	Danshen	Danshen Root	*Salvia miltiorrhiza* Bunge
#5	Shanzhuyu	Common Macrocarpium Fruit	*Cornus officinalis* Sieb. et Zucc.
#6	Baizhu	Largehead Atractylodes Rhizome	*Atractylodes macrocephala* Koidz.
#7	Zexie	Oriental Waterplantain Rhizome	*Alisma orientalis* (Sam.) Juzep.
#8	Danggui	Chinese Angelica	*Angelica sinensis* (Oliv.) Diels
#9	Dahuang	Rhubarb	*Rheum palmatum* L.
#10	Zhifuzi	Prepared Common Monkshood Daughter Root	*Aconitum carmichaelii* Debeaux
#11	Chuanxiong	Szechuan Lovage Rhizome	*Ligusticum chuanxiong* Hort.
#12	Dangshen	Pilose Asiabell Root /Moderate Asiabell Root/Szechwon Tangshen Root	*Codonopsis pilosula* (Franch.) Nannf.
#13	Mudanpi	Tree Peony Bark	*Paeonia suffruticosa* Andrews
#14	Gouqizi	Barbary Wolfberry Fruit	*Lycium barbarum* L.
#15	Yimucao	Motherwort Herb	*Leonurus heterophyllus* Sweet
#16	Chishao	Red Paeony Root	*Paeonia lactiflora* Pall.
#17	Maimendong	Dwarf Lilyturf Tuber	*Ophiopogon japonicus* (L.f.) Ker-Gawl.
#18	Zhuling	Agaric	*Polyporus umbellatus* (Pers) Fries
#19	Niuxi	Twotooth Achyranthes Root	*Achyranthes bidentata* Bl.
#20	Yinyanghuo	Epimedium Herb	*Epimedium brevicornum* Maxim.
Wang Mei 2011 [84]	Pinyin name	English name	Plant name
#1	Huangqi	Astragalus	*Astragalus membranaceus* (Fisch.) Bunge
#2	Danshen	Danshen Root	*Salvia miltiorrhiza* Bunge

(Table 1) cont.....

Wang Mei 2011 [84]	Pinyin name	English name	Plant name
#3	Shanzhuyu	Common Macrocarpium Fruit	*Cornus officinalis* Sieb. et Zucc.
#4	Fuling	Indian Buead	*Poria cocos* (Schw.) Wolf
#5	Shanyao	Common Yam Rhizome / Wingde Yan Rhizome	*Dioscorea opposita* Thunb.
#6	Shengdihuang	Rehmannia Root	*Rehmannia glutinosa* Libosch.
#7	Chuanxiong	Szechuan Lovage Rhizome	*Ligusticum chuanxiong* Hort.
#8	Dahuang	Rhubarb	*Rheum palmatum* L.
#9	Danggui	Chinese Angelica	*Angelica sinensis* (Oliv.) Diels
#10	Zexie	Oriental Waterplantain Rhizome	*Alisma orientalis* (Sam.) Juzep.
#11	Yimucao	Motherwort Herb	*Leonurus heterophyllus* Sweet
#12	Shudihuang	Rehmannia Root	*Rehmannia glutinosa* Libosch.
#13	Baizhu	Largehead Atractylodes Rhizome	*Atractylodes macrocephala* Koidz.
#14	Taizishen	Heterophylly Falsestarwort Root	*Pseudostellaria heterophylla* (Miq.) Pax ex Pax et Hoffm.
#15	Dangshen	Pilose Asiabell Root /Moderate Asiabell Root/Szechwon Tangshen Root	*Codonopsis pilosula* (Franch.) Nannf.
#16	Gouqizi	Barbary Wolfberry Fruit	*Lycium barbarum* L.
#17	Fuzi	Prepared Common Monkshood Daughter Root	*Aconitum carmichaelii* Debeaux
#18	Maidong	Dwarf Lilyturf Tuber	*Ophiopogon japonicus* (L.f.) Ker-Gawl.
#19	Shuizhi	Leech	*Whitmania pigra* Whitman
#20	Honghua	Safflower	*Carthamus tinctorius* L.

Formulae

Formulae are the most common prescription form used in the TCM arena. These various single TCM's possessing different pharmacological effects are carefully intermixed to produce more potent effects. There are many formulae used for the management of DN and some of them have been tested in terms of inflammation and cytokines.

Buyang Huanwu Decoction (BHD)

BHD was first recorded in "Correction of Errors in Medical Classics" and was a formula used for chronic kidney disease. Two small sample size RCT reported that BHD could significantly reduce the serum TGF-β_1 level and improve proteinuria in early stage DN patients [85, 86]. Another RCT reported the reduction of proteinuria was associated with the down-regulation of serum IL-6 and TNF-α [87]. The TCM ingredients of BHD are listed in Table **2**. The active ingredients of BHD contain Paeoniflorin (Fig. **8**), Calycosin7-O-β-D-Glucopyranoside (Fig. **9**) and Quercetin (Fig. **10**) assayed by HPLC [88, 89].

Table 2. The TCM ingredients of BHD.

Pinyinname	English name	Plant/Animal name
Huangqi	Astragalus	Astragalus membranaceus (Fisch.) Bunge
Chishao	Red Paeony Root	Paeonia lactiflora Pall.
Danggui	Chinese Angelica	Angelica sinensis (Oliv.) Diels
Chuanxiong	Szechuan Lovage Rhizome	Ligusticum chuanxiong Hort.
Honghua	Safflower	Carthamus tinctorius L.
Taoren	Peach Seed	Prunus persica (L.) Batsch
Dilong	Earthworm	Pheretima aspergillum (E. Perrier)

The Pinyin name and English name of TCM are referenced from the Chinese Clinical Trial Registry (http://f1.clinical-trialecrf.org/doc/2011/8/15/115111957 30792805.pdf). The plant names correspond to the latest revision in "The Plant List" (www.theplantlist.org).

Fig. (8). Paeoniflorin.

Fig. (9). Calycosin7-O-β-D-Glucopyranoside.

Fig. (10). Quercetin.

Fufang Danshen Diwan (FDD)

FDD consists of three TCM which are listed in Table **3**. FDD was reported to be able to reduce the proteinuria and the levels of serum TNF-α, IL-6, hs-CRP in patients with DN [90, 91]. This indicated the anti-proteinuric effect of FDD was associated with its anti-inflammatory property. The active ingredient of FDD contains Danshensu (Fig. **11**) assayed by HPLC [49].

Table 3. The TCM ingredients of FDD.

Pinyin name	English name	Plant name
Danshen	Danshen Root	Salvia miltiorrhiza Bunge
Sanqi	Sanchi	Panax notoginseng (Burkill) F.H.Chen
Bingpian	Borneol	Dryobalanops aromatica Gaertn. f.

The Pinyin name and English name of TCM are referenced from the Chinese Clinical Trial Registry (http://f1.clinicaltrialecrf.org/doc/2011/8/15/115111957 30792805.pdf). The plant names correspond to the latest revision in "The Plant List" (www.theplantlist.org).

Fig. (11). Danshensu.

Bushen Tongluo Formula (BTF)

Min and his co-workers [92, 93] carried out two RCTs to assess the efficacy of BTF and the underlying mechanisms. The TCM used in BTF is listed in Table (**4**). BTF could improve proteinuria. It was also able to significantly reduce not only the levels of connective tissue growth factor (CTGF), matrix metalloproteinase (MMP)-9, but also the levels of CRP, TNF-α, and MCP-1. This indicates the clinical anti-proteinuric benefit of BTF is associated with its anti-inflammatory activity.

Table 4. The TCM ingredients of BTF.

Pinyin name	English name	Plant/Animal name
Shanzhuyu	Common Macrocarpium Fruit	*Cornus officinalis* Sieb. et Zucc.
Huangqi	Astragalus	*Astragalus membranaceus* (Fisch.) Bunge
Mudanpi	Tree Peony Bark	*Paeonia suffruticosa* Andrews
Quanxie	Scorpion	*Buthus martensii* Karsch
Dihuang	Rehmannia Root	*Rehmannia glutinosa* Libosch.

The Pinyin name and English name of TCM are referenced from the Chinese Clinical Trial Registry (http://f1.clinicaltrialecrf.org/doc/2011/8/15/115111957 30792805.pdf). The plant names correspond to the latest revision in "The Plant List" (www.theplantlist.org).

QidanYishenJiangtang Capsule (QYJC)

The individual TCM ingredients which make-up QYJC are listed in Table 5. An experimental study reported that QYJC exerted reno-protective effects by significantly down-regulating the expression of AGEs in the renal tissues of diabetic rats [94]. Another experimental study reported that the renal benefits conferred by QYJC was associated with the down-regulation of TGF-β_1 and VEGF [95]. Jia *et al.* [96], examined the effects of QYJCs in patients with DN. 88 patients were randomly assigned to the QYJC group or another TCM product (JiangtangShu capsule, JSC) group. The authors declared their study was a double blind, double dummy trial, but they did not report more methodological details. Now that both the QYJC arm and the control used capsules, why was the dummy necessary? Their results showed that the serum levels of TGF-β and IGF-1 were significantly reduced in the QYJC group compared with the JSC group.

Table 5. The TCM ingredients of QYJC.

Pinyin name	English name	Plant name
Huangqi	Astragalus	*Astragalus membranaceus* (Fisch.) Bunge
Dihuang	Rehmannia Root	*Rehmannia glutinosa* Libosch.
Shanzhuyu	Common Macrocarpium Fruit	*Cornus officinalis* Sieb. et Zucc.
Shanyao	Common Yam Rhizome / Wingde Yan Rhizome	*Dioscorea opposita* Thunb.
Fuling	Indian Buead	*Poria cocos* (Schw.) Wolf
Danshen	Danshen Root	*Salvia miltiorrhiza* Bunge
Chuanxiong	Szechuan Lovage Rhizome	*Ligusticum chuanxiong* Hort.
Sanleng	Common Burreed Rhizome	*Sparganium stoloniferum* (Buch.-Ham. ex Graebn.) Buch.-Ham. ex Juz.

(Table 5) cont.....

Pinyin name	English name	Plant name
Yimucao	Motherwort Herb	*Leonurus heterophyllus* Sweet
Fuzi	Prepared Common Monkshood Daughter Root	*Aconitum carmichaelii* Debeaux

The Pinyin name and English name of TCM are referenced from the Chinese Clinical Trial Registry (http://f1.clinicaltrialecrf.org/doc/2011/8/15/115111957 30792805.pdf). The plant names correspond to the latest revision in "The Plant List" (www.theplantlist.org).

In addition to the above-mentioned TCM formulae, a large number of articles reported encouraging results concerning TCM formulae for DN by regulating inflammatory cytokines, such as: Bushen Huoxue Qufeng Formula [97]; Jianpi Yishen Tongluo Yin [98]; Shenyan Kangfu Pian [99]; Gepi Jianji [100, 101]; Modified Lianpu Yin [102]; Shenwei Ning Granules [103]; Danggui Shaoyao San [104]; Liuwei Dihuang Decoction [105]; Huanglian Jiedu Decoction [106]; Yangyin Qingre Formula [107]; Jianpi Yishen Decoction [108]; Tangmai Kang [109]; Jiangtang Bushen Formula [110]; and Jianpi Bushen Decoction [111]; as well as others. Obviously, TCM formulae hold promise in this field.

CONCLUSIONS

The literature published in the past few decades demonstrates that when used alone or in combination with conventional western medicine, TCM may confer laboratory and/or clinical benefits to DN. The renal benefits of TCM are usually associated with their property of regulating inflammatory cytokines. Although some research-concerning TCM for inflammatory cytokines in DN are in-depth, most research, if not all, are still very superficial. The signaling pathways and underlying mechanisms of TCM affecting inflammatory cytokines remain unclear. These preliminary discoveries promise expanding conceptualizations of TCM interactive qualities. The methodological shortcomings obviously weaken the validity of current literature. For example, in most RCTs, it is unclear if the allocation sequence was concealed or not, which could result in selection bias. Although there are methodological shortcomings in this field - fortunately, the experimental reno-benefits of TCM can be translated into immediate clinical treatment and this indicates new, effective and promising therapeutic drugs.

ABBREVIATIONS

ACEI   angiotensin-converting enzyme inhibitors
ACR   albumin to creatinine ratio
AGE   advanced glycosylated end product
APS   astragalus polysaccharides
AR   aldose reductase

ASI	astragalus Saponin I
AS-IV	astragaloside IV
BHD	buyang huanwu decoction
BMP	bone morphogenetic protein
BRE	breviscapine
BTF	bushen tongluo formular
BUN	blood urea nitrogen
CCr	creatinine clearance rate
CCL5	chemokine (C-C motif) ligand 5
COX	cyclooxygenase
CRP	creactive protein
CTGF	connective tissue growth factor
DM	diabetes mellitus
DN	diabetic nephropathy
ECM	extracellular matrix
ELISA	enzyme-linked immuno sorbent assay
ER	endoplasmic reticulum
ET	endothelin
FDD	Fufang Danshen Diwan
GFR	glomerular filtration rate
GSH-Px	glutathione peroxidase
HbA1c	hemoglobin A1c
hs-CRP	high sensitivityc reactive protein
ICAM	intercellular adhesion molecule
IFN	interferon
IGF	insulin-like growth factor
IL	interleukin
JAK	janus kinase
JSC	jiangtang shu capsule
MAPK	mitogenactivated protein kinase
MCP	monocyte chemoattractant protein
MCs	mesangial cells
MMP	matrix metalloproteinase
NECA	(N-ethylcarboxamido)-adenosine
NF-κB	nuclear factor-κB

OPN	osteopontin
PKC	protein kinase C
PNS	panax notoginseng saponins
p-p38MAPK	phosphorylated p38mitogenactivated protein kinase
QYJC	qidan yishen jiangtang capsule
RCT	randomized controlled trial
RNA	ribonucleic acid
ROS	reactive oxygen species
RT-PCR	reverse transcription-polymerase chain reaction
SCr	serum creatinine
SIRT	Silent information regulator
SOCS	suppressors of cytokine signaling
SOD	superoxide dismutase
STAT	signal transduction and activator of transcription
sTNFR1	soluble tumor necrosis factor receptors 1
sTNFR2	soluble tumor necrosis factor receptors 2
STZ	streptozocin
TC	cholesterol
TCM	traditional chinese medicine
TG	triglyceride
TGF	transforming growth factor
TLR4	toll like receptor 4
TNF	tumor necrosis factor
Tyk	tyrosine kinase
T1DM	type 1 diabetes mellitus
UAE	urinary albumin excretion
UAER	urinary albumin excretion rate
uVDBP	urinary vitamin D-binding protein
VEGF	vascular endothelial growth factor

CONSENT FOR PUBLICATION

Not applicable.

CONFLICT OF INTEREST

The authors confirm that they have no conflict of interest to declare for this publication.

ACKNOWLEDGEMENTS

The present chapter is funded by Demonstration Project Chengdu City - 2014: Improvement of Livelihood through Science & Technology (2014-HM02-00014-SF) and the Research Project for Practice Development of National TCM Clinical Research Bases (JDZX2015219). The present chapter is also supported by the TCM Science & Technology: A Research Project of Sichuan Province (2016Z002).

REFERENCES

[1] Menke A, Casagrande S, Geiss L, Cowie CC. Prevalence of and Trends in Diabetes Among Adults in the United States, 1988-2012. JAMA 2015; 314(10): 1021-9.
[http://dx.doi.org/10.1001/jama.2015.10029] [PMID: 26348752]

[2] Danaei G, Finucane MM, Lu Y, et al. National, regional, and global trends in fasting plasma glucose and diabetes prevalence since 1980: systematic analysis of health examination surveys and epidemiological studies with 370 country-years and 2·7 million participants. Lancet 2011; 378(9785): 31-40.
[http://dx.doi.org/10.1016/S0140-6736(11)60679-X] [PMID: 21705069]

[3] Mogensen CE, Christensen CK, Vittinghus E. The stages in diabetic renal disease. With emphasis on the stage of incipient diabetic nephropathy. Diabetes 1983; 32 (Suppl. 2): 64-78.
[http://dx.doi.org/10.2337/diab.32.2.S64] [PMID: 6400670]

[4] Tervaert TW, Mooyaart AL, Amann K, et al. Pathologic classification of diabetic nephropathy. J Am Soc Nephrol 2010; 21(4): 556-63.
[http://dx.doi.org/10.1681/ASN.2010010010] [PMID: 20167701]

[5] Satirapoj B. Nephropathy in diabetes. Adv Exp Med Biol 2012; 771: 107-22.
[PMID: 23393675]

[6] Raptis AE, Viberti G. Pathogenesis of diabetic nephropathy. Exp Clin Endocrinol Diabetes 2001; 109 (Suppl. 2): S424-37.
[http://dx.doi.org/10.1055/s-2001-18600] [PMID: 11460589]

[7] García-García PM, Getino-Melián MA, Domínguez-Pimentel V, Navarro-González JF. Inflammation in diabetic kidney disease. World J Diabetes 2014; 5(4): 431-43.
[http://dx.doi.org/10.4239/wjd.v5.i4.431] [PMID: 25126391]

[8] Donate-Correa J, Martín-Núñez E, Muros-de-Fuentes M, Mora-Fernández C, Navarro-González JF. Inflammatory cytokines in diabetic nephropathy. J Diabetes Res 2015; 2015: 948417.
[http://dx.doi.org/10.1155/2015/948417] [PMID: 25785280]

[9] Navarro-González JF, Mora-Fernández C. The role of inflammatory cytokines in diabetic nephropathy. J Am Soc Nephrol 2008; 19(3): 433-42.
[http://dx.doi.org/10.1681/ASN.2007091048] [PMID: 18256353]

[10] Lan HY. Diverse roles of TGF-β/Smads in renal fibrosis and inflammation. Int J Biol Sci 2011; 7(7): 1056-67.
[http://dx.doi.org/10.7150/ijbs.7.1056] [PMID: 21927575]

[11] Wang W, Koka V, Lan HY. Transforming growth factor-beta and Smad signalling in kidney diseases. Nephrology (Carlton) 2005; 10(1): 48-56.
[http://dx.doi.org/10.1111/j.1440-1797.2005.00334.x] [PMID: 15705182]

[12] Lan HY. Transforming growth factor-β/Smad signalling in diabetic nephropathy. Clin Exp Pharmacol Physiol 2012; 39(8): 731-8.
[http://dx.doi.org/10.1111/j.1440-1681.2011.05663.x] [PMID: 22211842]

[13] Lan HY, Chung AC. Transforming growth factor-β and Smads. Contrib Nephrol 2011; 170: 75-82.
[http://dx.doi.org/10.1159/000324949] [PMID: 21659760]

[14] Wang X, Shaw S, Amiri F, Eaton DC, Marrero MB. Inhibition of the Jak/STAT signaling pathway prevents the high glucose-induced increase in tgf-beta and fibronectin synthesis in mesangial cells. Diabetes 2002; 51(12): 3505-9.
[http://dx.doi.org/10.2337/diabetes.51.12.3505] [PMID: 12453907]

[15] Hebenstreit D, Horejs-Hoeck J, Duschl A. JAK/STAT-dependent gene regulation by cytokines. Drug News Perspect 2005; 18(4): 243-9.
[http://dx.doi.org/10.1358/dnp.2005.18.4.908658] [PMID: 16034480]

[16] Shi Y, Zhang Y, Wang C, et al. Suppressor of cytokine signaling-1 reduces high glucose-induced TGF-beta1 and fibronectin synthesis in human mesangial cells. FEBS Lett 2008; 582(23-24): 3484-8.
[http://dx.doi.org/10.1016/j.febslet.2008.09.014] [PMID: 18801363]

[17] Sun L, Kanwar YS. Relevance of TNF-α in the context of other inflammatory cytokines in the progression of diabetic nephropathy. Kidney Int 2015; 88(4): 662-5.
[http://dx.doi.org/10.1038/ki.2015.250] [PMID: 26422621]

[18] Carlsson AC, Östgren CJ, Nystrom FH, et al. Association of soluble tumor necrosis factor receptors 1 and 2 with nephropathy, cardiovascular events, and total mortality in type 2 diabetes. Cardiovasc Diabetol 2016; 15(1): 40.
[http://dx.doi.org/10.1186/s12933-016-0359-8] [PMID: 26928194]

[19] Awad AS, You H, Gao T, et al. Macrophage-derived tumor necrosis factor-α mediates diabetic renal injury. Kidney Int 2015; 88(4): 722-33.
[http://dx.doi.org/10.1038/ki.2015.162] [PMID: 26061548]

[20] Navarro JF, Mora-Fernández C. The role of TNF-alpha in diabetic nephropathy: pathogenic and therapeutic implications. Cytokine Growth Factor Rev 2006; 17(6): 441-50.
[http://dx.doi.org/10.1016/j.cytogfr.2006.09.011] [PMID: 17113815]

[21] Navarro-González JF, Jarque A, Muros M, Mora C, García J. Tumor necrosis factor-alpha as a therapeutic target for diabetic nephropathy. Cytokine Growth Factor Rev 2009; 20(2): 165-73.
[http://dx.doi.org/10.1016/j.cytogfr.2009.02.005] [PMID: 19251467]

[22] Yang X, Wang Y, Gao G. High glucose induces rat mesangial cells proliferation and MCP-1 expression via ROS-mediated activation of NF-κB pathway, which is inhibited by eleutheroside E. J Recept Signal Transduct Res 2016; 36(2): 152-7.
[http://dx.doi.org/10.3109/10799893.2015.1061002] [PMID: 26644089]

[23] Wei M, Li Z, Xiao L, Yang Z. Effects of ROS-relative NF-κB signaling on high glucose-induced TLR4 and MCP-1 expression in podocyte injury. Mol Immunol 2015; 68 (2 Pt A): 261-71.
[http://dx.doi.org/10.1016/j.molimm.2015.09.002] [PMID: 26364141]

[24] Chung CH, Fan J, Lee EY, et al. Effects of Tumor Necrosis Factor-α on Podocyte Expression of Monocyte Chemoattractant Protein-1 and in Diabetic Nephropathy. Nephron Extra 2015; 5(1): 1-18.
[http://dx.doi.org/10.1159/000369576] [PMID: 25852733]

[25] Shoukry A, Bdeer Sel-A, El-Sokkary RH. Urinary monocyte chemoattractant protein-1 and vitamin D-binding protein as biomarkers for early detection of diabetic nephropathy in type 2 diabetes mellitus. Mol Cell Biochem 2015; 408(1-2): 25-35.
[http://dx.doi.org/10.1007/s11010-015-2479-y] [PMID: 26104579]

[26] Fufaa GD, Weil EJ, Nelson RG, et al. Urinary monocyte chemoattractant protein-1 and hepcidin and early diabetic nephropathy lesions in type 1 diabetes mellitus. Nephrol Dial Transplant 2015; 30(4): 599-606.

[http://dx.doi.org/10.1093/ndt/gfv012] [PMID: 25648911]

[27] Nazir N, Siddiqui K, Al-Qasim S, Al-Naqeb D. Meta-analysis of diabetic nephropathy associated genetic variants in inflammation and angiogenesis involved in different biochemical pathways. BMC Med Genet 2014; 15: 103.
[http://dx.doi.org/10.1186/s12881-014-0103-8] [PMID: 25280384]

[28] Stefanidis I, Kreuer K, Dardiotis E, et al. Association between the interleukin-1β Gene (IL1B) C-511T polymorphism and the risk of diabetic nephropathy in type 2 diabetes: a candidate-gene association study. DNA Cell Biol 2014; 33(7): 463-8.
[http://dx.doi.org/10.1089/dna.2013.2204] [PMID: 24839897]

[29] Shikano M, Sobajima H, Yoshikawa H, et al. Usefulness of a highly sensitive urinary and serum IL-6 assay in patients with diabetic nephropathy. Nephron 2000; 85(1): 81-5.
[http://dx.doi.org/10.1159/000045634] [PMID: 10773760]

[30] Rakitianskaia IA, Riabov SI, Dubrova AG, Azanchevskaia SV, Riabova TS, Gurkov AS. [The role of IL-6 in the development of morphological changes in renal tissue in elderly patients with type 2 diabetes complicated by diabetic nephropathy]. Adv Gerontol 2012; 25(4): 632-7.
[PMID: 23734508]

[31] Elsherbiny NM, Al-Gayyar MM. The role of IL-18 in type 1 diabetic nephropathy: The problem and future treatment. Cytokine 2016; 81: 15-22.
[http://dx.doi.org/10.1016/j.cyto.2016.01.014] [PMID: 26836949]

[32] Luo C, Li T, Zhang C, et al. Therapeutic effect of alprostadil in diabetic nephropathy: possible roles of angiopoietin-2 and IL-18. Cell Physiol Biochem 2014; 34(3): 916-28.
[http://dx.doi.org/10.1159/000366309] [PMID: 25200363]

[33] Elsherbiny NM, Abd El Galil KH, Gabr MM, Al-Gayyar MM, Eissa LA, El-Shishtawy MM. Reno-protective effect of NECA in diabetic nephropathy: implication of IL-18 and ICAM-1. Eur Cytokine Netw 2012; 23(3): 78-86.
[PMID: 22995127]

[34] Domingueti CP, Dusse LM, Carvalho MD, de Sousa LP, Gomes KB, Fernandes AP. Diabetes mellitus: The linkage between oxidative stress, inflammation, hypercoagulability and vascular complication J Diabetes Complications 2015 Dec 18; (S1056-8727)(15): 00507-3.

[35] Navarro-González JF, Mora-Fernández C, Muros de Fuentes M, García-Pérez J. Inflammatory molecules and pathways in the pathogenesis of diabetic nephropathy. Nat Rev Nephrol 2011; 7(6): 327-40.
[http://dx.doi.org/10.1038/nrneph.2011.51] [PMID: 21537349]

[36] Liu JY, Chen XX, Tang SC, et al. Chinese medicines in the treatment of experimental diabetic nephropathy. Chin Med 2016; 11: 6.
[http://dx.doi.org/10.1186/s13020-016-0075-z] [PMID: 26913057]

[37] Tu X, Ye X, Xie C, Chen J, Wang F, Zhong S. Combination Therapy with Chinese Medicine and ACEI/ARB for the Management of Diabetic Nephropathy: The Promise in Research Fragments. Curr Vasc Pharmacol 2015; 13(4): 526-39.
[http://dx.doi.org/10.2174/1570161112666141014153410] [PMID: 25360835]

[38] Kim J, Moon E, Kwon S. Effect of Astragalus membranaceus extract on diabetic nephropathy. Endocrinol Diabetes Metab Case Rep 2014; 2014: 140063.
[http://dx.doi.org/10.1530/EDM-14-0063] [PMID: 25298884]

[39] Li M, Wang W, Xue J, Gu Y, Lin S. Meta-analysis of the clinical value of Astragalus membranaceus in diabetic nephropathy. J Ethnopharmacol 2011; 133(2): 412-9.
[http://dx.doi.org/10.1016/j.jep.2010.10.012] [PMID: 20951192]

[40] Zhang J, Xie X, Li C, Fu P. Systematic review of the renal protective effect of Astragalus membranaceus (root) on diabetic nephropathy in animal models. J Ethnopharmacol 2009; 126(2): 189-

96.
[http://dx.doi.org/10.1016/j.jep.2009.08.046] [PMID: 19735713]

[41]   Liu YH, Yang L, Liu J. [Clinical observation on treatment of early diabetic nephropathy by milkvetch injection combined with captopril]. Zhongguo Zhong Xi Yi Jie He Za Zhi 2005; 25(11): 993-5.
[PMID: 16355614]

[42]   Chen YQ, Zeng JS, Long L. Effect of astragalus on expression of NF-κB and MCP-1 in renal tissues of diabetic rats. Guizhou Medical Journal 2009; 33(2): 102-5.

[43]   Peng G. OuYang JP, Mao XQ, Zhou Feng. The effects of astragalus polysaccharides on TGF-β1 in renal tissues of STZ-induced type 2 DM rats. Chinese Journal of Microcirculation 2007; 17(2): 8-10.

[44]   Sun JH, Jing Y, Li SK, Yin Z, Jiang GR, Yin ZC. Effects of astragalus on expression of Cyclooxygenase-2 and NF-κB in diabetic nephropathy rats. Chinese Journal of Integrated Traditional and Western Nephrology 2014; 15(9): 770-2.

[45]   Yin X, Zhang Y, Wu H, *et al.* Protective effects of Astragalus saponin I on early stage of diabetic nephropathy in rats. J Pharmacol Sci 2004; 95(2): 256-66.
[http://dx.doi.org/10.1254/jphs.FP0030597] [PMID: 15215651]

[46]   Deng HO, Kai L, Li YL, Zhi XM, Wen W. The effects of astragalus polysaccharides on TNF-α, IL-6 and immunologic function in old patients with early stage diabetic nephropathy. Zhong Yao Cai 2014; 37(4): 713-6.

[47]   Wang ZS, Xiong F, Xie XH, Chen D, Pan JH, Cheng L. Astragaloside IV attenuates proteinuria in streptozotocin-induced diabetic nephropathy *via* the inhibition of endoplasmic reticulum stress. BMC Nephrol 2015; 16: 44.
[http://dx.doi.org/10.1186/s12882-015-0031-7] [PMID: 25886386]

[48]   Yin X, Zhang Y, Yu J, *et al.* The antioxidative effects of astragalus saponin I protect against development of early diabetic nephropathy. J Pharmacol Sci 2006; 101(2): 166-73.
[http://dx.doi.org/10.1254/jphs.FP0050041] [PMID: 16766854]

[49]   Pharmacopeia of People's Republic of China. 2015 Edition., Beijing: China Medical Science Press 2015.

[50]   Lee SH, Kim YS, Lee SJ, Lee BC. The protective effect of Salvia miltiorrhiza in an animal model of early experimentally induced diabetic nephropathy. J Ethnopharmacol 2011; 137(3): 1409-14.
[http://dx.doi.org/10.1016/j.jep.2011.08.007] [PMID: 21856399]

[51]   Yin D, Yin J, Yang Y, Chen S, Gao X. Renoprotection of Danshen Injection on streptozotocin-induced diabetic rats, associated with tubular function and structure. J Ethnopharmacol 2014; 151(1): 667-74.
[http://dx.doi.org/10.1016/j.jep.2013.11.025] [PMID: 24269771]

[52]   Kim SK, Jung KH, Lee BC. Protective effect of Tanshinone IIA on the early stage of experimental diabetic nephropathy. Biol Pharm Bull 2009; 32(2): 220-4.
[http://dx.doi.org/10.1248/bpb.32.220] [PMID: 19182379]

[53]   Chen GY, Tang SF, Su BL, *et al.* Effect of Tanshinone IIA on Renal Tumor Growth Factor-beta 1 and Nuclear Factor-kappa B in Diabetic Nephropathy Rats. Journal of Guangzhou University of Traditional Chinese Medicine 2015; 32(5): 891-5.

[54]   Lu WB, Yang PJ, Li SM, Lv YP, Huang ZY, Hong H. Effect of Salvianolate on Inflammatory Cytokines and Renal Vascular Endothelial Function in Early Diabetic Nephropathy. Zhongguo Shiyan Fangjixue Zazhi 2014; 20(2): 184-7.

[55]   Wang B, Ni Q, Wang X, Lin L. Meta-analysis of the clinical effect of ligustrazine on diabetic nephropathy. Am J Chin Med 2012; 40(1): 25-37.
[http://dx.doi.org/10.1142/S0192415X12500036] [PMID: 22298446]

[56]   Su BL, Jing L, Chen GY. The effects of ligustrazine on inflammatory cytokines and TGF-β1 in patients with DN. Shaanxi Journal of Traditional Chinese Medicine 2012; 33(8): 976-7.

[57] Zhen L. The effects of injection of Danshen Root and ligustrazine on inflammatory cytokines in patients with early stage of DN. Lishizhen Medicine and Materia Medica Research 2013; 24(7): 1693-4.

[58] Wang B, Chen S, Yan X, *et al.* The therapeutic effect and possible harm of puerarin for treatment of stage III diabetic nephropathy: a meta-analysis. Altern Ther Health Med 2015; 21(1): 36-44.
[PMID: 25599431]

[59] Pan X, Wang J, Pu Y, Yao J, Wang H. Effect of Puerarin on Expression of ICAM-1 and TNF-α in Kidneys of Diabetic Rats. Med Sci Monit 2015; 21: 2134-40.
[http://dx.doi.org/10.12659/MSM.893714] [PMID: 26201474]

[60] Cui XL, Wang YZ, Liu XJ. Effect of puerarin on the expression of NF-κB65 and TNF-αin kidney of diabetic rats Medical Journal of Chinese People's Liberation Anny 2010; 35(6): 679-81.

[61] Tao X, Lipsky PE. The Chinese anti-inflammatory and immunosuppressive herbal remedy Tripterygium wilfordii Hook F. Rheum Dis Clin North Am 2000; 26(1): 29-50, viii. [viii.].
[http://dx.doi.org/10.1016/S0889-857X(05)70118-6] [PMID: 10680192]

[62] Liu G, Shen Y, You L, Song W, Lu K, Li Y. [Tripterygium wilfordii polyglycoside suppresses inflammatory cytokine expression in rats with diabetic nephropathy]. Xibao Yu Fenzi Mianyixue Zazhi 2014; 30(7): 721-4.
[PMID: 25001937]

[63] Kun L, Liu GL, Shen YJ, Wei S, Li YC. Protective effect of Tripterygium wilfordii polyglycoside on inflammatory injury kidney of rats with diabetic nephropathy. Pharmacology and Clinics of Chinese Mareria Medica 2015; 31(3): 93-6.

[64] Huang YR, Wan YG, Sun W, *et al.* [Effects and mechanisms of multi-glycoside of Tripterygium wilfordii improving glomerular inflammatory injury by regulating p38MAPK signaling activation in diabetic nephropathy rats]. Zhongguo Zhongyao Zazhi 2014; 39(21): 4102-9.
[PMID: 25775776]

[65] Song CD, Yang XL, Xue LM, Ren RY, Hou XJ. The effects of Tripterygium Glycosides on the expression of TGF-β1/p38MAPK in the kidney tissues of early stage DN rats. Chinese Journal of Basic Medicine in Traditional Chinese Medicine 2012; 18(12): 1348-50.

[66] Chen BJ. Triptolide, a novel immunosuppressive and anti-inflammatory agent purified from a Chinese herb Tripterygium wilfordii Hook F. Leuk Lymphoma 2001; 42(3): 253-65.
[http://dx.doi.org/10.3109/10428190109064582] [PMID: 11699390]

[67] Gao Q, Shen W, Qin W, *et al.* Treatment of db/db diabetic mice with triptolide: a novel therapy for diabetic nephropathy. Nephrol Dial Transplant 2010; 25(11): 3539-47.
[http://dx.doi.org/10.1093/ndt/gfq245] [PMID: 20483955]

[68] Ma RX, Liu LQ, Yan X, Wei J. Protective effect of triptolide on renal tissues in type2 diabetic rats. Chinese Journal of Hypertension 2008; 16(12): 1120-4.

[69] Ge Y, Xie H, Li S, *et al.* Treatment of diabetic nephropathy with Tripterygium wilfordii Hook F extract: a prospective, randomized, controlled clinical trial. J Transl Med 2013; 11: 134.
[http://dx.doi.org/10.1186/1479-5876-11-134] [PMID: 23725518]

[70] Liu YX, Wang HY. Mechanism of Herba Houttuyniae on relieving renal impairment in STZ-induced diabetic rats. Zhongyao Xinyao Yu Linchuang Yaoli 2010; 21(2): 107-10.

[71] Fang W, Lu FE, Xu LJ. Influence of Houttuynia Cordata Thunb on expression of BMP-7 and TGF-β1 in kidney tissue of diabetic rats. Tianjin Journal of Traditional Chinese Medicine 2006; 23(4): 334-7.

[72] Zheng YP, Kang HY. Effect of Radix Paeoniae Rubra on TNF-α, MCP-1, and ICAM-1 expression and macrophage infiltration in the kidney of diabetic nephropathy rats. Chinese Journal of Basic Medicine in Traditional Chinese Medicine 2013; 19(6): 637-9.

[73] Chen L. Effect of fructus corni on diabetic on Nephropathy in rats. Medical Journal of Qilu 2015; 30(

6): 674-6.

[74] Du SY, Xin Z, Lou LM, Qian YK. Effect of water extract of wolfberry fruit on expresions of interleukin-6 and tumor necrosis factor. Chin J Immunol 1994; 10(6): 356-8.

[75] Zhang LN, Xie XS, Zuo C, Fan JM. [Effect of ginsenoside Rg1 on the expression of TNF-α and MCP-1 in rats with diabetic nephropathy]. Sichuan Da Xue Xue Bao Yi Xue Ban 2009; 40(3): 466-71. [Medical Science Edition].
[PMID: 19627007]

[76] Wei HF, Wei YH, Li YQ, Miao CS, Cai L. Inhibitory effects of tannic acid on inflammatory factor expressions of renal tissues in diabetic rats. Journal of JiLin University 2011; 37(3): 393-7. [Medicine Edition].

[77] Wei HF, Cai L, Lei L, *et al*. Ameliorating effect of tannic acid on oxidative stress and micro-inflammatory state in diabetic rats. Chinas Med 2014; 9(10): 1479-84.

[78] Wang XG, Wei HF, Wei YH, Li YZ. The effect of tannic acid on the expression of C-reactive protein in serum and kidney of diabetic rats. Chinese Journal of Laboratory Diagnosis 2010; 12(14): 1910-2.

[79] Jun L, Wu LY, Wang CY. Influence of breviscapine injection to correlated inflammatory cytokines in patients with type 2 early diabetic nephropathy. Pharmacology and Clinics of Chinese MateriaMedica 2011; 27(3): 110-2.

[80] Cui XL, Liu XJ. Effects of breviscapine on expressions of PKCa,TNF-αin kindey of diabetic rats. Chin Hosp Pharm 2012; 32(6): 417-21.

[81] Xie X, Chang X, Chen L, *et al*. Berberine ameliorates experimental diabetes-induced renal inflammation and fibronectin by inhibiting the activation of RhoA/ROCK signaling. Mol Cell Endocrinol 2013; 381(1-2): 56-65.
[http://dx.doi.org/10.1016/j.mce.2013.07.019] [PMID: 23896433]

[82] Tang LQ, Ni WJ, Cai M, Ding HH, Liu S, Zhang ST. Renoprotective effects of berberine and its potential impact on the expression of β-arrestins and ICAM-1/VCAM-1 in streptozocin induced-diabetic nephropathy rats. J Diabetes 2015; 8(5): 693-700.
[http://dx.doi.org/10.1111/1753-0407.12349] [PMID: 26531813]

[83] Wang SF, Tang S, Cao KG. Analysis on medication rule of traditional Chinese medicine in treating diabetic nephropathy. Tianjin J Trad Chin Med 2009; 26(2): 167-8.

[84] Mei W, Yu JN, Yan W. Preliminary exploration in the therapeutic strategies and prescription laws of TCM for the management of diabetic nephropathy in modern literature. Chin J Integr Trad Western Nephrol 2011; 12(3): 259-60.

[85] Ying L, Chen F, Wang HY, Song HP. Effects of Buyanghuanwu Decoction(BHD) on TGF-β1 in Early Diabetic Nephropathy. Chinese Journal of Medicinal Guide 2011; 13(4): 656-7.

[86] Zou ZX, Liu GH, Heng XP, *et al*. Flavored tonifying Yang also five decoction on diabetic nephropathy TGF - beta-1 and express the influence of Smad7. Fujian. J Tradit Chin Med 2013; 06(44): 20-2.

[87] Ye RQ, Lin GB, Deng SL, *et al*. Effect of Buyanghuawu decoction on IL-6 and TNF-α in serum of patient of early diabetic nephropathy. Hebei Journal of Traditional Chinese Medicine 2011; 33(3): 383-4.

[88] Liang Z, Peng LX, Luo JY, He FY, Gang Z. Fingerprint analysis of the Chinese Traditional Medicine "Buyang Huanwu Decoction" by HPLC. Journal of Southwest University 2009; 31(4): 48-51. [Natural Science Edition].

[89] Xing Z, Lu TL, Mao CQ, *et al*. Quality standard for Buyang Huanwu Decoction. Zhongchengyao 2013; 35(3): 529-33.

[90] Zhen Z. Effect of compound dan-shen dropping pills on serum adiponectin TNF-a☐IL-6 of early diabetic nephropathy patients. Clinical Journal of Chinese Medicine 2014; 06(09): 5-6.

[91] Wu DL, Gong JH, Yan C, Yang YM. Effect of compound Dan-shen dropping pills on microalbuminuria and micro-inflammatory state in type 2 diabetic patients with diabetic nephropathy. Hainan Medical Journal 2015; 26(05): 687-9.

[92] Min CY, Fu TT. Du Yu, Wang CJ, Zhan Feng. Kidney T2DM CTGF in patients with diabetic nephropathy, the effect of MMP. Jorunal of Chinese Medicinal Materials 2012; 35(3): 507-9.

[93] Min CY, Chen YS, Fu TT, et al. Kidney in T2DM with the influence of inflammatory factors in patients with diabetic nephropathy. Jorunal of Chinese Medicinal Materials 2013; 36(02): 334-7.

[94] Gong CX, Wang XP, Gao YF, Jing L, Wang ZP. Qidan Yishen Jiangtang Capsule on diabetic nephropathy which express the effects of rat kidneys. Journal of Emergency in Traditional Chinese Medicine 2012; 21(10): 1607-8.

[95] Zhou FW, Li YQ, Gong CX, et al. The effects of Qidan Yishen Jiangtang Capsule on TGF - beta 1 and VEGF in rat model with diabetic nephropathy. Chinese Journal of Ethnomedicine and Ethnopharmacy 2009; 18(22): 15-6.

[96] Jia FX, Li YQ, Gui L, Song CH, Wang ZP. Study of T L ym phocyte Subgroup and Changes of TG F - B IGF - 1 in Patients w ith D iab etic N ephroph thy on Qidan Yishen Jiang tang Capsule. Zhonghua Zhongyiyao Xuekan 2008; 26(10): 2201-3.

[97] Zhou JJ, Guo HW. Clinical Effects of Kidney-invigorating,Blood-promoting and Wind-dispelling Therapy on Diabetic Nephropathy in IV and V Stages and Its Influence on MCP-1. Acta Universitatis Traditionis Medicalis Sinensis Pharmacologiaeque Shanghai 2012; 26(2): 42-5.

[98] Peng SL. Impacts on Inflammatory Factor of Diabetic Nephropathy Treated with the Therapy for Strengthening the Spleen and Kidney and Activating Blood Circulation. World Journal of Integrated Traditional and Western Medicine 2015; 10(5): 686-7.

[99] Zhu LQ, He LJ, Lin YB, Yu Y, Zhang YJ. Effects of nephritis rehabilitation tablets combined with irbesartan on expresions of inflammatory cytokines IL-6 and TNF-α in Early Diabetic Nephropathy. Chinese Journal of Integrated Traditional and Western Nephrology 2014; 15(6): 539-40. [http://dx.doi.org/10.1016/S0254-6272(15)30059-5]

[100] Yao MX, Qian X, Shang YF, et al. Study of Ge Pi Decoction on regulation of monocyte chemoattractant protein-1 level to reverse early diabetic nephropathy. Zhongguo Shiyan Fangjixue Zazhi 2010; 16(4): 145-8.

[101] Yao MX, Zhang FH, Wang HL, Xu AM, Ma AG. Effects of Gepi decoction on level of TNF-α in rats with type 2 diabetic mellitus China Tropical Medicine 2009; 9(3): 423-5.

[102] Yao Z, Ning Z, Gao QH, Duan YJ, Lv WL. The Study of the Effects of Anti-inflammatory Factor in Experimental Type 2 Diabetic Rats Drank by Lianpu Drink. Hubei Journal of Traditional Chinese Medicine 2013; 35(03): 6-7.

[103] Jian Y, Ting L, Zheng BL, et al. Effects of Shenweining Granules in Early and Middle staged diabetic Nephropathy of Inflammatory Cytokines. New Journal of Traditional Chinese Medicine 2015; 47(4): 105-6.

[104] Zhu MQ, Xia JY. Effect of Modified Angelica Peony Powder Combined with Irbesartan on Inflammatory Cytokine in Patients with Early Diabetic Nephropathy. Zhongguo Jiceng Yiyao 2014; 21(16): 2546-7.

[105] Lan S, Yu LQ, Xiong PH. Protective effect and mechanism of Liuwei Dihuang combined with HuangLianJiedu decoction on diabetic nephropathy rats. Journal of NanJing University of TCM 2013; 29(6): 553-7.

[106] Yi T, Lu FE, Xu LJ, Leng SH, Wang KF. Effect of Antidotal Decoction of Coptis on Cytokines IL-4 and IL-10 in Type2 Diabetic Rats. Chinese Journal of Microcirculation 2015; 15(3): 44-5.

[107] Yu LQ, Liao GH. ShenLan. Effect of Yangyin Qingre Decoction on Inflammatory Cytokine in Serum of Type 2 Diabetic Rats. Chinese Journal of Integrated Traditional and Western Nephrology 2013; 14

(11): 995-7.

[108] Zhang ML, Liao WY. Clinical Effect of Jianpi Yishen Decoction on Patients with Early Diabetic Nephropathy and the Study of Its Influence on Level of C-reactive protein and TNF-αin serum. New Journal of Traditional Chinese Medicine 2006; 38(12): 26-7.

[109] Juan Lv. Effects of Tangmaikang (TMK) on expressions of inflammatory cytokines in patient of early diabetic nephropathy. Journal of Practical Diabetology 2014; 10(6): 18-9.

[110] Fan GJ, Tang XY, Lu S, *et al.* Effect of Jiangtang Bushen Prescription on Expressions of TNF-α,I--6,CRP in serum of Diabetic rats. Lishizhen Medicine and Materia Medica Research 2011; 22(11): 2721-2.

[111] Wei F, Jun Y, Du YL. Clinical observation Jianpi Bushen decoction therapy on 54 patients the inflammation of diahetic nephropathy in early phase. Chinese Journal of Integrated Traditional and Western Nephrology 2010; 11(12): 1102-3.

CHAPTER 5

Through the Perspective of Histology - The Alzheimer's Disease Promotion by Obesity and Glucose Metabolism: Type 3 Diabetes

Cigdem Elmas* and Cemile Merve Seymen

Department of Histology and Embryology, Faculty of Medicine, Gazi University, Ankara, Turkey

Abstract: Since the 1990s The World Health Organisation (WHO) has stated that "... obesity should now be regarded as one of the greatest neglected public health problems of our time..." and defined the global epidemic of being overweight and obese as "globesity". A positive energy balance, which consists of an imbalance between energy intake and calorie expenditure is the main cause of obesity, however, genetic, environmental, socioeconomical, behavioral and psychological factors may also be the inducing factors when it comes to obesity. The excess of adiposity has an enhancing effect on the development of hypertension, cardiovascular diseases and type 2 diabetes mellitus (T2DM) as a result of the resistance to insulin-mediated glucose disposal. T2DM which represents a common disease associated with obesity and often aging is characterized by high blood glucose levels, impaired insulin production and peripheral insulin resistance. Homeostatic degradation of glucose affects the cerebral functions directly or indirectly because glucose is a significant metabolic substrate for all cells and also for the cells of the brain. Insulin has a key effect on the regulation of energy metabolism of neurons and neuronal recovery, which acts as a growth factor on all cells including neurons in the central nervous system. Therefore, simply put, impairment in neuronal homeostasis which occurs as a result of insulin deficiency. These are considered to be a risk factor for Alzheimer's disease (AD) development. Indeed, many studies have shown that glucose intolerance and impairment of insulin secretion are associated with a higher risk to develop dementia or AD. It is worth remembering that AD is associated with brain insulin resistance and deficiency, whereas T2DM is associated with peripheral insulin resistance. In short, it can be said that T2DM causes AD-type neurodegeneration in the brain. T2DM and AD share several molecular processes that underlie the degenerative developments. Dysregulated glucose metabolism, abnormalities in insulin signaling, the formation of advanced glycation end products, oxidative stress, the activation of inflammatory pathways and abnormal protein processing are the common characteristics of T2DM and AD. The misfolding of proteins plays an important role in both diseases, so as the aggregation of amyloid peptides. AD is characterized by the deposition of amyloid within neurons and amyloid plaques.

---

* **Corresponding author Cigdem Elmas:** Department of Histology and Embryology, Faculty of Medicine, Gazi University, 06500 Besevler, Ankara, Turkey; Tel: +90-312-202-4648, 05336550394; Fax:+903122124647; E-mail: 00cigdem@gmail.com

Atta-ur-Rahman (Ed.)
All rights reserved-© 2019 Bentham Science Publishers

Also in AD, the formation of amyloid fibers could be the product of ubiquitin-mediated protein degradation defects induced by a dysfunction of the proteasome. According to one study which was conducted on T2DM rats, T2DM-dependent decreases in p62 (a known cargo molecule that transports polyubiquitinated tau to proteosomal and autophagic degradation systems) transcription which is a primary mechanism underlying increased AD-like pathology. In some studies, brain amyloid deposition occurs as a result of increased blood-brain barrier permeability in case of diabetes conditions. In the recent years, according to some members of the diabetes community, AD is seen as a neuroendocrine disorder and the term "Type 3 Diabetes" defines the insulin deficiency and resistance in the brain of those with AD. In the context of these information, in this chapter, we propose a study about "Type 3 Diabetes" with the underlying mechanisms through the perspective of histology.

Keywords: Advanced Glycation Products, Alzheimer's Disease, Amyloid Beta Deposition, Calcium Dysregulation, Ceramides, Endoplasmic Reticulum Stress, Glucose Metabolism, Inflammatory Response, Insulin, Insulin Degrading Enzyme, Insulin Resistance, Mitochondrial Dysfunction, Neurofibrillary Tangles, Neuronal Homeostasis, Neurodegeneration, Obesity, Oxidative Stress, Tau Hyperphosphorylation, The Apolipoprotein E, Toll-Like Receptors, Type 2 Diabetes Mellitus, Type 3 Diabetes, Ubiquitin Proteosome System.

THE PATHOPHYSIOLOGY OF ALZHEIMER'S DISEASE

Alzheimer's Disease (AD) is the most common reason of dementia charts with 60-80% rate. It occurs due to losses of neurons and synapses in various parts of the Central Nervous System (CNS). It is a progressive, chronic neurodegenerative disease and characterized with a decrease in cognitive functions, self-care deficiencies, variety neuropsychiatric and behavioral disorders [1, 2]. It is estimated that 4.6 million new cases of AD occur in the world every year [3]. It was originally defined with its clinical and neuropathological features by a German doctor Alois Alzheimer in 1907 [4, 5]. The risk of incidence of this disease increases depending logarithmically on age [6, 7]. While frequency of incidence in the population between 60-65 ages is approximately 0.1%, the frequency of incidence in population over 85 years extends to 47% [7, 8]. It is predicted that more than 25% of the world population will be over 65 years by 2050. These numbers demonstrate that AD, which increases logarithmically with an increasing old age, will be one of the most significant health problems worldwide [9].

The macroscopic pathological findings of AD are atrophy in the brain, and stenosis in gyrus, extension in sulcus and ventricles [10]. Neurofibrillary tangles accumulating in neuron and amyloid plaques accumulating extracellularly and neuron losses are accepted as basic microscopic changes of the disease [11]. It is reported that in AD, neurofibrillary tangle pathology shapes the progress parallel to clinical symptoms in the brain and that amyloid plaques are not directly connected with clinical symptoms [12].

Neurofibrillary tangles consist of neurofibrillary tangles fibril clusters which involve the hyperphosphorylated shape of tau protein connected with microtubule and accumulated in cell body and dendrites [13]. It is alleged that hyperphosphorylation of tau could occur as a result of imbalance between various kinases arranging phosphorylation and phosphatase activities [14]. And these hyperphosphorylated tau proteins cause to disrupt the cell functions [15].

There is extracellular accumulation of amyloid beta (Aβ) deriving from APP (Amyloid Precursor Protein) in senile plaques in AD [13]. APP is metabolized by two ways. First, the non-amyloidogenic way is a metabolized pathway, in which alpha secretase (α-secretase) and gamma secretase (γ-secretase) enzymes digesting APP, exist and consist of products which are nontoxic. In amyiodogenic way, Aβ1-40 and Aβ1-42 occur *via* beta secretase (β-secretase) and gamma secretase (γ-secretase) enzymes [16, 17]; approximately, 90% of Aβ is Aβ1-40 [18]. Neurotoxic Aβ piling up in cellular compartment impairs cellular functions by causing mitochondrial and synaptic damages and hyperphosphorylation of tau protein [15, 19].

In AD, the rise in neurotransmitter and Acetylcholine (ACh) levels increase caution of physiological conditions [20]. It is stated that changes in other neuromediator levels are lesser than changes in ACh levels. It is reported that in AD, decrease in ACh synthesis depends on a decrease in the amount and function of Choline Acetyltransferase (CAT) enzyme and a decrease in re-uptake of choline. And this situation on damages and losses the cholinergic neuron and axon, which are projected to the cortex and hippocampus. Losses have been observed in nicotinic receptors whose effects on learning and memory are obvious, and in presynaptic M2 muscarinic receptors but the concentration of postsynaptic M1 muscarinic receptors did not change in AD [21].

It has been reported that cholinergic loss which occurred in AD is associated with various behavioral and psychiatric symptoms such as depression, agitation, anxiety and psychosis [22]. It is assumed that irregularities and neuron losses occurred in serotonergic and dopaminergic neurotransmission as well as cholinergic dysfunction [22, 23].

According to some studies that were conducted to explain the physiopathology of AD, intracellular Calcium (Ca^{+2}) concentration which increase as a result of stimulating of glutamatergic system, chronically causes neuronal excitotoxicity, thus results in neuronal dysfunction and cell death [24]. The increased levels of Ca^{+2} concentrations in the brain enhance neutral proteinases (calpains) and causes amyloid plaques and neurofibrillary tangles according to some studies [25].

One of the hypothesis associated with etiopathogenesis of AD is oxidative stress hypothesis. It is pinpointed that in AD, oxidative damage increases and that this increase could cause neuronal degeneration and death. Among the findings showing contribution of oxidative stress to etiopathogenesis of the disease, an increase in iron, aluminum and mercury concentrations stimulating formation of free radicals in Alzheimer's patients, an increase in lipid peroxidation, protein and DNA oxidation, disruption in energy metabolism, decrease in amount of cytochrome oxidase C, existence of peroxynitrite, heme oxygenase-I and superoxide dismutase-I in neurofibrillary tangles and lastly a production of amyloid plaques [26, 27].

The results of some AD studies report that arginine-NO pathway is associated with Alzheimer's pathogenesis [28], and that nitric oxide (NO) levels which increased during the disease process ruin the brain homeostatis and cause brain lesion and pathology [29]. It has been suggested that over production of NO could play a role in the neurotoxicity resulting from NMDA receptor [30]. Reactive oxygen types occurring as a result of over production NO activates various neurotoxic mechanism and these mechanisms could cause neuron and memory losses in Alzheimer's patients [31].

Some studies suggest that unstable distribution of neurotrophic factors, inflammatory processes, energy metabolism disorders, mitochondrial damage, viral diseases and neurotoxicity are associated with Alzheimer ethiopathology [26, 31].

Neuronal Loss

Neuronal loss in AD starts from entorhinal cortex. Following the limbic system it is determined in superior temporal sulcus. Advance style of neuronal loss in time and anatomical susceptibility type of neuronal loss generally resembles a type of neurofibrillary tangles. There is a negative correlation found between neurofibrillary tangles and the number of neurons. In addition to this, neurofibrillary tangles cannot be regarded as being responsible for the death of neurone on its own. Neuronal loss is not required in sections where neurofibrillary tangles exist as well as subcortical cores. On the other hand, heavy neuronal loss may occur in sections where neurofibrillary tangles exist in small numbers. Amyloid neurotoxicity and trans-synaptic degeneration are the other factors that are regarded to play a role in cell death. As another mechanism of cell death, apoptosis or programmed cell death is also referred [32].

Synapse Loss

Synapse loss is a basic structural change correlated with clinical dementia weight in cortical biopsy samples. Synapse loss is mostly explained with secondary effect resulting from anterograde Wallerian degeneration of neurofibrillary tangle and neuron weight. However, it is possible that primary damage can be found in synapses and this causes neurofibrillary tangle, and finally cell death [32].

INTERACTIONS BETWEEN DIABETES MELLITUS AND ALZHEIMER'S DISEASE

In the last few years, AD generally seems to be related to a neuroendocrine disorder and is defined as Type 3 Diabetes (T3D). T3D defines the insulin deficiency and resistant in the brain of those with AD. The risk of having Alzheimer doubly increases in patients with insulin resistance, metabolic syndrome and Type 2 Diabetes (T2D) diseases [33].

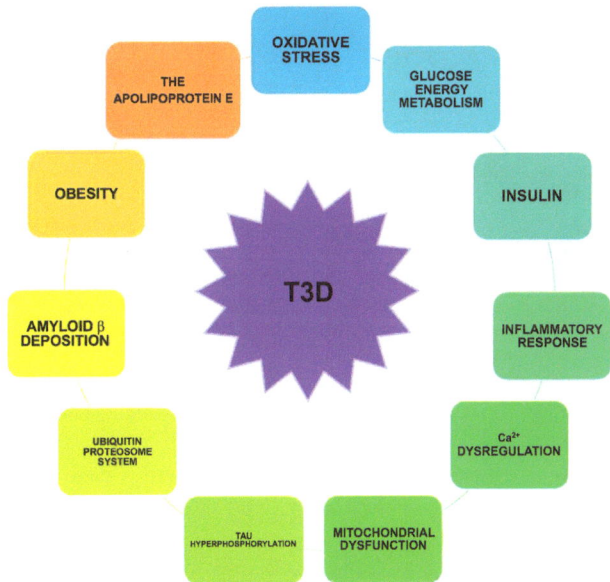

Fig. (1). Some factors responsible for development of T3D pathology.

There have been attempts to explain the relationship between DM and AD through findings. Proteins glycated in the highest degree can enhance amyloid toxicity and facilitate neurofibrillary tangle formation. Hyperglycemia represses acetyl choline production by affecting the use of glucose in the brain [34].

Now, we will examine the pathophysiological similarities of DM and AD and also see the possible mechanisms of T3D or AD that is triggered by DM through different headings:

OXIDATIVE STRESS

The speed of free radical formation in organism and the speed of destroying of them stand in balance. This condition is called oxidative balance. If oxidative balance is supplied, the organism is not influenced by free radicals. Increase in formation of these radicals or decrease in destroying their speed causes a breakdown in the balance. Briefly, this condition called "Oxidative Stress" demonstrates imbalance between free radical formation and antioxidant defense mechanism and it causes tissue damage [35].

Diabetes and Oxidative Stress

The role of reactive oxygen types in diabetes has been an issue of discussion since 1980's [36]. Many studies point out the relationships between diabetes, complications and reactive oxygen types. Tissue damage occurring as a result of metabolic stress, sorbitol way activity, hypoxia and ischemia-reperfusion which result from changes in nonenzymatic glycation and energy metabolism, plus the enhancement of free radical production [37] and changes in the antioxidant defense system [38 - 42]. It is recognized that expressions and antioxidant capacity of antioxidative enzymes such as superoxide dismutase, catalase, and glutathione peroxidase have the lowest levels in pancreas islet cells in comparison to other tissues such as liver, kidney, skeletal muscle and adipose tissues [43, 44]. It is thought that damage observed in beta cells known to be one of sensitive structures against oxidative stress results from toxic effects of hyperglycemia [45]. Researchers think that hydrogen peroxide has an influence upon insulin signal receptor system after transformed OH radical, one of ROS production having reactivity, and can play a key role in signal transduction arranged *via* receptor by insulin [46]. Findings indicate that free radical production mediated glycation causes beta cell apoptosis also support this view [46, 47]. It is thought that T and B lymphocytes produce toxic effects of inflammatory cells like macrophages to beta cells *via* free radicals [48]. The increase was observed in 8-OHdG (8-hydroxy deoxy guanosine) levels considered as an oxidative stress marker in rat experiment models for diabetes [49]. As well as the studies supporting the idea that free radical formation is a direct result of hyperglycemia [50], it was observed that free radical formation begins when endothelial and smooth muscle cells are incubated in an ambiance containing glucose in high concentration [51, 52]. The idea that there is a close relationship between hyperglycemia and oxidative stress is supported *in vivo* studies [53]. It is thought that streptocolysin, which is derivative of N-nitroso compounds used for constituting diabetes similar to humans in experimental animal studies [54], destroys langerhans islets selectively by occurring oxidant matters and initiates diabetes by giving inappropriate NO answers [55, 56].

As well as hyperglycemia, the view related to this suggests that there is a risk for complications of diabetes in hypertriglyceridemia which in itself is significant [57 - 59]. Studies have shown that diabetic cases enhance lipid peroxydation in plasma lipoproteins, erythrocyte membrane lipids and various tissues. However, it is unknown whether this increase results from more enzymatical (arachidonic acid path) lipid peroxydation or nonenzymatic lipid peroxydation. Lipid peroxydation originates nonenzymatically from both prostaglandins with lipooxygenase becoming active as a result of common vascular inflammation and from lipids in membranes of endothelial and phagocytic cells with effects of free radicals and transition metals. Then, it is reported that products belonging to both sides enhance lipid peroxydation by activating each other mutually. Epidemiology demonstrates that an increase in plasma lipid peroxydation is associated with vascular disease and hypertriglyceridemia instead of diabetes [55, 60, 61]. Findings of researchers reveal that there are increases in both LDL oxidation and nonenzymatic glycation depending on hyperglycemia in diabetic patients with vascular complication. In diabetic cases protein oxidation increases addition to lipids. Especially as a result of oxidation of extracellular proteins in collagen, elastin and myelin, diabetic complications such as cataract, microangiopathy, atherosclerosis and nephropathy develop in tissues like lens, vessel and basal membrane [55, 62, 63].

Oxidative stress is a link between T2DM and also AD, hence it plays an important role in the pathogenesis of T3D [64]. The accumulation and the insufficient detoxification of oxidative damaged proteins, nucleic acids or lipids are the possible same mechanism of two disorders [65]. Advanced Glycation Products (AGEs) and Advanced Lipid Peroxidation Products (ALEs) are produced as a result of oxidative stress with slow and complex reactions and proinflammatory responses in case of T2DM, and moreover in AD. Hence, they are the potential common markers of the two disorders [66].

Advanced Glycation Products (AGEs)

Glycation products occurring as a result of high plasma diabetes in diabetic patients cause neuronal damage associated with oxidative stress. Also, these glycation products can activate fundamental inflammatory pathways such as nuclear factor cappa B by connecting their receptors [67 - 69].

DM is a chronic disease characterized with hyperglycemia forming at the end of insulin effects in insulin secretion or defect in both of them. Deficiency of insulin effects in target tissues underlies at the base of anomalies in carbohydrate, fat and protein metabolisms observed in DM [70]. As well as the metabolic problems of diabetes, exposing these patients to hyperglycemia for a long time can cause microvascular and macrovascular complications such as nephropathy, neuropathy and atherosclerosis. Complications appear mostly in tissues like cardia, nervous system and small blood vessels when glucose enters into a cell independently from insulin [71]. AGEs play a role in developing these complications due to high diabetes level in the blood, and appear in connective tissue diseases such as rheumatoid arthritis, and in neurologic diseases such as AD and in many pathological diseases such as at the end of stage renal disease (ESRD) [72, 73].

AGEs (Advanced Glycation End-Products) Formation Mechanism

AGEs consist of heterogeneous components which occur as a result of nonenzymatic glycation of nitrogen groups in proteins, lipoproteins and/or nucleic acids with carbonyl groups of reductive glucose. AGE products were originally described in 1912 by Louis Camille Maillard. AGEs were used in food chemistry at first and then started to be researched with exploration of HbA1c in diabetic patients in 1968 [74]. Protein glycation starts when Schiff alcali is formed by carbonyl group of glucose and free amino groups of protein. Formation of Schiff alcali occurs within hours and then transforms Amadori products within days. Amadori products then transform into dicarbonyl compounds and into AGEs within weeks. Although Amadori products are partially recyclable, future stages are not recyclable as glycation depends on concentration. In addition, dependency is more prominent in the early stages and for this reason its production increases in diabetes [72].

At the end of glucose and lipids oxidate, other mechanism in AGEs formation is the formation of low molecular weight dicarbonyl compounds such as 3-deoxiglucosone, glyoxal and metilglyoxal whose reactivities are high, as intermediate products. Dicarbonyl compounds can occur from generally glucose intermediate products, degradation of protein exposed to glycation and peroxydation of lipids. In addition, they can occur from metilglyoxal, ketone bodies metabolisms and threonine catabolism in fewer amounts. Dicarbonyl compounds have high chemical activity and can lead to AGEs formation by entering reaction with terminal amino acids directly. Another fundamental mechanism which plays an important role in AGEs formation is poliol pathway. A part of glucose occurring in high amount depending on diabetes, transforms into sorbitol and then into 3-deoxiglucosone, one of AGE product [75].

However, these reactions lead to indirect increases in oxidative stress because they lead to the consumption of NADPH and glutation. Because many agents play a role in AGEs formation through this way, AGEs have a heterogeneous structure. Foods, which we take in our diet exclude these, and tobacco products contain reactive AGE precursors. When we look at AGEs in foods in order to compare cooking foods in high level (grilling, frying) the AGE level increases significantly and cooking with plenty of water in lower level and in minimum time reduces AGEs [76]. These AGEs in heterogeneous structure are divided into 3 parts according to their chemical structure [74]:

1. Fluorochrome cross linked AGEs; 'pentosidine' and 'crossline'
2. Nonfluorochrome cross linked AGEs; 'glucosepane' and 'MOLD'
3. Non cross linked AGEs; 'N- carboxymethyl lysine (CML) and 'Pyrraline'

General Features of AGEs and AGE Receptors

Effective factors in AGEs formation are the production and destruction the speed of proteins, hyperglycemia levels and the amount of environmental oxidative stress and its prevalence [77]. The fact is that AGEs formation takes a long time, and as such it affects proteins in the long term. Lysine, histidine and arginine amino acids in proteins are more sensitive towards glycation. In case of AGEs formation in high level such as the end stage of renal failure, nucleic acid and lipid compounds participate in AGEs formation because the amount of glycation depends on hyperglycemia level and it can occur both in the cell and out of the cell. While glucose has the lowest producing glycation among reductive glucoses, such as fructose, threose and gliseraldehit-3-phosphate; it has a higher rate of causing glycation [78].

Fundamental diabetes complications in AGEs play two different roles. Firstly, it ruins matrix structure and functions by forming cross ligaments between proteins especially through extracellular matrix structures. Secondly AGEs cause many metabolic changes by building a way for various transcription factors and synthesis of cytokines by activating various signals [78].

AGEs mostly connect to receptors for end products of advanced glycation (RAG E), to scavenger receptors (Class A, CD36, class B tip1, LOX-1, FEEL-1, FEEL- 2), and to receptors named AGE-R1 (oligosaccharide transferase-48), AGE-R2, AGE-R3 (Galactin-3). The most examined of these receptors is RAGEA and it is a member of IgG super family. It mainly exists in mononuclear phagocyte, endothelial cell, smooth muscle cell and astrocyte although they are expressed in little amounts in normal vessels and tissues [39]. RAGE can be stimulated by inflammatory cytokine, amphoter, amyloid- β and other fibrillary proteins, apart from AGEs. In RAGE expression, diabetes and inflammation increases. Extracellular section of RAGE consists of two sections including type V section and type C section. While type V section is responsible for ligand bonding, type C section provides stability of type V section. Transmembrane section which follows this section provides anchoring of receptor to the membrane. At the end of transmembrane, there is a small intracytoplasmic tail, this section provides intracellular signalization [79]. Different RAGE isoforms exist. These are full-length RAGE, soluble RAGE (sRAGE) and dominant negative RAGE (DNRAG E). The basic difference between these 3 receptors is that all of them except full-length RAGE have intracytoplasmic tail. Hence, full-length RAGE can perform intracellular signalization. Due to this feature, DNRAGE and sRAGE AGE effects are effective in repressive style [80].

With bonding of AGE to its receptor, GTPs (Guanosine triphosphate) such as protein kinases (MAPKs) is activated with NAD(P)H oxidase, p21ras and mitogen, such as kinases/2, p38, Cdc42 ve Rac is regulated with extracellular signal, activates NF κB to warn intracellular signal pathways. Activation of NF κB provides expression of inflammatory cytokines, adhesion molecules and various mediators. With activation of NF κB, scavenger receptors bind AGE-R1, AGE-R2, AGE-R3 receptors to AGE, however do not initiate signal conduction with RAGE and provide detoxification and clearance of AGE [73].

AGEs accumulation is seen as normal aging within the various cell types but this accumulation process significantly increases in DM and also in AD [66]. In the case of extracellular AGEs accumulation, neurofibrillary tangles and Aβ plaques are modified by it, so the production rate of AGEs is very important for the detection of T2DM and also for AD [81].

GLUCOSE/ENERGY METABOLISM IN ALZHEIMER'S DISEASE

The brain is a very active organ in terms of metabolism energy and it is used for regulating events such as ion concentration in synaptic conduction, formation of electrical potentials, active uptake of excitatory neurotransmitter and syntheses operations. While performing these tasks the brain uses glucose, which is an important metabolic substrate, as a basic energy resource. Glucose metabolism in the brain is oxidative and most of the glucose oxidizes to carbon dioxide and water. Complete oxidation of glucose to carbon dioxide and water cannot always occur. This case is called glycogen pathway and it is important for astrocytic glycogen stores [82]. Glycogen is stored especially in astrocytes. It is stated that most of the glycogens in the brain are biologically usable and can be snowed under glucose, phosphate and lactate [83]. This subject will be detailed in the following sections under the title of "Glucose Transporters (GLUTs)". At this stage carrying glucose to the brain fulfills *via* high molecular weight glucose transporter protein Glucose Transporter-1 (GLUT-1), which is independent of Blood-Brain obstacle and sodium. Other transporter proteins exist in the membranes of different cell types. Astrocyte is rich in terms of molecule weight made glucose, especially GLUT-1. GLUT-1 and GLUT-4 are reported in hippocampus, cerebellum, olfactory bulbus and lateral hippocampus and in less amount globus pallidus. GLUT-3 and GLUT-8 exist in neuronal cells, GLUT-5 exists in microglial cells. GLUT-4 and GLUT-8 are sensitive when it comes to insulin and glucose transporters in the brain [84]. The passing of glucose to the brain from a tissue depends on extracellular glucose concentration. Therefore, intracellular glucose level is abnormally high in hyperglycemic cases. As in the short term, hyperglycemia and DM occur as a result of the increase of glucose in the blood plasma according to normal values, long term hyperglycemia causes blood brain obstacle to change *via* a variety of metabolic ways and neuronal damage [85]. Some of these metabolic ways form an indirect way where passive responses exist independently of neuronal activity. It has also been stated that neuronal responses, when hyperglycemia constitutes in gene expression in neurons in the central nervous system, have a very important place in neuronal damage [86].

When we look at the passive response where hyperglycemia constitutes in the central nervous system, DM is considered to be a disease which spreads all over the world due to people 's life conditions at present, and has a high mortality and morbidity risks. In completed studies, it is stated that free radicals and lipid peroxydation increase significantly in diabetic patients and rats in which diabetes was created experientially and that oxidative stress has a role in diabetes etiology and advance [87]. In addition to this, it is pointed out that changes appearing in continuing oxidative stress and antioxidant capacity can be associated with the formation of chronic complications of diabetes [88]. Nonenzymatic glycation, changes in energy metabolism, sorbitol pathways activity, enhance production of free radicals. Changes of antioxidant defense system can be listed as some of the underlying causes of diabetes complications like tissue damage, metabolic stress, hypoxia and ischemia-reperfusion due to reactive oxygen species [89]. For example it is observed that when endothelial and smooth muscle cells are incubated in ambiance containing glucose in high concentrations, free radical formation starts [37]. Formation of reactive oxygen type is explained with three mechanisms:

Glucose Autoxidation

Glucose autoxidation in oxidative phosphorylation found in electron transporter chain provides prosuction of reduced Nicotinamide Dinucleotide (NADH), which takes part in energy production, and Flavin Adenin Dinucleotide (FAD). Electrochemical proton gradient formed *via* mitochondrial electron transformer chain lead to the production of intracellular superoxide radicals in two main sections: NADH dehydrogenase complex (Complex I) and ubiquinone (Complex II) interface. Mitochondrial superoxide production initiates protein kinase C by enhancing in high glucose concentration. Activated protein kinase C is an important molecule playing a key role in pathogenesis of vascular complications by participating in significant changes such as vascular disorders and increase in cytokine production [90].

Glycation

Advanced glycation products are heterogeneous compounds forming as a result of nonenzymatic glycation between nitrogen groups in proteins, lipoproteins and / or nucleic acids and carbonyl groups of reduced glucose. Protein glycation begins with this particular glucose carbonyl group and free amino group of proteins to making Schiff alkali. Formation of Schiff alcali occurs within hours and then transforms Amadori products within days after which they transform into dicarbonyl compounds and then into advanced glycation products within weeks. By connecting private receptors of end products of advanced glycation, this activates NF-κB, AP-1, MAP kinase, PI3-kinase, Akt, Ras, Stat3, JAK ve PKA signals pathways and enhances cytokine expression such as vascular cell adhesion molecule type-1, interleukin-1, tumor necrosis factor-alpha and free radical production. It was stated that end products of advanced glycation cause an increase in production of protein C kinase [91]. This was observed in experimental animal models where end products of advanced glycation enhance diabetic neuropathy by reducing sensomotor nerve transferring speed and blood flowing toward peripheral nerves. Also it was stated that high hemoglobin levels exposed to glycation cause a decrease in psychomotor activity and motor speed [92].

Poliol Pathway

NADPH is required for nitric oxide (NO) and operation of transforming of oxidized glutathione, which is an important cell function to fulfill into reduced form. More glucose can cause sorbitol production *via* poliol by causing the decrease in intracellular NADPH levels, required for glucose aldose reductase activation. As a result of this, antioxidant levels diminish. Sorbitol is also a compound which can oxide to fructose. Fructose and glucose production causes the decrease in myoinositol levels in cells and therefore reduction in Na-K ATP-az enzyme activation. End products of advanced glycation were found in senile plaques and neurofibrillary tangles of Alzheimer's patients [93, 94]. End products of advanced glycation in cortical neurons and cerebral vessels of old cerebrovascular patient were associated with severity of cognitive deterioration. AD risk is high in diabetic patients. Accumulation of end products of advance glycation in diabetic patients who have AD, and "up regulation" of their receptors were found [95]. It is stated that hyperglycemia enhances lactate production and tissue acidosis, increases cerebral blood flow, ruins cerebral auto regulation by constituting hyperthrombotic case and vasogenic edema [96].

When we look at the active answer which hyperglycemia constitutes in the central nervous system, changes which hyperglycemia makes in various neurons of the central nervous system in gene level are demonstrated in studies conducted in the field.

Changes in Hypothalamic Neurons

It was observed that hyperglycemia enhances vasopressin emission of mango cellular neuro secretory cells placing in hypothalamic supraoptic and paraventricular nucleus against hyperosmolarity increasing and as a result, chronic glucose rises [97]. These cells contain sodium channels with mechanosensitive voltage gate perceiving extracellular osmolarity changes. Consistently the continuation of hyperosmolar case causes "up-regulation" of sodium channels and reduction in threshold value for formation of action potentially together increasing sodium channels. Molecular changes in these cells contribute to the degeneration of cells. "Up-regulation" of mangocellular neurosecretory cells causes the increase in neuronal nitric oxide synthases enzyme and increase in N-methyl-D-aspartate (NMDA) receptors, and these increases cause intracellular calcium accumulation [98]. Even though nitric acid in high concentration is protective against reactive oxygen types, glutamate toxicity develops as a result of over activation of neurons.

Changes in Hippocampal Neurons

Encephalopathy, which shows itself in cognitive and behavioral disorders, can appear as one of the late completions of diabetes. Progressive dementia incidence arise in people with diabetes depending on hyperglycemia [99]. It is stated that in diabetes there are specific changes in 30 gene transcription such as inhibitor neuropeptide galanin, synuclein gamma and non conjugated protein 2 in hippocampus and 22 gene transcription such as protein kinase C gamma or epsilon, ABCA1 (ATP-Bonding Cassette A1). These genes take part in significant duties such as neurotransmission, lipid metabolism, neural development, oxidative damage and DNA repair. *In vitro* electrophysiological, measurements show that long term potentiation hyperglycemia is impaired depending on its severity in CA1 and CA3 sections in hippocampus of diabetic patients. It is observed that long term depression is uncomplicated in CA1 hippocampal neurons [100]. In diabetics through experimental studies it is noted that intracellular calcium level increases. It is thought that changes in calcium levels contribute to an increase in long term depression and decrease in long term potentiation [101]. A growing factor like insulin in hippocampal dentate gyrus in chronic experimental diabetic patient and expression of receptor were "down regulated". This case is significant in neuronal apoptosis tendency. Another change in CA1 and CA3 neurons is that nitric oxide synthase mRNA and protein concentrations decrease. This case reduces long term potentiation and causes disorders in cognitive memory and learning [102].

Changes in Neruronal and Glial Cells in Occipital and Frontal Cortex

Myosin-Va is an important protein included in mRNA transporting, membrane intensiveness and synaptic plasticity. It is reported that hyperglycemia causes the decrease in occipital and frontal cortex in motor protein miyosin-Va and protein expression. A decrease in protein expression brings a hereto decrease in conduction [103].

INSULIN AND ALZHEIMER'S DISEASE

In postmodern studies it is reported that there is a decrease in insulin receptors within the Alzheimer's patients, The Insulin like Growth Factors (IGF)-1 and 2 insulin messenger ribonucleic acid (m-RNA). According to the studies that were made on laboratory animals, streptolysin given to cerebrospinal fluid causes the increase in insulin, IGF and brain volume [104]. As a result of these effects, disorders occur in the memory of laboratory animals [105].

Streptolysin is alkylating chemotherapeutic and is used to develop T1D model in laboratory animals. The basic effect mechanism of streptolysin is to destroy islet-cell *via* DNA and protein damage by entering into glucose carrier 2 and the beta island of pancreas [106]. It is acknowledged that when streptolysin is given to cerebrospinal fluid, it destroys neuronal systems producing insulin and decrease the amount of insulin in the brain. In laboratory animals in which streptolysin is given their cerebrospinal fluid, this cerebrospinal fluid again provides healing in memory functions [107, 108].

Insulin passes across Blood Brain Barrier (BBB) and is also synthesized internally by a group of neurons. Insulin gene expression exists in neuronal cells of mature and immature mammals [107].

Insulin m-RNA and insulin exists in hippocampal pyramidal neurons and neurons of medial prefrontal cortex, enthorinal cortex, perprinal cortex, thalamus, olfactory bulbus, hypothalamus. Insulin m-RNA and insulin do not exist in glia cells. Insulin receptor exists in glia and neurons and has wide distribution in the brain [109, 110].

Insulin receptor causes apoptosis inhibition *via* tyrosine kinase, phosphatidylinositol-3-kinase (akt-protein kinase B), mitogen activated protein kinase. Also, they cause insulin *via* these kinase systems and cause modulation of channel *via* phosphorylation of NMDA [111, 112]. As a result of modulation in NMDA channel, increase of calcium flow takes charge in memory formation. In canal modulation, kinase systems and insulin receptor substrate (IRS) [113].

In the brain volume of rats whose insulin receptor substrate gene was cleaned the proliferation decreases in their hippocampus neurons and neurofibrillary tangle accumulation increases. Rapid cognition and recovery in motor abilities were monitored in patients with stroke and those using insulin [114].

Cognitive functions enhanced without variance in intranasal plasma glucose and insulin concentration was applied over twenty one days [115 - 117]. Insulin acts as a growth factor in the brain. It reduces neuronal apoptotic process and facilitates neuronal repair [118]. Also determining energy metabolism in insulin neurons affects neuronal survival significantly. Glucose uptake occurring during insulin deficiency and the production of adenosine triphosphate cause deterioration of neuronal homeostatis. Insulin stimulates choline acetyltransferase which is a fundamental enzyme in synthases of acetylcholine. Therefore, decreasing insulin or insulin resistance decreases the amount of choline acetyltransferase [119].

Other effect of insulin is to be regulated *via* phosphorylation of intracellular skeleton. Its accumulation in form of neurofibrillary tangle, depending on faulty phosphorylation of tau protein in AD. Tau protein is portrayed in microtubule formation, performs adhesion duty in the organization of proteins called a tubulin and in the formation of microtubule polymers [120]. The small amount of insulin prepares the ground for an increase in precursor protein and accumulation of Aβ [121, 122].

Insulin Resistance

Ceramides occurs with a combination of sphingosine and fatty acids. In obesity induced with a diet, they cause over emission of ceramide from fatty tissues. Ceramides occurs with a combination of sphingosine and fatty acids. In obesity induced with diet, they also cause over emission of ceramide from fatty tissue. Ceramide leads to insulin resistance by activating inflammatory cytokine [123]. Ceramide toxicity exists in hepatosteatosis and metabolic syndrome [124].

Giving ceramide intraperitoneally causes decreased learning in laboratory animals [125]. In addition to this, intestinal flora has fundamental functions in insulin activity and resistance. This changes cause the increase in mucosal transmittance and a change of distribution of Toll Like Receptor (TLR) [126]. Carbohydrate diet causes expression of TLR-4 type receptors in mucosal cells. TLRs is-involved in the recognition of antigen and determining of severity of immune answers [127]. TLR-4 type receptors cause the increase in inflammation by generating immune response, and the increase in Alzheimer risk *via* fatty liver, metabolic syndrome and insulin resistance [128]. It was shown that TLR-4 damages the effects of insulin *via* variety kinases on macrophage, adipocyte and liver cells. It is thought that fatty acids and TLR-4 activation impair the effects of insulin in the last organ by scaling up levels of tumor necrosis factor-alpha and interleukin-4 [129]. It was shown that cytokine is produced from adipocytes of obese persons and that it warns inflammatory situation like macrophages [130].

Insulin Like Growth Factors (IGFs)

Growing factors like insulin derives from a family of single chained serum protein in various cells, significant regulator of cell reproduction and cell differentiation [131]. Two different polypeptides were determined for this family: Insulin Like Growth Factor-1 (IGF-1) and Insulin Like Growth Factor-2 (IGF-2). These are growing factors resembling insulin both biochemically and functionally. IGFs are synthesized by many tissues (liver, smooth muscle, and placenta) and carry bondings to specific bonding proteins in plasma (IGD) Bonding Proteins) [132].

IGFs are one chained polypeptides. IGF-1 is alkaline peptides including 70 amino acids and its molecular weight is 7471 kDa. IGF-2 is slight acidic peptide containing 67 amino acids and its molecular weight is 7471 kDa. These two IGF molecules have A and B chains similarly proinsulin and these chains are connected with each other *via* disulfide bonds named peptide C. Aminoacids ranges of IGF-1 and IGF-2 show homology with proinsulin in a rate of 43% and 41%.

In contradistinction to proinsulin, IGFs contain section D in carboxy terminal. This structural similarity with proinsulin explains the bonding of both IGF molecules to insulin receptors in lower affinity. On the other hand, structural differences prohibit the bonding of insulin to IGF binding proteins [133].

IGFs are secreted from liver with stimulation of growth hormones. Equally, Growth Hormone (GH) changes autocrine and paracrine effects of hormones by regulating IGF synthases and its emission from many tissues. IGF concentration of blood is 300-600 ng/ml. This mechanism is warned by GH, and activation of GH receptors enhances IGF synthases and its emission. It was observed that when GH in circulation is eliminated in animals in which hyphophysectomy was performed, lGF level in circulation decreases significantly. Although IGF-I depends on growth hormone, it changed depending on age. IGF-I level is lower in babyhood and childhood periods, and it is high in puberty and decreases with age [134].

IGFs has metabolic effects like insulin. By inhibiting proteolysis in skeleton and muscle tissue, they play a role as a mediator in protein synthesis and metabolic effects on GH [135]. IGFs are strong simulators of cell differentiation such as myoblast, osteoblast, oligodendrocyte and adipocyte. They enhance lipolysis inhibition, glucose oxidation in fatty tissue, transportation of glucose and amino acid to diaphragm and cardiac muscle, collagen and proteoglycane synthases. They have positive effects in calcium, magnesium and potassium homeostasis. At the same time, by stimulating production and function of IGF-I and IGF-II lymphocyte, they play an effective role on the immune system [136].

Insulin Receptors

Bonding of insulin to target cell surface initiated the biological answer. Many cells have private surface insulin receptors [137]. An insulin receptor is a member of the receptor tyrosine kinase family and is a glycoprotein which is formed by 2α and 2β subunits connected with disulfide binds. Unit α is completely extracellular and includes the place of insulin bonding. B unit consists of intracellular sections having extracellular, transmembrane and intrinsic tyrosine kinase activation. The evidences show that two isoforms resulting from different cutting of insulin receptor exon 11 and recognised as A and B are different in terms of insulin susceptibility. With bonding insulin to α unit, autophosphorylation appears in tyrosine residues in cytoplasmic section of unit β. Activated β unit provides phosphoylation of intracellular substrates. Among these, there are insulin receptor substrate (IRS) family members Shc homology and collagen (Shc) protein, binder protein-1 associated with growth factor receptor-2 and others. IRS protein phosphorylation activates phosphatidylinositol-3 kinase (PI3K), tyrosine kinase, tyrosine protein phosphatase and many small proteins. Activated PI3 kinase produces lipid phosphatidylinositol-3,4,5 triphosphate. Increasing PIP3, initiates protein kinase cascade namely protein kinase B (PKB), serine/threonine kinases and protein kinase C (PKC) having different isoforms [138]. This complex pathway pursues 2 ways. The first is the mitogenic pathway which transmits the effects of insulin in GH and the other is metabolic pathway regulating food metabolism. This causes motion of vesicles containing glucose transporter protein-4 (GLUT-4) to cell membrane in signal pathway, PI3 kinase, skeletal muscle and adipocyte, and an increase in glycogen and lipid synthases and stimulates other metabolic ways [117, 119]. PI3 kinase is an enzyme playing a key role in revealing metabolic effects of insulin. PKB plays roles in many effects of insulin such as glucose involvement, glycogen synthases and stimulation of protein synthases [139]. Because PI3 kinase and PKB are the central molecules in many effects of insulin, the activity of this molecule, expression levels and possible gene mutation play a role in insulin resistance [140].

The number of insulin receptors and their susceptibility are important in insulin effect. If the insulin level is chronically high, the numbers of receptor decreases. Situations associated with high insulin level and decreased bonding to receptor are linked to obesity and over uptaking of carbohydrate and the usage of insulin in high dose for a long time. Situations associated with the decrease of insulin level and over bonding are hunger and exercise. High cortisol level reduces bonding of insulin to the receptor [137].

Insulin sensitive insulin receptors are unquestionable in neuronal cells and also at the BBB, and if the uptake of glucose increases in the brain, insulin receptors will be upregulated. In AD conditions, this process (receptors upregulation) is compromised hence, glucose metabolism rates decrease in the brain. As such, insulin plays an important role for the cognitive functions [66].

Additionally, decreased immunreactivity of Choline Acetyl Transferase (ChAT), insulin or IGF-1 receptors are seen within the cortical neurons in AD. According to experimental analyses, tau protein and ChAT gene expression are controlled by insulin as well as IGF-1 receptors [141]. Impaired insulin mechanisms like insulin resistance or deficiency cause brain plasticity, synaptic dysfunction and neurodegeneration. As a result, it is possible that ChAT establishes a connection between DM and AD [142].

INFLAMMATORY RESPONSE

Inflammation can be defined as the part of the body's mechanisms of defence against infections and injury, and specialized cells that are mobilized to neutralize such risks and restore normal body physiology. Inflammation also includes some soluble factors [143]. Inflammation is an important characteristic of diabetes and AD, and play critical roles in the pathogenesis of both disorders [144] (Fig. 2).

Cytokines and Other Bioactive Substances

The single most important risk factor for AD is aging, but it is often associated with chronic state of low-grade inflammation resulting from deregulated levels of pro- and anti-inflammatory cytokines, the condition called as "inflamm-aging" [145].

The raised levels of inflammatory mediators are connected with the dysfunctioning of immunological response that results in insulin resistance [146]. The key advice of T2DM is insulin resistance which is related with the raised level of these inflammatory mediators [147 - 149]. Furthermore, the accumulation of inflammatory products at different rates in Alzheimer's brain can be explained in a simple way [66]. Some studies have stated the presence of inflammatory mediators in the AD brain, including elevated levels of cytokines / chemokines and gliosis (especially microgliosis) [150]. Cytokines may originate from peripheral immune cells and then reach the central nervous system by crossing the blood-brain barrier or may be directly produced within the central nervous system by neurons and glial cells. Additionally, cytokines are very significant because they mediate both immune response and also neuron functions/neuronal survival by binding to spesific cytokine receptors [151, 152]. In other words, AD is associated with inflammatory response [66].

Macrophage activation or infiltration into adipose tissue and overproduction of proinflammatory cytokines, like tumor necrosis factor-alpha (TNF- α) are key characteristics of the pathophysiology of metabolic disorders because overexpression of TNF- α in adipose tissue of obese individuals causes peripheral insulin resistance [153]. Adipocytes and macrophages are closely related and share many functions, such as to secrete the cytokines. Besides, both can be activated by pathogen-associated components, such as lipopolysaccharides [154]. Also, it is stated that macrophages and pre-adipocytes are genetically related, and preadipocytes have been shown to transdifferentiate into macrophages [155]. According to a study, the brain TNF-α levels were found significantly higher in the saline-treated streptozotocin (STZ) rats group, but exenatide administration regulates the serum glucose level, reduces the TNF- α expression and improves hippocampal neuron numbers and memory within these STZ-treated rats [156].

The brain and also peripheral tissues illustrate similar inflammatory processes. Hence, uncontrolled chronic inflammation generate the advance of tissue damage in degenerative diseases both in the brain and also in peripheral tissues [66]. For example, blood concentrations of inflammatory mediators; TNF-α, interleukin 6 (IL-6) and IL-1b are found at high levels in AD patients [157]. Inflammatory mediators IL-6, interleukin 1a (IL-1a) and plasminogen activator inhibitor type 1 (PAI-1) are upregulated during obesity and they are potential participants to insulin resistance/the metabolic syndrome [158]. In addition to this, PAI gene-knockout mice demonstrate decreased adiposity and an improved metabolic profile in case of increased calories intake [159, 160]. Moreover, some studies have shown that C-reactive protein (CRP) and α1-antichymotrypsin are associated with the risk of AD [146]. Acute-phase reactants and IL-6 are the major cytokine mediators of acute-phase responses. Also, they are a strong identifier of T2DM development [161]. Substantially, T2D causes some changes in BBB permeability [162]. In addition, increased levels of IL-6 were found compared with nondiabetic AD brains in postmortem examinations of diabetic AD brains [163]. Proliferator-activated receptor-G (PPARG) agonists class of antidiabetic drugs are shown to have anti-inflammatory effects because these drugs decrease the IL-6 levels, and also, other inflammatory mediators. Interestingly, it is seen that these drugs not only reduce insulin resistance but also reduce the incidence of AD [164, 165].

Ceramides

Ceramides are generally cytotoxic, lipid signaling molecules [166] and they are formed *via* 3 main pathways [167]. They occur from the composition of fatty acids and sphingosine [168] and may form under conditions of cell stress and promote cellular signaling, including the regulation of apoptosis [169]. They can also activate inflammatory pathways including JNK and NF-κB [170]. To sum up, in general, ceramides cause insulin resistance by activating pro-inflammatory cytokines and protein phosphatase 2A (PP2A), also inhibiting transmission of signals through phosphatidyl-inositol-3 kinase (PI3K), Akt. They can stimulate apoptosis by activating protein kinase C, protein phosphatase 1 (PP1), caspases and cathepsin D [167]. Diet-induced obesity causes the excessive ceramide release from adipose tissue and these excessive ceramides cause the insulin resistance by activating inflammatory cytokines [168].

To summarize briefly, particular types of sphingolipids, ceramides and its metabolites, cause inflammation and are increasingly connected with T2DM [166, 171]. Ceramides which are generated peripherally, cross the blood-brain barrier and can contribute to AD pathogenesis by changing the microenvironment of lipid mass, therefore favoring amyloid-β generation or by inducing the central inflammation and disruption in neuronal insulin signaling [166, 172]. Similarly, it was shown that, intraperitoneal ceramide administrations cause the decreased learning activity within the experimental animals [168].

Endoplasmic Reticulum (ER) Stress

Increased levels of ceramide production can cause the endoplasmic reticulum (ER) stress and as a result, contribute to the progress of cellular degeneration [167]. Additionally, the accumulation of protein aggregates and also other genetic or environmental alterations are responsible to produce chronic stress beyond the capacity of ER, leading to deleterious consequences [173].

Recent studies from experimental models show that ER stress plays a critical role in the initiation and integration of pathways of inflammation and insulin action in obesity and T2DM [174]. A common cause of cell abrasion in diseases is characterized by oligomerisation and misfolding of amyloidogenic proteins may be said to be ER stress–induced apoptosis [175]. Components of the unfolded proteins can be used as indicators to measure the ER stress [176, 177]. Some morphological changes related to ER stress from T2DM patients have been reported, distention of the ER [173], induction of ER stress-associated death response markers like C/EBP homologous protein (CHOP), and activation of caspases in islet β cells [175]. Also it has been reported that these pathological changes are spesific for T2DM patients [173].

Protein misfolding through ER dysfunction could be the final common pathway for many neurological diseases [178]. Additionally, ER dysfunction could instead be the result of an impaired cell metabolism such as modified calcium homeostasis [179].

Toll-like Receptors (TLR)

Toll-like receptor (TLR) family is a class of proteins [180] which gives rise to metabolically or nutritionally induced inflammatory responses [174]. TLRs are transmembrane proteins and the expression of pro-inflammatory cytokines occurs by activation of these receptors, particularly TLR-2 and TLR-4 [126]. Especially, TLR4 plays an important role in many inflammatory disorders among all the TLR's [181].

Carbohydrate-rich diets cause some changes in the intestinal flora, also these changes cause an increase in mucosal permeability and force to change the distribution of the TLRs subtypes [126]. These type of diets give rise to expression of TLR-4 type receptors in the mucosal cells [127]. This type of receptors generate an exaggerated immune response and this leads to an increase in inflammation. This reults in a fatty liver, metabolic syndrome, insulin resistance and AD [148]. It was shown that, TLR-4s disrupt the effects of insulin *via* some kinases which are located in macrophages, adipocytes and liver cells [168]. Also it was believed that acid oils and TLR-4 activation cause an increase TNF-α and IL-6 levels [129]. Lately, the association between the TLR-4 polymorphism gene and AD was investigated by some researchers. According to other studies, patients with AD and animal models of AD show an increased expression of CD14 (a co-receptor for TLR-4), TLR-2 and TLR-4 which are thought to occur autonomously in response to the presence of amyloid-β [128]. Besides, according to some studies it was concluded that TLR-4 is not involved in the trigger of amyloid-β deposition, and as amyloid-β deposits start, microglia are activated *via* TLR-4 signaling to decrease amyloid-β deposits and protect cognitive functions from amyloid-β -mediated neurotoxicity [182]. According to these observations, activation of microglia *via* TLR-4 is neuroprotective is seen in the early stages of AD pathogenesis, so TLR-4 signaling pathways may provide potential therapeutic targets [128].

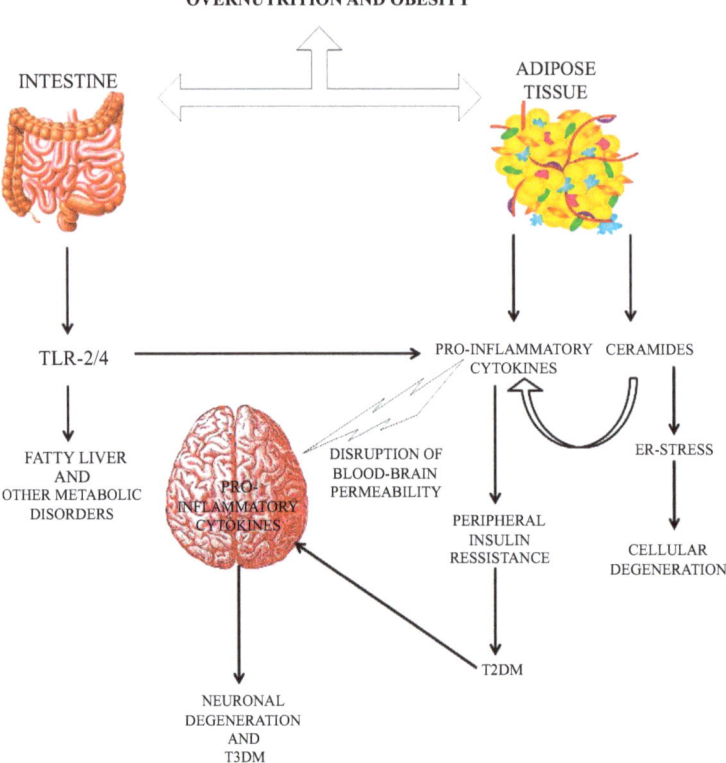

Fig. (2). General diagram of inflammatory response in T3D.

NEURONAL CALCIUM (CA²⁺) DYSREGULATION

The Ca^{2+} hypothesis of brain aging and dementia may be thought to play a role on the pathogenesis of dementia and certain neurodegenerative disorders [183]. The homeostasis of cytosolic calcium concentration ($[Ca^{2+}]_i$) plays a critical role in aging [184]. Additionally, Ca^{2+} mediated signal transduction cascades and Ca^{2+} homeostasis are part of the final common pathways for the neuropathological changes associated with AD like neuronal dysfunction and cell death [183]. Also it is known that, there is a close relationship between Ca^{2+} homeostasis and the production of reactive oxygen species, ischaemia, and (brain) cell death [184, 185]. Synaptic plasticity is considered to be a player in forming the neuronal basis of learning and memory. Long-term depression (LTD) and Long-term potentiation (LTP) are two experimentally stimulated forms of activity-dependent synaptic plasticity that have been studied considerably. Both are triggered by an increase in the level of post-synaptic intracellular Ca^{2+} concentration [183].

Several studies have been done when it comes to the examination on the effects of experimental diabetes on synaptic plasticity in the hippocampus, and as a result, learning deficits have been found in STZ-diabetic rats dependent on diabetes duration and severity [183]. Also it has been shown that, insulin treatment prevents the development of the changes in LTP, but it is less effective on LTP deficits [186].

According to experimental diabetes models, learning disorders are associated with the changes in the Ca^{2+}-dependent hippocampal synaptic plasticity. Dysregulation in the homeostasis of cytosolic free Ca^{2+} may be a pathway for the multifactorial pathogenesis which involves vascular changes, oxidative stress and non-enzymatic protein glycation in the case of neurological complications of diabetes. It has been shown that some of Ca^{2+} channel blockers like nimodipine, improve the experimental peripheral neuropathy, through a vascular mechanism by using direct neuronal effects. It also improves Ca^{2+}-dependent forms of synaptic plasticity in the hippocampus of diabetic rats according to some preliminary studies [183].

According to another study, pathological conditions in diabetes generate a reduction in the capacity of nociceptive neurons to eliminate Ca^{2+} from the cytosol by mitochondria or ER [187]. The excess of glucose is converted into sorbitol and fructose that is linked to decrease levels of myoinositol. High glucose concentrations cause the formation of advanced glycation end-products in high levels. Therefore, these reasons may be listed for the accumulation of Ca^{2+} in diabetic conditions [188, 189].

MITOCHONDRIAL DYSFUNCTION

The most complex and metabolically active organelle in the cell is mitochondria and the cell may die by apoptosis or necrosis if the severe mitochondrial dysfunction occurs [190]. Mitochondrial energy–transducing capacity is necessary for the repair of cellular function, and impaired mitochondrial energy metabolism or redox homeostasis is a feature of both brain and peripheral aging [191, 192]. The brain, particularly neurons, have a high requirement for energy. Energy metabolism is the most important function of mitochondria for all cell types [190].

Moreover, mitochondrial dysfunction is one possible mechanism responsible for the energy deficiency in AD [190]. Cytochrome c oxidase (complex IV) which is the type of electron transport complexes, is an intracellular measure of oxidative energy metabolic capacity and respiration [193]. Through the AD patients, this type of complex was found in decreased levels in the mitochondria so it can be said that this is the potential molecular candidate contributor to the energy hypometabolism [190]. To take the arguement further, it can be said that impaired mitochondrial function in brain endothelial cells may also contribute to the microvasculature distortion [194].

The normal mitochondrial metabolism products are ROS. They are cytotoxic and generated by monoelectronic reduction of oxygen in the respiratory chain. Transient ROS generation is responsible to physiological stimuli such as insulin signaling in peripheral tissues [144]. However, high levels of ROS in the brain can be detrimental [164] and include the molecular etiology of AD, markers of oxidative stress like oxidized forms of lipids, proteins and DNA present in an AD brain [21]. An overall decrease in mitochondrial mass, an increase in cytoplasmic mitochondrial DNA and cytochrome oxidase in lipofuscin-containing vacuoles are seen in the neurons of affected by AD pathology [195].

Mitochondrial dysfunction relate with pathological changes seen in Alzheimer's disease [196]. Compromised energy metabolism like low ATP, oxidative stress and damage like DNA oxidation and metabolic outcomes like low regulatory heme are the three key consequences of mitochondrial dysfunction relevant to AD [197]. Deficiency of heme is the possible mechanism for decreased levels of complex IV and the resulting mitochondrial dysfunction in the case of AD [190]. Transient production of ROS is implicated in synaptic signaling and facilitates LTP and memory-related mechanisms in the brain [198].

As will be described more in detail at the following sections, amyloid-β (Aβ) peptides are seen in AD brain, and also these peptides are found in the mitochondria [190]. The amyloid precursor protein (APP) is known to be associated with the outer mitochondrial membrane [199], and also, β-secretase and Aβ are present in mitochondria which inhibits cytochrome oxidase in the presence of copper. As such, Aβ could negatively affect mitochondrial electron transport [200] and cause the dysfunction of mitochondrial electron transport proteins in AD [201]. Heme is synthesized in the mitochondria and Aβ also binds to it to form the complex Aβ-heme which exhibits peroxidase activity. Thus, these findings may provide a direct connection between key cytopathologies of AD like mitochondrial dysfunction and Aβ [190].

Abnormal stimulation of excitatory NMDA-Rs has been suggested as a key mechanism leading to high ROS production in AD, likely to be a result of Ca^{2+} related mitochondrial dysfunction by amyloid-β oligomers (AβOs) [202].

Mitochondrial dysfunction which occurs as a result of chronically elevated reactive oxygen species (ROS) levels or an imbalance between ROS production and intracellular levels of antioxidant defenses lead to oxidative stress. This condition has been associated with both T2DM and also AD [21]. Increased levels of mitochondrial ROS production and oxidative stress produce an outcome of high-fat diet consumption in skeletal muscle, leading to the development of peripheral insulin resistance in T2DM [203]. In other words, scientific evidence indicates that increased oxidative stress and decreased antioxidant capacity are seen in the case of DM in humans and also in experimental diabetes. It is also accepted that increased oxidative stress has been involved in the pathology of several diseases like diabetes and AD [204].

According to a study of STZ-induced type 1 diabetic rat models, the isolated mitochondria from the brain had a lower content of coenzyme Q9 (CoQ9) and this finding indicates a deficit in antioxidant defences in DM [205].

The importance of the intracellular Ca^{2+} concentration in DM and AD conditions was previously mentioned. If we look through a different perspective, rising levels of intracellular Ca^{2+} concentration occurs in oxidative stress conditions, and the role of the cytosolic Ca^{2+} levels in the modulation of several intracellular signalling pathways like protein kinase C and Calmodulin dependent signalling pathway is very important for apoptotic processes [183, 206]. Increase concentration of calcium was found in the brain tissues of patients with AD. Intracellular calcium levels are higher in neurons including intracellular neurofibrillary tangles than in neurons from healthy control patients [207]. In addition it was observed that, the diabetes decreases the accumulation Ca^{2+} capacity of mitochondria, so the convenient intracellular environment for the mitochondrial permeability transition opening occurs [204, 205].

Mitochondrial dysfunction triggers neuronal degeneration and cell death, and it is thought that Amyloid-β oligomers (AβOs)-induced neuronal oxidative stress is blocked by insulin so brain insulin signalling and oxidative stress seem to be closely linked [208, 209]. Additionally, neuronal oxidative stress blockage or protection by insulin seems to involve the prevention of abnormal NMDA-R activation [210]. Metabolic disorders and AD have been related with JNK activation and inhibition of insulin/IGF-1 and their receptors [203]. In summary, mitochondria, insulin and JNK signalling play key roles in neuronal dysfunction in brain aging and AD [211].

Suppressing the production of cellular energy leads to the reduction or elimination of insulin secretion or function [212]. Again, the decrease of oxidative phosphorylation efficiency cause the disruption of glucose homeostasis and as a result, by the change in glucose tolerance, tissue and blood lactate levels increase [213]. About 90% of the ATP requirement of the normal functioning for neurons is provided by mitochondria, so the maintenance of oxidative phosphorylation capacity is incredibly important for the brain. The inhibition of oxidative phosphorylation will affect the brain system before any other system. This is due to the high levels of ATP requirement of neurons, and also, the central nervous system such as the transmission of impulses along the neuronal pathway [204].

TAU HYPERPHOSPHORYLATION

Neuronal microtubule-associated protein (MAP) tau stimulates tubulin polymerisation and stabilizes microtubules in the brain [214, 215]. Tau is found predominantly in axons [216]. Tau gene includes 16 exons and locates on the long arm of chromosome 17 (position 17q21) in humans [217]. The N-terminal end of the tau protein is highly acidic and the repeats in the C-terminal half of the tau protein are the domains where tau binds to microtubules [218 - 220]. In an adult human brain, one single tau gene encodes six tau isoforms [221], also it is noted that, only the shortest tau isoform is expressed in the fetal brain, though all six isoforms are seen in an adult human brain [222, 223]. Each of the six tau isoforms are differentially expressed during different development stages to stimulate microtubule assembly with different efficiencies, so each of the six tau isoforms has its different biological activities and physiological role [222 - 224].

It is believed that tau's ability to bind microtubules can have positive and negative outcomes. Stabilizing to microtubules and permitting neurites' extension can be considered as positive effects of tau. Furthermore, tau might compete with the motor protein kinesin during microtubule binding, and as a result, may cause to decrease axonal transport. This is the possible negative effect of tau protein. Several amyotrophic lateral sclerosis-transgenic mouse models show neurofilament accumulation in motor neurons and tau overexpression. This finding supports the negative effect [224], but according to a recent study, no defects are found in axonal transport through this kind of mice [225].

Hyperphosphorylated tau has been detected in neurofibrillary lesions and also other central nervous system disorders. The microtubule binding function of tau is impaired by hyperphosphorylation but paired helical filaments (PHF)-tau which is the hyperphosphorylated form of tau does not bind to microtubules unless it is dephosphorylated. The reduced binding ability of PHF-tau to microtubules cause to destabilize the microtubules, and this results in the disruption of vital cellular processes, like rapid axonal transport, which results in the degeneration of affected neurons [120].

Tau abnormal hyperphosphorylation leads to neurodegeneration in AD and in the so-called tauopathies [226] microtubule assembly is negatively regulated in the case of phosphorylation of tau [227, 224]. It has been found that isolated from

autopsied AD brain includes 3- to 4-fold higher phosphorylation level of tau than the normal human brains [224].

Tau phosphorylation at different sites has a different effect on its biological function and also pathogenic role [224]. More than 40 phosphorylation sites have been identified in tau protein isolated from AD brain until today [228]. For example, phosphorylation of tau at Ser262, Thr231 and Ser235 inhibits its binding to microtubules by ~35%, ~25%, and ~10% accordingly a quantitive *in vitro* study [229]. Also it has been found that, Ser199/Ser202/Thr205, Thr212, Thr231/Ser235, Ser262/Ser356 and Ser422 are the critical phosphorylation sites which convert tau to an inhibitory molecule accordingly *in vitro* kinetic studies of the binding between hyperphosphorylated tau and normal tau [230]. Furthermore, phosphorylation of Thr231, Ser396 and Ser422 sites stimulates self-aggregation of tau into filaments, and also the mutation of Ser396 and Ser404 sites into Glu generate more fibrillogenic tau by pretending phosphorine [231]. In addition, the mutation of Ser422 site into Glu shows a significantly increased propensity to aggregate tau [224]. According to these informations, the abnormal hyperphosphorylation of tau does not cause neurodegeneration by loss of normal activity that can be compensated for by other MAPs, but also causes neurodegeneration by the gain of toxic functions [224].

Neurofibrillary Tangles (NFT)

Senile amyloid plaques (also known as neuritic plaques) (SPs), neurofibrillary tangles (NFT), synaptic neurons loss and marked atrophy in the brain are histopathological changes in AD [6]. The basic component of NFTs are hyperphosphorylated form of tau proteins. The unconnected phosphorylated tau is polymerized into unsolved double helical filaments and these become intraneuronal NFT overtime. It is believed that this process occurs as a result of a long period of years. To support, also it is believed that tau-targeted therapeutic strategies may reduce the form of unsolved NFTs, and as a result, incidence of AD may be reduced in parallel to this. Eventually, the NFTs cause to disrupt the integrity of cytoskeleton and also axonal transport and as a result, give rise to cell death. The cell death results in extracellular NFTs which are called "ghost tangle s". In light of all this information we can say that the severity of cognitive deterioration in AD is correlated with NFT quantity instead of amyloid deposition [232 - 234]. Some of genetic engineering studies in mice suggest that primary neurotoxic formations occur as a result of mutant tau proteins rather than NFTs [235, 236]. So it appears that, both amyloid plaques and also NFTs are not the mediators of AD pathogenesis as they may be the products of AD. Being different from all of them, the familial AD increases the amyloid β not the mutant tau proteins.

However, it has been shown that, when tau hyperphosphorylation occurs in the brain, this formation can cause neurodegeneration [237].

Determination of SPs and NFTs are necessary but not sufficient for AD diagnosis because they are seen in other neurodegenerative diseases and also in normal aging. As such, for the definitve diagnosis of AD, they are shown in certain quantities and in certain neuroanatomical distribution. For example, from the age of sixty, SPs start to occur in neocortex and NFTs start to occur in limbic system, but if NFTs accumulate in the limbic system and pass to the neocortex. In other words if NFTs are seen in neocortex, also, SPs are seen in limbic system;this shows 100% sensivity and specifity for AD [232].

According to some studies, neurofibrillary degeneration has been seen in transgenic mice which express both human amyloid precursor protein (APP) and mutant tau, Also several studies have observed that Aβ-fibrils induce tau hyperphosphorylation *in vitro* and that Aβ-fibrils do not cause degeneration of hippocampal neurons within tau knock-out mice, so it was suggested that tau is one of the major downstream targets of toxic Aβ [238].

Tau Hyperphosphorylation in Diabetes and Alzheimer's Disease

O-GlcNAcylation is a unique type of O-linked glycosylation of nucleocytoplasmic proteins by a monosaccharide β-N-acetylglucosamine (O-GlcNAc) and recently it was found that altered O-GlcNAcylation is regulated by glucose metabolism. In another words it links the impairment of brain glucose metabolism to hyperphosphorylation of tau [239]. Thanks to this, some researchers hypothesize that the impaired brain glucose uptake or metabolism can cause neurodegeneration *via* down-regulation of O-GlcNAcylation and deregulation of tau phosphatases, and as a result, abnormal hyperphosphorylation of tau occurs [230, 240]. Therefore it can be said that, T2DM may contribute to the formation of AD by the same or similar mechanism [239].

A well-known insulin sensitiser "rosiglitazone" is generally used for the treatment of T2DM. Also, it has been shown to decrease tau hyperphosphorylation in cultured cells and in obese rats [224, 241]. It is possible that this sensitizer performs its effectiveness *via* restoring brain glucose metabolism, increasing brain insulin sensitivity and tau O-GlcNAcylation. As a result, it inhibits tau hyperphosphorylation, so these effective ways can be presented to support the mechanism which is described above [224].

Tau O-GlcNAcylation is dynamically regulated by two enzymes which are O-GlcNAc transferase and β-N-acetylglucosaminidase [242]. So, as an alternative to the above treatment, tau hyperphosphorylation may be altered by targeting these two enzymes [224].

According to a study that supports T2DM linked AD by tau hyperphosphorylation;

a long-term administration of STZ by intracereb-roventricular injection generates tau hyperphosphorylation, neuronal damage, cell loss and the accumulation of Aβ in the brain [243]. Other studies can be found about the overproduction of tau protein and the accumulation of Aβ in leptomeningal vessels in the case of intracerebroventricular STZ injection [244].

The expression of APP and tau mRNAs and also immunoreactive bands have been found in normal pancreas, T2D pancreas and insulinoma beta cells by RT-PCR and Western blotting. These findings indicate that APP and tau are present in the pancreatic tissue and also in islet beta cells [245]. In T2D patients, slight upregulation of tau expression is also detected at the gene level in pancreatic islets compared to controls [246].

All these results show that tau phosphorylation and Aβ formation are the characteristics of T2D [161]. According to a study, insulin stimulates tau phosphorylation at Ser202 site directly and this site is responsible for predispositione for tangle formation [247]. This study also summarized the mechanism of insulin signalling and tau phosphorylation in the brain as follows:

1. The signal can be quickly activated by insulin in the brain, even in regions separated from the circulation by a tight BBB such as hippocampus.
2. Insulin, which is administered peripherally, evokes tau phosphorylation at Ser202 site and this not only occurs rapidly after the insulin injection, but also occurs upon prolonged stimulation during clamp studies.
3. Insulin-stimulated tau phosphorylation at Ser202 site and also insulin action depend on neuronal insulin receptors because the absence of insulin receptors such as IGF-1 receptor cannot be compensated by other receptors.

Therefore, neurodegeneration through activation of tau phosphorylation is directly connected to insulin action in the brain [247]. According to another study, when the site-specific tau phosphorylation levels between AD and T2DM-AD have been compared it has been found that, even if tau is abnormally hyperphosphorylated in both groups, there have been clear site-specific differences between pure AD and T2DM-AD groups. In the T2DM-AD group, decreased hyper phosphorylation at the N-terminal half and aggravated hyperphosphorylation at the C-terminal half in the tau molecule as compared to AD group [248]. Thus, T2DM might not only elevate the risk of AD by promoting tau phosphorylation, but also quicken the AD by aggravating tau hyperphosphorylation at the critical abnormal phosphorylation sites [239].

Glucose Transporters (GLUTs)

Glucose transporters (GLUTs) are responsible for glucose transport across the blood-brain barrier because neurons are unable to synthesize or store glucose in the brain so they are fully dependent on GLUTs. Different GLUTs are encoded but in the mammalian brain, GLUT1 and GLUT3 are the preponderant GLUTs responsible for glucose transport [249]. GLUT1 is responsible for transporting glucose from the blood into the extracellular space of the brain and it is highly expressed in the endothelial cells of the BBB [239]. Also, GLUT3 helps transport glucose from the extracellular space into the neurons and it is the major neuronal GLUTs [250]. Additionally, GLUT2 is responsible to glucose transport into astrocytes and it is expressed in astrocytes of the brain. According to several studies, decreased levels of GLUT1 and GLUT3 have been observed in the AD brain [239] and this decreases associated with the hyperphosphorylation of tau [251]. Additionally, according to a study, GLUT3 and tau O-GlcNAcylation were found in decreased levels, however, tau phosphorylation was found in increased levels in T2DM brain [239]. O-GlcNAcylation negatively regulates the tau phosphorylation [239, 252]. In summary, T2DM contributes to the increased risk for AD by a decreased brain glocose uptake / metabolism because of the down-regulation of the global protein O-GlcNAcylation, also, GLUT3 and hyperphosphorylation of tau mechanism. Additionally, the decrease in global protein O-GlcNAcylation levels were found to be similar in T2DM and T2DM-AD brains [239] (Fig. **3**).

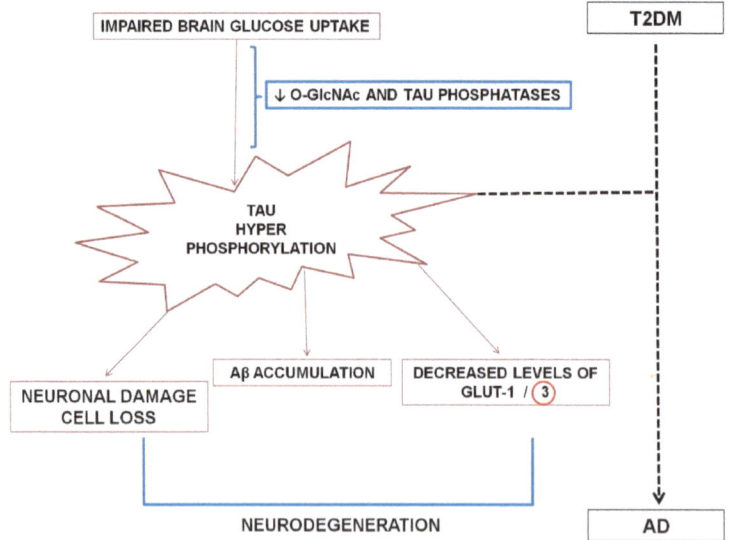

Fig. (3). General diagram of tau hyperphosphorylation in T3D.

Glycogen-Synthase Kinase-3 (GSK-3)

Tau phosphorylation is promoted by several serine/threonine protein kinases *in vitro* including MAPK (Mitogen-Activated Protein Kinase) [253, 254], GSK-3 (Glycogen-Synthase Kinase-3) [120], cdk-5 (Cyclin-dependent Kinase 5), PKA (cAMP-dependent Protein Kinase) [120, 224], Ca^{2+} / Calmodulin-dependent Protein kinase II [255] and stress-activated protein kinases [224]. Studies in neuronal cell culture models, it has been found that tau phosphorylation is promoted by Erk family kinases (The Extracellular Regulated Kinase) and GSK-3 [247].

Two different genes encode the two isoforms of GSK-3, which are GSK-3α (51 kDa) and GSK-3β (46 kDa). According to some studies on mouse models, GSK-3β knock-out mice die in utero [256], while GSK-3α knock-out mice are viable but show improved glucose tolerance in response to glucose load and elevated hepatic glycogen storage and insulin sensitivity [257]. These two isoforms share 85% homology at the amino acid level [258]. The activities of both isoforms can be down regulated in response to insulin and growth factors like IGF-1 (Insulin-like Growth Factor) [120]. According to some studies that have been done on cultured cerebellar neurons, IGF-1 and insulin promote survival by signalling through the PI(3)K / PKB (Phosphoinositide 3-Kinase / Protein Kinase B) pathway [120, 259]. So it appears that through the activation of IGF-1, insulin or PKB may induce inhibition of GSK-3 and affect tau phosphorylation in neuronal cells [120].

Except for the hyperphosphorylation of tau, GSK-3 is closely related in the increased production of Aβ, inflammatory responses and also memory impairment. In addition, GSK-3 reduces acetylcholine synthesis [260]. Moreover, GSK-3 might directly contribute to neuronal loss in AD because it is a key mediator of apoptosis [261].

Inhibition of Glycogen-Synthase Kinase-3 (GSK-3) as a Therapeutic Strategy

Kinase inhibitors are more preferred than tau phosphatase activators in tau phosphorylation because enzyme inhibition by small molecules is much more easily realized than activation according to general pharmacologic experiences. The studies are focused on GSK-3β inhibitors for the tau kinase–targeting drugs, because of GSK-3β activity may increase with aging and it plays an important role on AD pathogenesis [224, 262]. According to *in vitro* and *in vivo* mice-models studies pharmacological or genetic inhibition of GSK-3β can inverse hyperphosphorylation of tau and prevent behavioural impairments in mice. So it appears that GSK-3β inhibition is a good therapeutic target for AD. At the same time GSK-3β inhibition is well tolerated from the human body and this feature marks the importance of GSK-3β inhibition in this process [224].

In addition, GSK-3β plays a critical role in the pathogenesis of several other diseases like T2DM and cancer and inhibition of GSK-3β. It may also inhibit Aβ production and helps cell survival [224].

In tandem, tau hyperphosphorylation occupies a very important place within the several therapeutic strategies for treating AD. These strategies can be summarized as follows:

1. Inhibition of GSK-3β, cdk5 and other tau kinases,
2. Renovation of the most important and major tau phosphatase PP2A (Protein phosphatase-2A) activity, because unlike protein kinases, protein phosphatases have broad substrate specificities and targeting these enzymes is a logical consideration for reversing abnormal
    hyperphosphorylation of tau.
3. Targeting tau O-GlcNAcylation [224].

UBIQUITIN / PROTEOSOME SYSTEM (UPS)

Proteins can be attached to small molecules such as phosphate, methyl, or acetyl groups, or they can be covalently modified; the array of proteins in a eukaryotic cell or organism depend on the variety of post-translational covalent modifications; these changes greatly extend the functional diversity and dynamics of the proteome, or typically only transiently, by certain other proteins. Among these protein modifiers, the first defined and mostly understood is ubiquitin. Ubiquitin is a small protein and it can be transiently attached to thousands of different proteins [263].

Ubiquitin system is a non-lysosomal proteolytic way and found in all eukaryotes. This system is responsible for selectively removing or destroying incorrect and unwanted proteins [264, 265]. There are two important ways which are seen in the eukaryotes for the degradation of intracellular proteins: Long-lived intracellular proteins are destroyed in the lysosomes by ATP-independent way, while short-lived or incorrect proteins are destroyed in the cytosol in an ATP-dependant manner using UPS [265].

Ubiquitin was isolated from the thymus firstly in 1975 and it was thought that it was a type of thymic hormone, but then it was found in all tissues and organisms [266]. Ubiquitin is heat stable, 76 amino acid polypeptide and approximately 8.5 kDa weight [267, 268].

The attachment of ubiquitin to a substrate protein is called ubiquitination or ubiquitylation, and this attachment can tag the substrate protein for their degradation *via* the proteasome, alter their cellular location, affect their activity, and promote to prevent protein interactions [269 - 271].

Failures in the UPS system cause some pathological conditions [265]. Also, accumulation of ubiquitin or ubiquitin conjugates are found in neurodegenerative and muscular diseases, brain ischemia and cancer [272]. Many neurodegenerative diseases include the same pathological characteristic like the presence of distinctive ubiquitin-positive, intra or extracellular, inclusion bodies in affected regions of the brain [273]. In addition, serum ubiquitin levels increase in the case of parasitic and allergic diseases, AD and patients with chronic kidney failure during hemodialysis treatment [272].

Most cellular processes such as cell cycling, DNA repair, gene transcription and apoptosis need the UPS-mediated post-translational modification and degradation of proteins [273]. Mutually, UPS is responsible for the regulation of multiple cellular processes like proliferation, differentiation, signal transduction, transcriptional regulation and stress response [274, 275]. The part of the system, ubiquitin is a small protein that is covalently linked to other cellular proteins, hence targeting them for degradation in the proteasome [276].

The UPS cascade requires three kinds of enzymes: A ubiquitin-activating enzyme (E1), an ubiquitin conjugating enzyme (E2), and a ubiquitin protein ligase (E3) [274, 275]. Ubiquitin covalently attached to the target protein by using these enzymes and then polyubiquitinated proteins are degraded by 26S proteasome [272].

The degradation mechanisms of the cytosolic or nuclear proteins by ATP and UPS-dependent way can be summarized as follows:

1. The C-terminal end of ubiquitin is stimulated by E1 in the presence of ATP,
2. Activated ubiquitin is transferred to the E2,
3. E3 transfers the ubiquitin to the lysine group of target protein [272].

Ubiquitin/Proteosome System (UPS) in Diabetes and Alzheimer's Disease

Studies have shown that impaired UPS function cause an accumulation of polyubiquitinated proteins in neurodegenerative disorders associated with protein misfolding [277]. Also, it is acknowledged that exogenous administration of ubiquitin carboxyl-terminal hydrolase L1 (UCH-L1) which is a member of the deubiquinating enzyme family, restored memory deficit in an AD model [276, 278]. Downregulation of UCH-L1 has been associated with idiopathic

Parkinson's Disease (PD) and AD [279]. According to a study, in a high-fat diet conditions, UCH-L1-null mice displayed mild glucose intolerance [280].

In another study, the patients with AD showed significant decrease in the proteasome activity in the hippocampus, parahippocampal and middle temporal gyri, and in the inferior parietal lobule than the patients in control group [281]. Also, a mutant form of ubiquitin (Ub^{+1}), which is an efficient substrate for polyubiquitination *in vitro* and in transfected human cells, has been optionally seen in the brains of AD patients including nonfamilial AD. Hence, it seems that inhibition of the Ub-proteasome system is due to the expression of this mutant form in an aging brain and leads to neuropathologic consequences [282].

From another perspective, Aβ accumulation in the brain causes a disturbed proteasome function, hence facilitating tau accumulation, but this is reversible. For example if Aβ is cleared, normal proteasome function is re-established and early tau deposits can be removed. It follows that at the same time the UPS is a link between Aβ and tau interaction [283].

T2DM is a type of disease that activates the UPS and this system plays an important role in diabetic neuropathy. Neuronal protein ubiquitination protects the neurons from the chronic attacks of organelles and disrupted neuronal proteins. Also, pathological activated protein ubiquitination causes neuronal apoptosis/death because it is the part of irreversible catabolic action in the late term cellular damage. In the T2DM, patients with high levels of serum ubiquitin concentration, the reduction has been identified in the amplitude of peripheral nerve axons and inverse correlation has been found between the serum ubiquitin levels and the amplitude of axons [284].

AMYLOID BETA (AB) DEPOSITION

Identification of misfolded protein aggregates in the case of the neurofibrillary tangles composed of hyperphosphorylated tau protein and senile plaques composed of Aβ peptides. As well as extensive oxidative damage and neuronal loss are important findings for the AD diagnosis in the post mortem brain [285]. Aβ peptide is considered to be the second main neuropathological change in AD [286 - 288] also known as "amyloid β cascade hypothesis" [289]. The predominant component of the senile plaques that accumulates in various brain regions and pathologically represent the AD condition is the Aβ peptide [285]. The Aβ peptide consists of 40-42 amino acid residues and occurs as a result of the proteolytic way from the transmembrane protein APP and is encoded by chromosome 19 [286 - 288]. APP includes APP_{695}, APP_{751}, and APP_{770} and they are the three common isoforms of APP. APP_{695} is the most common in the neuronal cells among them [285]. In other words, Aβ is one of the metabolism products of APP [290]. Once Aβ is generated from the APPs, it is found in many different forms like monomers, dimers, higher oligomers, and polymers; after which these forms comprise about 8 nm amyloid fibrils that accumulate in the disease [291]. Hydrophilic N-terminus and hydrophobic C-terminus ends are the structure of Aβ monomer characteristic [285]. So the "amyloid" term refers to the latter form when masses of fibrils accumulate as deposits in the extracellular space and its microvasculature in the brain. For this reason, "the Aβ hypothesis of Alzheimer's disease" term is more accurate than "the amyloid hypothesis of Alzheimer's disease" [291].

The function of this protein is not fully understood [286 - 288]. APP knockout mice do not show significant mortality or morbidity but APPs may have neurotrophic and neuroprotective activities [290]. According to another study, 15–20% reduction in adult body weight was seen in APP knockout mice (APP–/–) but subtle neurological deficits including reduced forelimb grip strength, decreased locomotor activity and increased gliosis were observed in these mice [292]. Also, it has been suggested that APP could be involved in regulating metal ion transport and homeostasis because metal-binding domain is localized near APP residues. It has to be underlined that high concentrations of metals like iron, copper and zinc are also found within the mature plaques from diseased brains [285].

APP occurs from the intracellular carboxyl end, membrane section that includes 28 amino acid residues and extracellular amino end like all transmembrane proteins. APP is metabolized by cutting a number of proteolytic enzymes like alpha, beta and gamma secretase (α, β and γ-Secretase) [293]. α -secretase enzymes cut the Aβ through the middle part, thus, the number of Aβ fragments do not increase [293, 294] because the new formations are not solved forms. Indeed, they are in the extracellular protein form called as APP or sAPP [293]. This form can be taken into the cell and can be found in various forms in different cellular components [294]. In addition, it has been shown that these forms have neurotrophic positive effects on the neurons through the cell culture conditions. However, β-secretase enzymes cut the APP from the amino end, also γ-secretase enzymes cut the APP from the carboxyl end and as a result, Aβ occurs as the product [293]. At the end of this proteolysis, two predominant Aβ structures occur, these formations include 39/40 amino acid residues that accumulate in the vascular area and 42/43 amino acid residues that accumulate in the plaques [294]. Patients with early dementia show increased levels of Aβ-42 and Aβ oligomers which have been found in the brain tissues, also in the cerebrospinal fluid, and these levels are parallel with the cognitive deficit [295 - 297]. Hence, it appears that the mediators of the AD are not the amyloid plaques, but also, the small Aβ oligomers [298, 299].

Collectively in the case of AD α-secretase pathway is suppressed but it is suggested that the balance deviates to the β and γ-secretase pathways. Therefore, in the AD conditions, Aβ levels increase due to APP rising or β and γ-secretase enzymes activations [300].

GSK-3α regulates the production of Aβ peptides and the decreased phosphorylation of GSK-3β may potentiate γ-secretase activity and the amyloidogenic AβPP processing, resulting in increased intracellular levels of Aβ [301].

From the age of 60, Aβ that is the product of β and γ-secretase activity, starts to accumulate in the neocortex in case of loose plaques for almost everyone, but these accumulations do not mean that they always cause the development of dementia [300]. Simply put, Aβ is normally generated by the cells and circulates in extracellular fluids in all individuals throughout their life. This has been known since 1992 because Aβ is a normal metabolic product of intracellular APP processing but the accumulation of Aβ has the potential in pathogenesis of AD. So, the presence of Aβ in the extracellular or intracellular area is normal but the incompatibility of balance between its arrival and removal within these area are the cause to increase its levels and to generate the accumulation process. Put in another way, the accumulation of dimers or oligomers in intracellular area or brain interstitial fluid induce microscopically apparent deposits that produce mild, largely subclinical neuronal dysfunction [291]. Loose plaques are harmless but as a result of Aβ accumulations, oxidative tension and free radical formations occur. Additionally, in this case, gliosis and microglial activations cause inflammatory changes and all of these changes result in concentrating the plaques to create the solid plaques. These solid plaques are responsible for the local neurotoxicity, cell apoptosis / death, and neuritic degenerations [300].

In the early stages of AD the smaller Aβ formations like oligomers and Amyloid β-Derived Diffusible Ligands (ADDLs) (will be explained in the next subsection), reduce the synaptic function and cause impairment of learning and memory formation processes as well as interfere with neurotransmission and/or other signalling pathways in the cells [285].

Amyloid β-Derived Diffusible Ligands (ADDLs)

Amyloid β-Derived Diffusible Ligands (ADDLs) are neurotoxic, nonfibrillar (globular) and soluble Aβ oligomers [302, 303]. These oligomeric structures contain 3-24 peptide monomers according to atomic force microscopy and gel analysis [303]. ADDLs cause disruption of LTP rapidly at submicromolar doses and kill differentiated central nervous system neurons in prolonged incubations. In summary, ADDLs cause early-stage memory loss firstly, leading to synapse degeneration and nerve cell death. ADDLs might block the structural plasticity to induce the synapse loss by blocking synapse replacement [304] and hyperphosphorylation of tau [305]. More than 90% of ADDLs colocalize with the major component of postsynaptic densities and the reliable marker of postsynaptic terminals in mature hippocampal cultures, PSD-95 (Postsynaptic Density Protein-95), so it can be argued that ADDLs specifically target the synapses [306].

Hippocampal neurons are lost in AD and according to selective expression of high affinity ADDL-binding proteins, toxin receptors play a role in hippocampus. These soluble oligomers have been found in APP-transgenic mice an AD patients. The molecular basis of ADDLs is very important because of their toxicity and differential expression of ADDL binding proteins has been verified by immunofluorescence microscopy and also other techniques like the ligand blot assay. According to these researches, ADDL binding sites have been found on neurons from hippocampus and cortex but not in the cerebellum [304].

ADDLs are very small, hence easily diffusible within the tissues, and for this reason they are more harmful than Aβ. They also contribute to insulin deficits in AD patients that include insulin resistance in the brain, consequently ADDLs break the communication relation between the insulin and its receptor at the synapse. In the next stage, ADDLs bind to the synapse itself and changes its conformation for modification of the synaptic shape. The resulting scenario results in insulin resistance [66, 305]. Because of ADDLs effect on insulin signalling, it is suggested that they are the neurological form of the brain diabetes. An adequate amount of insulin receptors have been found in the neurons that do not have ADDLs, therefore, measuring the ADDLs levels may be a new methodology for the AD diagnosis [66].

Collectively, according to ADDLs hypothesis, in the case of AD, increased production or decreased clearance result in elevated levels of Aβ42 formations and these formations increase ADDLs levels but not deposition. In short, Aβ42 assembly into ADDLs. ADDLs bind with high affinity to neuronal receptors and trigger aberrant synaptic signalling. After this point, it divides into two paths, firstly, it causes LTP blockage and in continuation of this memory impairment and dementia, especially in very low ADDLs concentrations. Secondly, it causes a persistent downstream signaling like GSK3β and in continuation of this synaptic degeneration, cytoskeletal collapses and notably causes neuron death over many years [303].

Amyloid Beta (Aβ) Deposition in Diabetes and Alzheimer's Disease

The deposition of Aβ and β-islet cells of pancreas is the pathogenic similarity between AD and T2DM. These β-islet cells produce both amyloid and insulin.

According to some studies elevated levels of amyloid effect the functioning of β-islet cells and cause the disruption of the glucose homeostasis [307]. Indeed, there is a toxic cycle between continuous insulin exposure in the case of brain insulin resistance and Aβ accumulation in the neurons [308].

Impaired insulin signaling can disturb both APP processing and Aβ clearance [303]. Aβ oligomer toxicity is a direct consequence of neuronal insulin resistance in the AD brain [309]. In parallel with these findings, increased levels of neurofibrillary tangles and Aβ plaques have been seen in the diabetic patient's brain [66]. Additionally Aβ levels are reduced and memory performance is improved in the case of treating AD patients and transgenic animals with either insulin or insulin sensitizer [309].

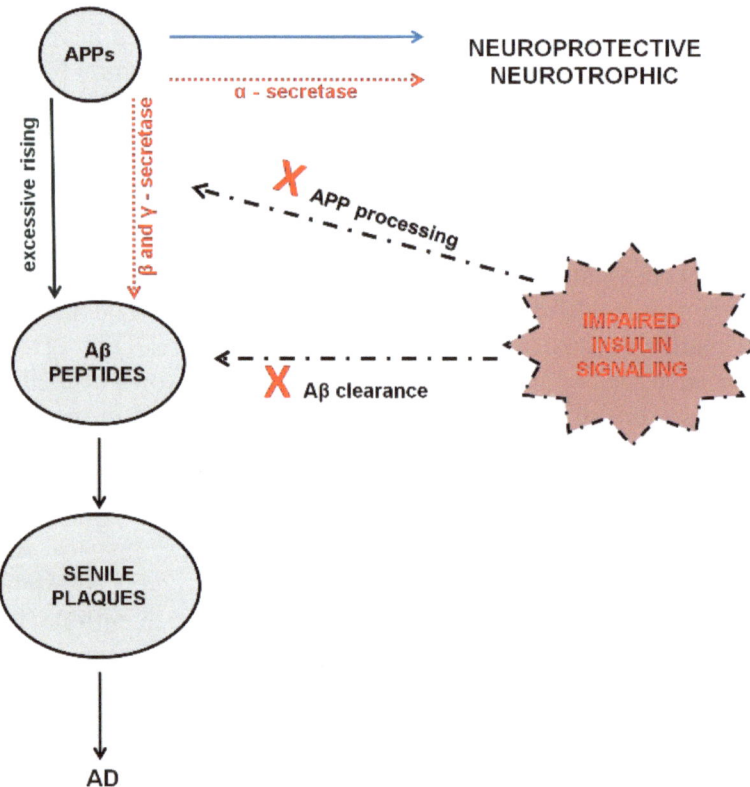

Fig. (4). General diagram of Aβ deposition in T3D.

In T3DM conditions, brain insulin resistance results in abnormal Aβ expression and protein processing [310]. However, Aβ toxicity may cause insulin resistance in the brain and disturb the insulin signalling by competing with insulin on its receptors. Also, it decreases the surface expression of insulin receptors and reduce the insulin affinity to its relative receptors [311]. The insulin-stimulated insulin receptor tyrosine phosphorylation and phosphorylation of several downstream kinases are inhibited by the Aβ oligomers in hippocampal neurons. According to another study, the insulin-stimulated insulin receptor tyrosine phosphorylation is inhibited by ADDLs at nanomolar concentrations in rat hippocampal primary neurons and in human insulin receptor expressing cells. A striking loss of insulin receptor α-subunit from the dendritic membranes occur as a result of ADDLs' binding to neurons. These types of ADDLs binding neurons show the insulin receptor immunoreactivity into the cell interior and nucleus due to translocation, however, neurons without ADDLs-binding show intact insulin receptor immunoreactivities in their dendrites [309].

To remain in DM and also hypercholesterolemia for a long time may cause deleterious effects on the functional integrity of the BBB, and as a result, coincident accumulation of $A\beta_{42}$ which has been seen in the neurons suggests that there is a relation between the BBB permeability and intraneuronal deposition of $A\beta_{42}$ in the brain [162] (Fig. **4**).

Insulin-Degrading Enzyme (IDE)

Insulin-degrading enzyme (IDE) is a metalloprotease and expressed throughout the body ubiquitously. This enzyme primarily localises to the cytoplasm and peroxisomes of cells and IDEs are responsible for degrading insulin after the connection of insulin and its receptor. Moreover, IDE is responsible for preventing over-accumulation of serum insulin levels [312].

The clearance of Aβ plaques occurs as a result of the participation of IDE as a catalyst in the brain [66]. Also IDE regulates the intracellular domains' levels of insulin, Aβ protein and APPs [308]. Both insulin and Aβ bind and compete with IDE and also compete in binding to the insulin receptor [309]. The affinity of IDE is very high for insulin which leads to the degradation of amyloid beta gets inhibited [66].

Microglial cells secrete the IDE to the extracellular area in the brain, where it degrades monomeric but not oligomeric Aβ peptide. Consequently, Aβ concentrations are reduced in the brain, as well as decreasing aggregation and plaque formations is decreased. In the AD patients, IDE levels have been found in decreased levels, especially in the hippocampus [312].

Mutant IDE rat model of T2DM show hyperinsulinaemia and glucose intolerance. Additionally, IDE (-/-) knockout mice show cerebral accumulation of endogenous Aβ in increased levels. Therefore, it can be said that IDE hypofunction cause hyperinsulinemia, diabetes, neurodegeneration and neuronal loss through the

T2DM or T3DM. Another example, reduced Aβ, regulated insulin and degraded APP have been seen *via* IDE within the normal subjects, so IDE, insulin and Aβ show a regulatory relationship [308]. Another fact is that the production of IDE declines with aging, and the levels of substrates increase [66].

Collectively, elevated insulin levels in the case of T2DM induce Aβ accumulation through competition between insulin and Aβ for IDE [313].

OBESITY

The risk of the vascular disease, sporadic dementia, AD, hypertension, diabetes, cardiovascular and cerebrovascular diseases are increased by overweight and obesity [314]. Obesity is also known as adiposity or fatness, and in addition to the above-mentioned diseases, is associated with the increases risk of insulin resistance, dyslipidemia, degenerative joint disease, cancer, and respiratory diseases. About 30% of society are obese, and the distribution of obesity is higher in women than men [315].

'Body memory'is related to food intake generally and there are numerous hypotheses relating to memory, hippocampal and hypothalamic functions, and control of energy intake [316].

In middle age elevated body mass index (BMI) (overweight and obesity), that can be grouped in the following overweight (BMI ≥ 25) and obesity (BMI ≥ 30), is associated with higher dementia risk and predicts higher dementia risk in older ages. However, there have been reports of no association with lower body mass index related to higher AD risk [315]. In other words, mid-life overweight and obesity have been linked to a higher risk of dementia in later life [314]. According to a population based studies, every 1% rise in BMI at the age of 70 could increase the risk of AD by 36% [66].

Cerebral (temporal) atrophy, white-matter lesions change and the disturbance of the BBB are the possible mechanisms of the brain pathology. In the case of dementia depend on BMI and body-fat distribution [314, 317].

Formation of vascular pathologies due to overweight and obesity, are the primary mechanism for increased dementia, AD and neurodegenerative pathologies. Alterations in adipose tissue hormone levels and subsequent feedback loops through peripheral overweight and obesity increase the metabolic risk and lead to direct neuronal toxicity possibly due to an altered BBB [314].

The renin-angiotensin system (RAS) is another potential link of adiposity, vascular disease and AD. Blood pressure regulation is the classical function of the

RAS but it may also provide a link between obesity, hypertension, vascular syndromes like T2DM and the health of the brain [316].

It is a known fact that, worldwide there is a concerning epidemic of obesity, insulin resistance and diabetes [315]. Obesity, hypertension, hypercholesterolemia and hyperlipidemia are all linked to T2DM. These types of diseases are also associated with the early cognitive changes in learning and memory, mental flexibility and mental speed [317].

Obesity may effect the brain directly or it can increase the risk of AD through insulin signaling pathway or hypertension [318] because it leads to the important precursors of T2DM, such as insulin resistance, impaired glucose metabolism, hypertension and dyslipidemia [66]. Elevated levels of BMI affect the insulin signaling pathway in the liver or muscle cells by increasing the proinflammatory response [319]. Experimental studies have shown that the developing risk of AD is increased by three fold in obese patients without T2DM. In obesity and T2DM cases, the risk of developing AD may be linked to hyperglycemia, insulin resistance, oxidative stress, enhanced accumulation of glycation end products mechanistically [66]. According to another view, recognition of the interactions between leptin and insulin in the brain may be key for the relationship between adiposity and diabetes in AD [314]. It has recently been suggested that leptin plays a significant role in the regulation of brain functions by studying the impaired leptin signaling pathway that can be used to propose a link between AD and T2DM [66]. Another view pinpoints that APPs upregulation have been seen in obesity, and there is a positive correlation between APPs expression, insulin resistance and adipocyte cytokine expression levels [314].

THE APOLIPOPROTEIN-E (APOE4)

Apolipoprotein E is a type of plasma protein and synthesized by the liver mainly, but different from other proteins, it is also synthesized by the central nervous system such as astrocytes, Schwann cells and oligodendrocytes. ApoE is one of the apolipoprotein type that directs the lipid metabolism [320]. It participates in the transport of cholesterol and lipoproteins within the circulatory system and so it affects the blood lipid levels [317]. Because dyslipidaemia is one of the features of diabetes along with insulin/glucose abnormalities and ApoE4 plays a key role in the processing of lipid, it is highly studied for its involvement in T2DM [66].

According to the findings of one study, there has been an interaction between diabetes and the apolipoprotein E ε4 allele (APOE ε4) [321]. ApoE4 is also referred to as "susceptibility gene" because it has shown a positive correlation with AD [66]. Aβ peptide binds to ApoE ε4 more easily than ApoE ε3 so ApoE ε4 has the most pronounced effect in promoting Aβ fibril formation [322]. In other words, ApoE ε4 responsible from the clearance of Aβ [317]. ApoE ε4 binds to extracellular senile plaques, neurofibrillary tangles and amyloid angiopathy regions to create a stable structures [323].

The risk of APOE ε4 is associated AD might be aggravated by diabetes because AD has seen two fold increase within the patients with diabetes who are ε4 carriers than individuals who harbor the ε4 allele but are not diabetic [317]. Also, ApoE ε4 co-localize with advanced glycation en products in the AD brain [322]. In ApoE ε4 heterozygotes, the risk of developing AD is increased by five folds and in the homozygous, the rate is about 50–90 percent [66]. Moreover, APOE ε4 affects the processing of tau because overexpression of APOE ε4 in transgenic mice seems to promote neuronal tau phosphorylation according to a study [317].

The lipid binding ability of ApoE is increased by cholesterol transporter ABCA1, so in the case of its deficiency, poor lipidation of ApoE occurs as a result in heavy deposition of Aβ and its overexpression occur [66].

According to a study, hyperinsulinaemia is associated with a high risk of AD among the subjects who do not carry ApoE ε4 allele, but no risk has been found for AD in the case when ApoE ε4 gene is present. Insulin-mediated energy metabolism is lower in the AD patients without ApoE ε4 than in those the AD patients with ApoE ε4 [324].

Collectively, elevated levels of insulin rather than T2DM alone leads to AD pathology, especially in populations without ApoE4 allele [324].

CONCLUSION

In recent years according to some members of the diabetes community, "Type 3 Diabetes" term has been introduced and it is a proposed condition for Alzheimer's disease resulting in an insulin deficiency and resistance in the brain. In this chapter we presented the information about "Type 3 Diabetes" with the underlying mechanisms through the perspective of histology.

Hopefully this chapter would provide some insight into the subject of "Type 3 Diabetes", also the underlying mechanisms, and the type of work that is researched by specialists in this field.

CONSENT FOR PUBLICATION

Not applicable.

CONFLICT OF INTEREST

The authors confirm that they have no conflict of interest to declare for this publication.

ACKNOWLEDGEMENTS

We would like to thank to Res. Assist. Pakize Nur AKKAYA for her assistance. Mr. Carmel Charles Sant English Lecturer at Anadolu University (Editing).

REFERENCES

[1] Gilman S. Alzheimer's disease. Perspect Biol Med 1997; 40(2): 230-45.
[http://dx.doi.org/10.1353/pbm.1997.0020] [PMID: 9058953]

[2] Lleó A, Greenberg SM, Growdon JH. Current pharmacotherapy for Alzheimer's disease. Annu Rev Med 2006; 57: 513-33.
[http://dx.doi.org/10.1146/annurev.med.57.121304.131442] [PMID: 16409164]

[3] Ferri CP, Prince M, Brayne C, et al. Global prevalence of dementia: a Delphi consensus study. Lancet 2005; 366: 2112-7.
[http://dx.doi.org/10.1016/S0140-6736(05)67889-0] [PMID: 16360788]

[4] Alzheimer A. Uber eine eigenartige Erkrankung der Hirnrinde. Allg Z Psychiat 1907; 64: 146-8.

[5] Alzheimer A. Über eigenartige Krankheitsfälle des späteren Alters. Z Gesamte Neurol Psychiatr 1911; 4: 356-85.
[http://dx.doi.org/10.1007/BF02866241]

[6] Katzman R, Saitoh T. Advances in Alzheimer's disease FASEB J 1991; 5: 278-86.
[http://dx.doi.org/10.1096/fasebj.5.3.2001787] [PMID: 2001787]

[7] Taneli B, Sivrioğlu Y, Taneli T. Alzheimer diease. In: İT Uzbay H Aydın, Ed. In: Gülhane Psychopharmacology Symposium; 1999; pp. 1. Printingt. Ankara: Gülhane Military Medical Academy Printing Office31-80.

[8] Evans DA, Funkenstein HH, Albert MS, et al. Prevalence of Alzheimer's disease in a community population of older persons. Higher than previously reported. JAMA 1989; 262(18): 2551-6.
[http://dx.doi.org/10.1001/jama.1989.03430180093036] [PMID: 2810583]

[9] Olshansky SJ, Carnes BA, Cassel CK. The aging of the human species. Sci Am 1993; 268(4): 46-52.
[http://dx.doi.org/10.1038/scientificamerican0493-46] [PMID: 8446881]

[10] Baysal AI, Yesilbudak Z. Clinical Findings Of Alzheimer's Disease. Turkiye Klinikleri J Neur 2003; 1(1): 1-5.

[11] Lopes JP, Oliveira CR, Agostinho P. Cell cycle re-entry in Alzheimer's disease: a major neuropathological characteristic? Curr Alzheimer Res 2009; 6(3): 205-12.
[http://dx.doi.org/10.2174/156720509788486590] [PMID: 19519302]

[12] Duyckaerts C, Panchal M, Delatour B, Potier MC. [Morphologic and molecular neuropathology of Alzheimer's disease]. Ann Pharm Fr 2009; 67(2): 127-35.
[http://dx.doi.org/10.1016/j.pharma.2009.01.001] [PMID: 19298896]

[13] Octave JN, Pierrot N. [Alzheimer's disease: cellular and molecular aspects]. Bull Acad Natl Med 2008; 192(2): 323-31.
[PMID: 18819686]

[14] Chung SH. Aberrant phosphorylation in the pathogenesis of Alzheimer's disease. BMB Rep 2009; 42 (8): 467-74.

[http://dx.doi.org/10.5483/BMBRep.2009.42.8.467] [PMID: 19712581]

[15] Mohandas E, Rajmohan V, Raghunath B. Neurobiology of Alzheimer's disease. Indian J Psychiatry 2009; 51(1): 55-61.
[http://dx.doi.org/10.4103/0019-5545.44908] [PMID: 19742193]

[16] Hooper NM. Roles of proteolysis and lipid rafts in the processing of the amyloid precursor protein and prion protein. Biochem Soc Trans 2005; 33(Pt 2): 335-8.
[http://dx.doi.org/10.1042/BST0330335] [PMID: 15787600]

[17] Kolev MV, Ruseva MM, Harris CL, Morgan BP, Donev RM. Implication of complement system and its regulators in Alzheimer's disease. Curr Neuropharmacol 2009; 7(1): 1-8.
[http://dx.doi.org/10.2174/157015909787602805] [PMID: 19721814]

[18] Silvestrelli G, Lanari A, Parnetti L, Tomassoni D, Amenta F. Treatment of Alzheimer's disease: from pharmacology to a better understanding of disease pathophysiology. Mech Ageing Dev 2006; 127(2): 148-57.
[http://dx.doi.org/10.1016/j.mad.2005.09.018] [PMID: 16278007]

[19] Reddy PH, Beal MF. Amyloid beta, mitochondrial dysfunction and synaptic damage: implications for cognitive decline in aging and Alzheimer's disease. Trends Mol Med 2008; 14(2): 45-53.
[http://dx.doi.org/10.1016/j.molmed.2007.12.002] [PMID: 18218341]

[20] Fisher A. Cholinergic treatments with emphasis on m1 muscarinic agonists as potential disease-modifying agents for Alzheimer's disease. Neurotherapeutics 2008; 5(3): 433-42.
[http://dx.doi.org/10.1016/j.nurt.2008.05.002] [PMID: 18625455]

[21] Mattson MP. Pathways towards and away from Alzheimer's disease. Nature 2004; 430(7000): 631-9.
[http://dx.doi.org/10.1038/nature02621] [PMID: 15295589]

[22] Grossberg GT. The ABC of Alzheimer's disease: behavioral symptoms and their treatment. Int Psychogeriatr 2002; 14 (Suppl. 1): 27-49.
[http://dx.doi.org/10.1017/S1041610203008652] [PMID: 12636179]

[23] Tanaka Y, Meguro K, Yamaguchi S, et al. Decreased striatal D2 receptor density associated with severe behavioral abnormality in Alzheimer's disease. Ann Nucl Med 2003; 17(7): 567-73.
[http://dx.doi.org/10.1007/BF03006670] [PMID: 14651356]

[24] Hynd MR, Scott HL, Dodd PR. Glutamate-mediated excitotoxicity and neurodegeneration in Alzheimer's disease. Neurochem Int 2004; 45(5): 583-95.
[http://dx.doi.org/10.1016/j.neuint.2004.03.007] [PMID: 15234100]

[25] Small DH. Network dysfunction in Alzheimer's disease: does synaptic scaling drive disease progression? Trends Mol Med 2008; 14(3): 103-8.
[http://dx.doi.org/10.1016/j.molmed.2007.12.006] [PMID: 18262842]

[26] Markesbery WR. Oxidative stres hypothesis in AD. Free Radic Biol Med 1997; 23(1): 134-47.
[http://dx.doi.org/10.1016/S0891-5849(96)00629-6] [PMID: 9165306]

[27] Reddy VP, Zhu X, Perry G, Smith MA. Oxidative stress in diabetes and Alzheimer's disease. J Alzheimers Dis 2009; 16(4): 763-74.
[http://dx.doi.org/10.3233/JAD-2009-1013] [PMID: 19387111]

[28] Vural H, Sirin B, Yilmaz N, Eren I, Delibas N. The role of arginine-nitric oxide pathway in patients with Alzheimer disease. Biol Trace Elem Res 2009; 129(1-3): 58-64.
[http://dx.doi.org/10.1007/s12011-008-8291-8] [PMID: 19099206]

[29] Aliyev A, Seyidova D, Rzayev N, et al. Is nitric oxide a key target in the pathogenesis of brain lesions during the development of Alzheimer's disease? Neurol Res 2004; 26(5): 547-53.
[http://dx.doi.org/10.1179/016164250176131] [PMID: 15265272]

[30] Snyder SH. Nitric oxide and neurons. Curr Opin Neurobiol 1992; 2(3): 323-7.
[http://dx.doi.org/10.1016/0959-4388(92)90123-3] [PMID: 1353698]

[31] Brion JP. The neurobiology of Alzheimer's disease. Acta Clin Belg 1996; 51(2): 80-90.
[http://dx.doi.org/10.1080/17843286.1996.11718490] [PMID: 8693872]

[32] Gürvit H. Degenerative Diseases Of The Nervous System. Dementia syndrome, Alzheimer's disease and non-Alzheimer's dementia.Neurology. İstanbul: Nobel Matbaacılık 2004; pp. 367-415.

[33] de la Monte SM, Neusner A, Chu J, Lawton M. Epidemilogical trends strongly suggest exposures as etiologic agents in the pathogenesis of sporadic Alzheimer's disease, diabetes mellitus, and non-alcoholic steatohepatitis. J Alzheimers Dis 2009; 17(3): 519-29.
[http://dx.doi.org/10.3233/JAD-2009-1070] [PMID: 19363256]

[34] Cankurtaran M, Arıoğul S. Risk Factors Of Alzheimer's Disease.Geriatrics and Gerontology 1 Press Ankara: MN Medial ve Nobel. 2006; pp. 953-68.

[35] Serafini M, Del Rio D. Understanding the association between dietary antioxidants, redox status and disease: is the Total Antioxidant Capacity the right tool? Redox Rep 2004; 9(3): 145-52.
[http://dx.doi.org/10.1179/135100004225004814] [PMID: 15327744]

[36] Baynes JW. Role of oxidative stress in development of complications in diabetes. Diabetes 1991; 40(4): 405-12.
[http://dx.doi.org/10.2337/diab.40.4.405] [PMID: 2010041]

[37] Baynes JW, Thorpe SR. Role of oxidative stress in diabetic complications: a new perspective on an old paradigm. Diabetes 1999; 48(1): 1-9.
[http://dx.doi.org/10.2337/diabetes.48.1.1] [PMID: 9892215]

[38] Altan N, Ongun CÖ, Hasanoğlu E, Engin A, Tuncer C, Sindel S. Effect of the sulfonylurea glyburide on superoxide dismutase activity in alloxan-induced diabetic rat hepatocytes. Diabetes Res Clin Pract 1994; 22(2-3): 95-8.
[http://dx.doi.org/10.1016/0168-8227(94)90041-8] [PMID: 8200301]

[39] Altan N, Ongun CÖ, Elmali E, Kiliç N, Yavuz O, Cayci B. Effect of the sulfonylurea glyburide on glutathione and glutathione peroxidase activity in alloxan-induced diabetic rat hepatocytes. Gen Pharmacol 1994; 25(5): 875-8.
[http://dx.doi.org/10.1016/0306-3623(94)90089-2] [PMID: 7835630]

[40] Saxena AK, Srivastava P, Kale RK, Baquer NZ. Impaired antioxidant status in diabetic rat liver. Effect of vanadate. Biochem Pharmacol 1993; 45(3): 539-42.
[http://dx.doi.org/10.1016/0006-2952(93)90124-F] [PMID: 8442752]

[41] Elmalí E, Altan N, Bukan N. Effect of the sulphonylurea glibenclamide on liver and kidney antioxidant enzymes in streptozocin-induced diabetic rats. Drugs R D 2004; 5(4): 203-8.
[http://dx.doi.org/10.2165/00126839-200405040-00003] [PMID: 15230625]

[42] Kiliç N, Malhatun E, Elmali E, Altan N. An investigation into the effect of sulfonylurea glyburide on glutathione peroxidase activity in streptozotocin-induced diabetic rat muscle tissue. Gen Pharmacol 1998; 30(3): 399-401.
[http://dx.doi.org/10.1016/S0306-3623(97)00277-2] [PMID: 9510093]

[43] Tiedge M, Lortz S, Drinkgern J, Lenzen S. Relation between antioxidant enzyme gene expression and antioxidative defense status of insulin-producing cells. Diabetes 1997; 46(11): 1733-42.
[http://dx.doi.org/10.2337/diab.46.11.1733] [PMID: 9356019]

[44] Tiedge M, Lortz S, Munday R, Lenzen S. Complementary action of antioxidant enzymes in the protection of bioengineered insulin-producing RINm5F cells against the toxicity of reactive oxygen species. Diabetes 1998; 47(10): 1578-85.
[http://dx.doi.org/10.2337/diabetes.47.10.1578] [PMID: 9753295]

[45] Robertson RP, Harmon J, Tran PO, Poitout V. β-cell glucose toxicity, lipotoxicity, and chronic oxidative street inn type 2 diabetes. Diabetes 2004; 53(1): 119-24.
[http://dx.doi.org/10.2337/diabetes.53.2007.S119]

[46] Houslay MD. 'Crosstalk': a pivotal role for protein kinase C in modulating relationships between signal transduction pathways. Eur J Biochem 1991; 195(1): 9-27.
[http://dx.doi.org/10.1111/j.1432-1033.1991.tb15671.x] [PMID: 1846812]

[47] Donath MY, Gross DJ, Cerasi E, Kaiser N. Hyperglycemia-induced β-cell apoptosis in pancreatic islets of Psammomys obesus during development of diabetes. Diabetes 1999; 48(4): 738-44.
[http://dx.doi.org/10.2337/diabetes.48.4.738] [PMID: 10102689]

[48] Eizirik DL, Flodström M, Karlsen AE, Welsh N. The harmony of the spheres: inducible nitric oxide synthase and related genes in pancreatic β cells. Diabetologia 1996; 39(8): 875-90.
[http://dx.doi.org/10.1007/BF00403906] [PMID: 8858209]

[49] Ihara Y, Toyokuni S, Uchida K, et al. Hyperglycemia causes oxidative stress in pancreatic β-cells of GK rats, a model of type 2 diabetes. Diabetes 1999; 48(4): 927-32.
[http://dx.doi.org/10.2337/diabetes.48.4.927] [PMID: 10102716]

[50] Giugliano D, Ceriello A, Paolisso G. Oxidative stress and diabetic vascular complications. Diabetes Care 1996; 19(3): 257-67.
[http://dx.doi.org/10.2337/diacare.19.3.257] [PMID: 8742574]

[51] Rösen P, Du X, Tschöpe D. Role of oxygen derived radicals for vascular dysfunction in the diabetic heart: prevention by α-tocopherol? Mol Cell Biochem 1998; 188(1-2): 103-11.
[http://dx.doi.org/10.1023/A:1006876607566] [PMID: 9823016]

[52] Du X, Stocklauser-Färber K, Rösen P. Generation of reactive oxygen intermediates, activation of NF-kappaB, and induction of apoptosis in human endothelial cells by glucose: role of nitric oxide synthase? Free Radic Biol Med 1999; 27(7-8): 752-63.
[http://dx.doi.org/10.1016/S0891-5849(99)00079-9] [PMID: 10515579]

[53] Ceriello A. Acute hyperglycaemia and oxidative stress generation. Diabet Med 1997; 14(3) (Suppl. 3): S45-9.
[http://dx.doi.org/10.1002/(SICI)1096-9136(199708)14:3+<S45::AID-DIA444>3.3.CO;2-I] [PMID: 9272613]

[54] Davidoff AJ, Rodgers RL. Insulin, thyroid hormone, and heart function of diabetic spontaneously hypertensive rat. Hypertension 1990; 15(6 Pt 1): 633-42.
[http://dx.doi.org/10.1161/01.HYP.15.6.633] [PMID: 2140815]

[55] Das K, Chainy GBN. Modulation of rat liver mitochondrial antioxidant defense system by thyroid hormone. Biochimica et Biophysica Acta (BBA)-. Molecular Basis of Disease 2001; 1537(1): 1-13.
[http://dx.doi.org/10.1016/S0925-4439(01)00048-5] [PMID: 11476958]

[56] Altan N, Buğdayci G, Tutkun FK, Sancak B, Nazaroğlu NK. The effect of the sulfonylurea glyburide on nitric oxide in streptozotocin-induced diabetic rat. Gen Pharmacol 1998; 31(2): 319-21.
[http://dx.doi.org/10.1016/S0306-3623(97)00430-8] [PMID: 9688480]

[57] Colhoun HM, Lee ET, Bennett PH, et al. Risk factors for renal failure: the WHO Mulinational Study of Vascular Disease in Diabetes. Diabetologia 2001; 44(2) (Suppl. 2): S46-53.
[http://dx.doi.org/10.1007/PL00002939] [PMID: 11587050]

[58] Battisti WP, Palmisano J, Keane WE. Dyslipidemia in patients with type 2 diabetes. relationships between lipids, kidney disease and cardiovascular disease. Clin Chem Lab Med 2003; 41(9): 1174-81.
[http://dx.doi.org/10.1515/CCLM.2003.181] [PMID: 14598867]

[59] Januszewski AS, Alderson NL, Metz TO, Thorpe SR, Baynes JW. Role of lipids in chemical modification of proteins and development of complications in diabetes. Biochem Soc Trans 2003; 31 (Pt 6): 1413-6.
[http://dx.doi.org/10.1042/bst0311413] [PMID: 14641077]

[60] Dean RT, Fu S, Stocker R, Davies MJ. Biochemistry and pathology of radical-mediated protein oxidation. Biochem J 1997; 324(Pt 1): 1-18.
[http://dx.doi.org/10.1042/bj3240001] [PMID: 9164834]

[61]	Dillard CJ, Downey JE, Tappel AL. Effect of antioxidants on lipid peroxidation in iron-loaded rats. Lipids 1984; 19(2): 127-33.
[http://dx.doi.org/10.1007/BF02534503] [PMID: 6708751]

[62]	Dillmann WH. Diabetes and thyroid-hormone-induced changes in cardiac function and their molecular basis. Annu Rev Med 1989; 40: 373-94.
[http://dx.doi.org/10.1146/annurev.me.40.020189.002105] [PMID: 2658757]

[63]	Diekman MJ, Romijn JA, Endert E, Sauerwein H, Wiersinga WM. Thyroid hormones modulate serum leptin levels: observations in thyrotoxic and hypothyroid women. Thyroid 1998; 8(12): 1081-6.
[http://dx.doi.org/10.1089/thy.1998.8.1081] [PMID: 9920361]

[64]	Nunomura A, Perry G, Aliev G, et al. Oxidative damage is the earliest event in Alzheimer disease. J Neuropathol Exp Neurol 2001; 60(8): 759-67.
[http://dx.doi.org/10.1093/jnen/60.8.759] [PMID: 11487050]

[65]	Stadtman ER. Protein oxidation in aging and age-related diseases. Ann N Y Acad Sci 2001; 928(1): 22-38.
[http://dx.doi.org/10.1111/j.1749-6632.2001.tb05632.x] [PMID: 11795513]

[66]	Mittal K, Katare DP. Shared links between type 2 diabetes mellitus and Alzheimer's disease: A review. Diabetes Metab Syndr 2016; 10(2) (Suppl. 1): S144-9.
[http://dx.doi.org/10.1016/j.dsx.2016.01.021] [PMID: 26907971]

[67]	Sima AA, Zhang W, Kreipke CW, Rafols JA, Hoffman WH. Inflammation in Diabetic Encephalopathy is Prevented by C-Peptide. Rev Diabet Stud 2009; 6(1): 37-42.
[http://dx.doi.org/10.1900/RDS.2009.6.37] [PMID: 19557294]

[68]	Vendrell J, Broch M, Fernandez-Real JM, et al. Tumour necrosis factor receptors (TNFRs) in Type 2 diabetes. Analysis of soluble plasma fractions and genetic variations of TNFR2 gene in a case-control study. Diabet Med 2005; 22(4): 387-92.
[http://dx.doi.org/10.1111/j.1464-5491.2004.01392.x] [PMID: 15787661]

[69]	Lechleitner M, Herold M, Dzien-Bischinger C, Hoppichler F, Dzien A. Tumour necrosis factor-alpha plasma levels in elderly patients with Type 2 diabetes mellitus-observations over 2 years. Diabet Med 2002; 19(11): 949-53.
[http://dx.doi.org/10.1046/j.1464-5491.2002.00846.x] [PMID: 12421433]

[70]	Diagnosis and classification of diabetes mellitus. Diabetes Care 2011; 34 (Suppl. 1): S62-9.
[http://dx.doi.org/10.2337/dc11-S062] [PMID: 21193628]

[71]	Ahmed N. Advanced glycation endproducts--role in pathology of diabetic complications. Diabetes Res Clin Pract 2005; 67(1): 3-21.
[http://dx.doi.org/10.1016/j.diabres.2004.09.004] [PMID: 15620429]

[72]	Singh R, Barden A, Mori T, Beilin L. Advanced glycation end-products: a review. Diabetologia 2001; 44(2): 129-46.
[http://dx.doi.org/10.1007/s001250051591] [PMID: 11270668]

[73]	Parmaksız İ. Advanced Glycation End-Products in Complications of Diabetes Mellitus. Marmara Medical Journal 2011; 24(3): 141-8.

[74]	Peyroux J, Sternberg M. Advanced glycation endproducts (AGEs): Pharmacological inhibition in diabetes. Pathol Biol (Paris) 2006; 54(7): 405-19.
[http://dx.doi.org/10.1016/j.patbio.2006.07.006] [PMID: 16978799]

[75]	Turk Z. Glycotoxines, carbonyl stress and relevance to diabetes and its complications. Physiol Res 2010; 59(2): 147-56.
[PMID: 19537931]

[76]	Huebschmann AG, Regensteiner JG, Vlassara H, Reusch JE. Diabetes and advanced glycoxidation end products. Diabetes Care 2006; 29(6): 1420-32.

[http://dx.doi.org/10.2337/dc05-2096] [PMID: 16732039]

[77] Goldin A, Beckman JA, Schmidt AM, Creager MA. Advanced glycation end products: sparking the development of diabetic vascular injury. Circulation 2006; 114(6): 597-605.
[http://dx.doi.org/10.1161/CIRCULATIONAHA.106.621854] [PMID: 16894049]

[78] Bierhaus A, Hofmann MA, Ziegler R, Nawroth PP. AGEs and their interaction with AGE-receptors in vascular disease and diabetes mellitus. I. The AGE concept. Cardiovasc Res 1998; 37(3): 586-600.
[http://dx.doi.org/10.1016/S0008-6363(97)00233-2] [PMID: 9659442]

[79] Schmidt AM, Yan SD, Yan SF, Stern DM. The multiligand receptor RAGE as a progression factor amplifying immune and inflammatory responses. J Clin Invest 2001; 108(7): 949-55.
[http://dx.doi.org/10.1172/JCI200114002] [PMID: 11581294]

[80] Ding Q, Keller JN. Evaluation of rage isoforms, ligands, and signaling in the brain. Biochim Biophys Acta 2005; 1746(1): 18-27.
[http://dx.doi.org/10.1016/j.bbamcr.2005.08.006] [PMID: 16214242]

[81] Takeuchi M, Sato T, Takino J, et al. Diagnostic utility of serum or cerebrospinal fluid levels of toxic advanced glycation end-products (TAGE) in early detection of Alzheimer's disease. Med Hypotheses 2007; 69(6): 1358-66.
[http://dx.doi.org/10.1016/j.mehy.2006.12.017] [PMID: 17888585]

[82] Clarke DD, Sokoloff L. Circulation and energy metabolism of the brain.Basic Neurochemistry. Philadelphia: Lippincott- Raven 1999; pp. 637-69.

[83] Bischof MG, Krssak M, Krebs M, et al. Effects of short-term improvement of insulin treatment and glycemia on hepatic glycogen metabolism in type 1 diabetes. Diabetes 2001; 50(2): 392-8.
[http://dx.doi.org/10.2337/diabetes.50.2.392] [PMID: 11272152]

[84] Reagan LP, Rosell DR, Alves SE, et al. GLUT8 glucose transporter is localized to excitatory and inhibitory neurons in the rat hippocampus. Brain Res 2002; 932(1-2): 129-34.
[http://dx.doi.org/10.1016/S0006-8993(02)02308-9] [PMID: 11911870]

[85] Horani MH, Mooradian AD. Effect of diabetes on the blood brain barrier. Curr Pharm Des 2003; 9(10): 833-40.
[http://dx.doi.org/10.2174/1381612033455314] [PMID: 12678883]

[86] Klein JP, Waxman SG. The brain in diabetes: molecular changes in neurons and their implications for end-organ damage. Lancet Neurol 2003; 2(9): 548-54.
[http://dx.doi.org/10.1016/S1474-4422(03)00503-9] [PMID: 12941577]

[87] Pitkänen OM, Martin JM, Hallman M, Akerblom HK, Sariola H, Andersson SM. Free radical activity during development of insulin-dependent diabetes mellitus in the rat. Life Sci 1992; 50(5): 335-9.
[http://dx.doi.org/10.1016/0024-3205(92)90434-Q] [PMID: 1531082]

[88] Van Dam PS, Van Asbeck BS, Erkelens DW, Marx JJ, Gispen WH, Bravenboer B. The role of oxidative stress in neuropathy and other diabetic complications. Diabetes Metab Rev 1995; 11(3): 181-92.
[http://dx.doi.org/10.1002/dmr.5610110303] [PMID: 8536540]

[89] Bukan N, Sancak B, Yavuz O, et al. Lipid peroxidation and scavenging enzyme levels in the liver of streptozotocin-induced diabetic rats. Indian J Biochem Biophys 2003; 40(6): 447-50.
[PMID: 22900374]

[90] Koya D, King GL. Protein kinase C activation and the development of diabetic complications. Diabetes 1998; 47(6): 859-66.
[http://dx.doi.org/10.2337/diabetes.47.6.859] [PMID: 9604860]

[91] Chappey O, Dosquet C, Wautier MP, Wautier JL. Advanced glycation end products, oxidant stress and vascular lesions. Eur J Clin Invest 1997; 27(2): 97-108.
[http://dx.doi.org/10.1046/j.1365-2362.1997.710624.x] [PMID: 9061302]

[92] Jacobson AM, Musen G, Ryan CM, *et al*. Long-term effect of diabetes and its treatment on cognitive function. N Engl J Med 2007; 356(18): 1842-52.
[http://dx.doi.org/10.1056/NEJMoa066397] [PMID: 17476010]

[93] Maritim AC, Sanders RA, Watkins JB III. Diabetes, oxidative stress, and antioxidants: a review. J Biochem Mol Toxicol 2003; 17(1): 24-38. [Review].
[http://dx.doi.org/10.1002/jbt.10058] [PMID: 12616644]

[94] Castellani RJ, Harris PLR, Sayre LM, *et al*. Active glycation in neurofibrillary pathology of Alzheimer disease: N(epsilon)-(carboxymethyl) lysine and hexitol-lysine. Free Radic Biol Med 2001; 31(2): 175-80.
[http://dx.doi.org/10.1016/S0891-5849(01)00570-6] [PMID: 11440829]

[95] Valente T, Gella A, Fernàndez-Busquets X, Unzeta M, Durany N. Immunohistochemical analysis of human brain suggests pathological synergism of Alzheimer's disease and diabetes mellitus. Neurobiol Dis 2010; 37(1): 67-76.
[http://dx.doi.org/10.1016/j.nbd.2009.09.008] [PMID: 19778613]

[96] McCormick M, Hadley D, McLean JR, Macfarlane JA, Condon B, Muir KW. Randomized, controlled trial of insulin for acute poststroke hyperglycemia. Ann Neurol 2010; 67(5): 570-8.
[PMID: 20437554]

[97] Dheen ST, Tay SS, Wong WC. Arginine vasopressin- and oxytocin-like immunoreactive neurons in the hypothalamic paraventricular and supraoptic nuclei of streptozotocin-induced diabetic rats. Arch Histol Cytol 1994; 57(5): 461-72.
[http://dx.doi.org/10.1679/aohc.57.461] [PMID: 7734175]

[98] Luo Y, Kaur C, Ling EA. Neuronal and glial response in the rat hypothalamus-neurohypophysis complex with streptozotocin-induced diabetes. Brain Res 2002; 925(1): 42-54.
[http://dx.doi.org/10.1016/S0006-8993(01)03258-9] [PMID: 11755899]

[99] Grober E, Hall CB, Hahn SR, Lipton RB. Memory impairment and executive dysfunction are associated with inadequately controlled diabetes in older adults. J Prim Care Community Health 2011; 2(4): 229-33.
[http://dx.doi.org/10.1177/2150131911409945] [PMID: 23804840]

[100] Biessels GJ, Kamal A, Ramakers GM, *et al*. Place learning and hippocampal synaptic plasticity in streptozotocin-induced diabetic rats. Diabetes 1996; 45(9): 1259-66.
[http://dx.doi.org/10.2337/diab.45.9.1259] [PMID: 8772732]

[101] Levy J, Gavin JR III, Sowers JR. Diabetes mellitus: a disease of abnormal cellular calcium metabolism? Am J Med 1994; 96(3): 260-73.
[http://dx.doi.org/10.1016/0002-9343(94)90152-X] [PMID: 8154515]

[102] Hölscher C. Nitric oxide, the enigmatic neuronal messenger: its role in synaptic plasticity. Trends Neurosci 1997; 20(7): 298-303.
[http://dx.doi.org/10.1016/S0166-2236(97)01065-5] [PMID: 9223222]

[103] da Costa AV, Calábria LK, Nascimento R, Carvalho WJ, Goulart LR, Espindola FS. The streptozotocin-induced rat model of diabetes mellitus evidences significant reduction of myosin-Va expression in the brain. Metab Brain Dis 2011; 26(4): 247-51.
[http://dx.doi.org/10.1007/s11011-011-9259-5] [PMID: 21842169]

[104] Shoham S, Bejar C, Kovalev E, Weinstock M. Intracerebroventricular injection of streptozotocin causes neurotoxicity to myelin that contributes to spatial memory deficits in rats. Exp Neurol 2003; 184(2): 1043-52.
[http://dx.doi.org/10.1016/j.expneurol.2003.08.015] [PMID: 14769399]

[105] Duelli R, Schröck H, Kuschinsky W, Hoyer S. Intracerebroventricular injection of streptozotocin induces discrete local changes in cerebral glucose utilization in rats. Int J Dev Neurosci 1994; 12(8): 737-43.

[http://dx.doi.org/10.1016/0736-5748(94)90053-1] [PMID: 7747600]

[106] Hoyer S, Lee SK, Löffler T, Schliebs R. Inhibition of the neuronal insulin receptor. An *in vivo* model for sporadic Alzheimer disease? Ann N Y Acad Sci 2000; 920: 256-8.
[http://dx.doi.org/10.1111/j.1749-6632.2000.tb06932.x] [PMID: 11193160]

[107] Lester-Coll N, Rivera EJ, Soscia SJ, Doiron K, Wands JR, de la Monte SM. Intracerebral streptozotocin model of type 3 diabetes: relevance to sporadic Alzheimer's disease. J Alzheimers Dis 2006; 9(1): 13-33.
[http://dx.doi.org/10.3233/JAD-2006-9102] [PMID: 16627931]

[108] Santos TO, Mazucanti CH, Xavier GF, Torrão AS. Early and late neurodegeneration and memory disruption after intracerebroventricular streptozotocin. Physiol Behav 2012; 107(3): 401-13.
[http://dx.doi.org/10.1016/j.physbeh.2012.06.019] [PMID: 22921433]

[109] Sakr HF. Effect of sitagliptin on the working memory and reference memory in type 2 diabetic Sprague-Dawley rats: possible role of adiponectin receptors 1. J Physiol Pharmacol 2013; 64(5): 613-23.
[PMID: 24304575]

[110] Trudeau F, Gagnon S, Massicotte G. Hippocampal synaptic plasticity and glutamate receptor regulation: influences of diabetes mellitus. Eur J Pharmacol 2004; 490(1-3): 177-86.
[http://dx.doi.org/10.1016/j.ejphar.2004.02.055] [PMID: 15094084]

[111] Skaper SD. Wnt-signalling: A new direction for alzheimer disease? CNS Neurol Disord Drug Targets 2014; 13(4): 556.
[http://dx.doi.org/10.2174/1871527313041407021048266] [PMID: 25133285]

[112] Shruster A, Eldar-Finkelman H, Melamed E, Offen D. Wnt signaling pathway overcomes the disruption of neuronal differentiation of neural progenitor cells induced by oligomeric amyloid β-peptide. J Neurochem 2011; 116(4): 522-9.
[http://dx.doi.org/10.1111/j.1471-4159.2010.07131.x] [PMID: 21138436]

[113] Schiöth HB, Craft S, Brooks SJ, Frey WH II, Benedict C. Brain insulin signaling and Alzheimer's disease: current evidence and future directions. Mol Neurobiol 2012; 46(1): 4-10.
[http://dx.doi.org/10.1007/s12035-011-8229-6] [PMID: 22205300]

[114] Grilli M, Ferrari Toninelli G, Uberti D, Spano P, Memo M. Alzheimer's disease linking neurodegeneration with neurodevelopment. Funct Neurol 2003; 18(3): 145-8.
[PMID: 14703895]

[115] Claxton A, Baker LD, Hanson A, *et al.* Long-acting intranasal insulin detemir improves cognition for adults with mild cognitive impairment or early-stage Alzheimer's disease dementia. J Alzheimers Dis 2015; 44(3): 897-906.
[http://dx.doi.org/10.3233/JAD-141791] [PMID: 25374101]

[116] Craft S, Baker LD, Montine TJ, *et al.* Intranasal insulin therapy for Alzheimer disease and amnestic mild cognitive impairment: a pilot clinical trial. Arch Neurol 2012; 69(1): 29-38.
[http://dx.doi.org/10.1001/archneurol.2011.233] [PMID: 21911655]

[117] Stockhorst U, de Fries D, Steingrueber HJ, Scherbaum WA. Insulin and the CNS: effects on food intake, memory, and endocrine parameters and the role of intranasal insulin administration in humans. Physiol Behav 2004; 83(1): 47-54.
[http://dx.doi.org/10.1016/S0031-9384(04)00348-8] [PMID: 15501490]

[118] Hoyer S. Brain glucose and energy metabolism abnormalities in sporadic Alzheimer disease. Causes and consequences: an update. Exp Gerontol 2000; 35(9-10): 1363-72.
[http://dx.doi.org/10.1016/S0531-5565(00)00156-X] [PMID: 11113614]

[119] Frölich L, Blum-Degen D, Bernstein HG, *et al.* Brain insulin and insulin receptors in aging and sporadic Alzheimer's disease. J Neural Transm (Vienna) 1998; 105(4-5): 423-38.
[http://dx.doi.org/10.1007/s007020050068] [PMID: 9720972]

[120] Hong M, Lee VM. Insulin and insulin-like growth factor-1 regulate tau phosphorylation in cultured human neurons. J Biol Chem 1997; 272(31): 19547-53.
[http://dx.doi.org/10.1074/jbc.272.31.19547] [PMID: 9235959]

[121] Messier C, Teutenberg K. The role of insulin, insulin growth factor, and insülin degrading enzyme in brain aging and Alzheimer's disease. Neural Plast. 2005;12:311-[122] Chen GJ, Xu J, Lahousse SA, Caggiano NL, de la Monte SM. Transient hypoxia causes Alzheimer-type molecular and biochemical abnormalities in cortical neurons: potential strategies for neuroprotection. J Alzheimers Dis 2003; 5: 209-28.
[PMID: 12897406]

[123] Lyn-Cook LE Jr, Lawton M, Tong M, et al. Hepatic ceramide may mediate brain insulin resistance and neurodegeneration in type 2 diabetes and non-alcoholic steatohepatitis. J Alzheimers Dis 2009; 16(4): 715-29.
[http://dx.doi.org/10.3233/JAD-2009-0984] [PMID: 19387108]

[124] Contreras C, González-García I, Martínez-Sánchez N, et al. Central ceramide-induced hypothalamic lipotoxicity and ER stress regulate energy balance. Cell Reports 2014; 9(1): 366-77.
[http://dx.doi.org/10.1016/j.celrep.2014.08.057] [PMID: 25284795]

[125] Ramirez T, Longato L, Dostalek M, Tong M, Wands JR, de la Monte SM. Insulin resistance, ceramide accumulation and endoplasmic reticulum stress in experimental chronic alcohol-induced steatohepatitis. Alcohol Alcohol 2013; 48(1): 39-52.
[http://dx.doi.org/10.1093/alcalc/ags106] [PMID: 22997409]

[126] Orsatti CL, Petri Nahas EA, Nahas-Neto J, Orsatti FL, Giorgi VI, Witkin SS. Evaluation of Toll-Like receptor 2 and 4 RNA expression and the cytokine profile in postmenopausal women with metabolic syndrome. PLoS One 2014; 9(10): e109259.
[http://dx.doi.org/10.1371/journal.pone.0109259] [PMID: 25329057]

[127] Ye D, Li FY, Lam KS, et al. Toll-like receptor-4 mediates obesity-induced non-alcoholic steatohepatitis through activation of X-box binding protein-1 in mice. Gut 2012; 61(7): 1058-67.
[http://dx.doi.org/10.1136/gutjnl-2011-300269] [PMID: 22253482]

[128] Trotta T, Porro C, Calvello R, Panaro MA. Biological role of Toll-like receptor-4 in the brain. J Neuroimmunol 2014; 268(1-2): 1-12.
[http://dx.doi.org/10.1016/j.jneuroim.2014.01.014] [PMID: 24529856]

[129] Brito BE, Zamora DO, Bonnah RA, Pan Y, Planck SR, Rosenbaum JT. Toll-like receptor 4 and CD14 expression in human ciliary body and TLR-4 in human iris endothelial cells. Exp Eye Res 2004; 79(2): 203-8.
[http://dx.doi.org/10.1016/j.exer.2004.03.012] [PMID: 15325567]

[130] Zhu T, Meng Q, Ji J, Lou X, Zhang L. Toll-like receptor 4 and tumor necrosis factor-alpha as diagnostic biomarkers for diabetic peripheral neuropathy. Neurosci Lett 2015; 585: 28-32.
[http://dx.doi.org/10.1016/j.neulet.2014.11.020] [PMID: 25445373]

[131] Caffesse R, Quinones CR. Polypeptide growth factors and attachment proteins in periodontal wound healing and regeneration. Periodontol 2000 1993; 1: 69-79.
[http://dx.doi.org/10.1111/j.1600-0757.1993.tb00208.x]

[132] McCauley LK, Somerman MJ. Biologic modifiers in periodontal regeneration. Dent Clin North Am 1998; 42(2): 361-87.
[PMID: 9597341]

[133] Daughaday WH, Rotwein P. Insulin-like growth factors I and II. Peptide, messenger ribonucleic acid and gene structures, serum, and tissue concentrations. Endocr Rev 1989; 10(1): 68-91.
[http://dx.doi.org/10.1210/edrv-10-1-68] [PMID: 2666112]

[134] Allen DL, Monke SR, Talmadge RJ, Roy RR, Edgerton VR. Plasticity of myonuclear number in hypertrophied and atrophied mammalian skeletal muscle fibers. J Appl Physiol 1995; 78(5): 1969-76.

[http://dx.doi.org/10.1152/jappl.1995.78.5.1969] [PMID: 7649936]

[135] Siddle K, Soos MA, Field CE, Navé BT. Hybrid and atypical insulin/insulin-like growth factor I receptors. Horm Res 1994; 41 (Suppl. 2): 56-64.
[http://dx.doi.org/10.1159/000183962] [PMID: 8088705]

[136] Whitler RJ, Meikle AW, Watts NB. Pituitary function.Tietz fundamentals of clinical chemistry. 4th ed. Philadelphia: WB saunders Company 1996; pp. 626-39.

[137] Greenspan FS, Gardner DG. Basic and Clinical Endocrinology. 7th. New York, Mc Graw Hill 2004; pp. 660-6.

[138] Sesti G. Pathophysiology of insulin resistance. Best Pract Res Clin Endocrinol Metab 2006; 20(4): 665-79.
[http://dx.doi.org/10.1016/j.beem.2006.09.007] [PMID: 17161338]

[139] Shepherd PR, Kahn BB. Glucose transporters and insulin action--implications for insulin resistance and diabetes mellitus. N Engl J Med 1999; 341(4): 248-57.
[http://dx.doi.org/10.1056/NEJM199907223410406] [PMID: 10413738]

[140] Bolu E. Mechanisms of insulin action and insulin resistance. 1. Metabolic Syndrome Symposium. Antalya. 2004; pp. 47-69.

[141] de la Monte SM, Tong M, Lester-Coll N, Plater M Jr, Wands JR. Therapeutic rescue of neurodegeneration in experimental type 3 diabetes: relevance to Alzheimer's disease. J Alzheimers Dis 2006; 10(1): 89-109.
[http://dx.doi.org/10.3233/JAD-2006-10113] [PMID: 16988486]

[142] Rivera EJ, Goldin A, Fulmer N, Tavares R, Wands JR, de la Monte SM. Insulin and insulin-like growth factor expression and function deteriorate with progression of Alzheimer's disease: link to brain reductions in acetylcholine. J Alzheimers Dis 2005; 8(3): 247-68.
[http://dx.doi.org/10.3233/JAD-2005-8304] [PMID: 16340083]

[143] Brown KL, Cosseau C, Gardy JL, Hancock REW. Complexities of targeting innate immunity to treat infection. Trends Immunol 2007; 28(6): 260-6.
[http://dx.doi.org/10.1016/j.it.2007.04.005] [PMID: 17468048]

[144] De Felice FG, Ferreira ST. Inflammation, defective insulin signaling, and mitochondrial dysfunction as common molecular denominators connecting type 2 diabetes to Alzheimer disease. Diabetes 2014; 63(7): 2262-72.
[http://dx.doi.org/10.2337/db13-1954] [PMID: 24931033]

[145] Vitale G, Salvioli S, Franceschi C. Oxidative stress and the ageing endocrine system. Nat Rev Endocrinol 2013; 9(4): 228-40.
[http://dx.doi.org/10.1038/nrendo.2013.29] [PMID: 23438835]

[146] Veurink G, Fuller SJ, Atwood CS, Martins RN. Genetics, lifestyle and the roles of amyloid beta and oxidative stress in Alzheimer's disease. Ann Hum Biol 2003; 30(6): 639-67.
[http://dx.doi.org/10.1080/03014460310001620144] [PMID: 14675907]

[147] Hak AE, Pols HA, Stehouwer CD, et al. Markers of inflammation and cellular adhesion molecules in relation to insulin resistance in nondiabetic elderly: the Rotterdam study. J Clin Endocrinol Metab 2001; 86(9): 4398-405.
[http://dx.doi.org/10.1210/jcem.86.9.7873] [PMID: 11549682]

[148] van de Ree MA, Huisman MV, Princen HM, Meinders AE, Kluft C. Strong decrease of high sensitivity C-reactive protein with high-dose atorvastatin in patients with type 2 diabetes mellitus. Atherosclerosis 2003; 166(1): 129-35.
[http://dx.doi.org/10.1016/S0021-9150(02)00316-7] [PMID: 12482559]

[149] Hotamisligil GS. Inflammatory pathways and insulin action. Int J Obes Relat Metab Disord 2003; 27 (Suppl. 3): S53-5.
[http://dx.doi.org/10.1038/sj.ijo.0802502] [PMID: 14704746]

[150] Perry VH, Nicoll JAR, Holmes C. Microglia in neurodegenerative disease. Nat Rev Neurol 2010; 6(4): 193-201.
[http://dx.doi.org/10.1038/nrneurol.2010.17] [PMID: 20234358]

[151] Szelényi J, Vizi ES. The catecholamine cytokine balance: interaction between the brain and the immune system. Ann N Y Acad Sci 2007; 1113: 311-24.
[http://dx.doi.org/10.1196/annals.1391.026] [PMID: 17584982]

[152] Watkins LR, Maier SF, Goehler LE. Cytokine-to-brain communication: a review & analysis of alternative mechanisms. Life Sci 1995; 57(11): 1011-26.
[http://dx.doi.org/10.1016/0024-3205(95)02047-M] [PMID: 7658909]

[153] Hotamisligil GS, Peraldi P, Budavari A, Ellis R, White MF, Spiegelman BM. IRS-1-mediated inhibition of insulin receptor tyrosine kinase activity in TNF-alpha- and obesity-induced insulin resistance. Science 1996; 271(5249): 665-8.
[http://dx.doi.org/10.1126/science.271.5249.665] [PMID: 8571133]

[154] Chung S, Lapoint K, Martinez K, Kennedy A, Boysen Sandberg M, McIntosh MK. Preadipocytes mediate lipopolysaccharide-induced inflammation and insulin resistance in primary cultures of newly differentiated human adipocytes. Endocrinology 2006; 147(11): 5340-51.
[http://dx.doi.org/10.1210/en.2006-0536] [PMID: 16873530]

[155] Charrière G, Cousin B, Arnaud E, et al. Preadipocyte conversion to macrophage. Evidence of plasticity. J Biol Chem 2003; 278(11): 9850-5.
[http://dx.doi.org/10.1074/jbc.M210811200] [PMID: 12519759]

[156] Solmaz V, Çınar BP, Yiğittürk G, Çavuşoğlu T, Taşkıran D, Erbaş O. Exenatide reduces TNF-α expression and improves hippocampal neuron numbers and memory in streptozotocin treated rats. Eur J Pharmacol 2015; 765(765): 482-7.
[http://dx.doi.org/10.1016/j.ejphar.2015.09.024] [PMID: 26386291]

[157] Swardfager W, Lanctôt K, Rothenburg L, Wong A, Cappell J, Herrmann N. A meta-analysis of cytokines in Alzheimer's disease. Biol Psychiatry 2010; 68(10): 930-41.
[http://dx.doi.org/10.1016/j.biopsych.2010.06.012] [PMID: 20692646]

[158] Guerre-Millo M. Adipose tissue and adipokines: for better or worse. Diabetes Metab 2004; 30(1): 13--.
[http://dx.doi.org/10.1016/S1262-3636(07)70084-8] [PMID: 15029093]

[159] Schäfer K, Fujisawa K, Konstantinides S, Loskutoff DJ. Disruption of the plasminogen activator inhibitor 1 gene reduces the adiposity and improves the metabolic profile of genetically obese and diabetic ob/ob mice. FASEB J 2001; 15(10): 1840-2.
[http://dx.doi.org/10.1096/fj.00-0750fje] [PMID: 11481248]

[160] Ma LJ, Mao SL, Taylor KL, et al. Prevention of obesity and insulin resistance in mice lacking plasminogen activator inhibitor 1. Diabetes 2004; 53(2): 336-46.
[http://dx.doi.org/10.2337/diabetes.53.2.336] [PMID: 14747283]

[161] Miklossy J, McGeer PL. Common mechanisms involved in Alzheimer's disease and type 2 diabetes: a key role of chronic bacterial infection and inflammation. Aging (Albany NY) 2016; 8(4): 575-88.
[http://dx.doi.org/10.18632/aging.100921] [PMID: 26961231]

[162] Acharya NK, Levin EC, Clifford PM, et al. Diabetes and hypercholesterolemia increase blood-brain barrier permeability and brain amyloid deposition: beneficial effects of the LpPLA2 inhibitor darapladib. J Alzheimers Dis 2013; 35(1): 179-98.
[http://dx.doi.org/10.3233/JAD-122254] [PMID: 23388174]

[163] Sonnen JA, Larson EB, Brickell K, et al. Different patterns of cerebral injury in dementia with or without diabetes. Arch Neurol 2009; 66(3): 315-22.
[http://dx.doi.org/10.1001/archneurol.2008.579] [PMID: 19139294]

[164] Lue LF, Walker DG, Rogers J. Modeling microglial activation in Alzheimer's disease with human postmortem microglial cultures. Neurobiol Aging 2001; 22(6): 945-56.

[http://dx.doi.org/10.1016/S0197-4580(01)00311-6] [PMID: 11755003]

[165] Hüll M, Strauss S, Berger M, Volk B, Bauer J. The participation of interleukin-6, a stress-inducible cytokine, in the pathogenesis of Alzheimer's disease. Behav Brain Res 1996; 78(1): 37-41.
[http://dx.doi.org/10.1016/0166-4328(95)00213-8] [PMID: 8793035]

[166] de la Monte SM. Triangulated mal-signaling in Alzheimer's disease: roles of neurotoxic ceramides, ER stress, and insulin resistance reviewed. J Alzheimers Dis 2012; 30(2) (Suppl. 2): S231-49.
[http://dx.doi.org/10.3233/JAD-2012-111727] [PMID: 22337830]

[167] de la Monte SM, Re E, Longato L, Tong M. Dysfunctional pro-ceramide, ER stress, and insulin/IGF signaling networks with progression of Alzheimer's disease. J Alzheimers Dis 2012; 30(2) (Suppl. 2): S217-29.
[http://dx.doi.org/10.3233/JAD-2012-111728] [PMID: 22297646]

[168] Erbaş O. Is Alzheimer's disease, type 3 diabetes? FNG&Bilim Tıp Dergisi 2015; 1(1): 48-51.

[169] Straczkowski M, Kowalska I, Nikolajuk A, et al. Relationship between insulin sensitivity and sphingomyelin signaling pathway in human skeletal muscle. Diabetes 2004; 53(5): 1215-21.
[http://dx.doi.org/10.2337/diabetes.53.5.1215] [PMID: 15111489]

[170] Summers SA. Ceramides in insulin resistance and lipotoxicity. Prog Lipid Res 2006; 45(1): 42-72.
[http://dx.doi.org/10.1016/j.plipres.2005.11.002] [PMID: 16445986]

[171] Holland WL, Bikman BT, Wang L-P, et al. Lipid-induced insulin resistance mediated by the proinflammatory receptor TLR4 requires saturated fatty acid-induced ceramide biosynthesis in mice. J Clin Invest 2011; 121(5): 1858-70.
[http://dx.doi.org/10.1172/JCI43378] [PMID: 21490391]

[172] Smith QR, Nagura H. Fatty acid uptake and incorporation in brain: studies with the perfusion model. J Mol Neurosci 2001; 16(2-3): 167-72.
[http://dx.doi.org/10.1385/JMN:16:2-3:167] [PMID: 11478371]

[173] Marchetti P, Bugliani M, Lupi R, et al. The endoplasmic reticulum in pancreatic beta cells of type 2 diabetes patients. Diabetologia 2007; 50(12): 2486-94.
[http://dx.doi.org/10.1007/s00125-007-0816-8] [PMID: 17906960]

[174] Hotamisligil GS. Inflammation and metabolic disorders. Nature 2006; 444(7121): 860-7.
[http://dx.doi.org/10.1038/nature05485] [PMID: 17167474]

[175] Huang CJ, Lin CY, Haataja L, et al. High expression rates of human islet amyloid polypeptide induce endoplasmic reticulum stress mediated beta-cell apoptosis, a characteristic of humans with type 2 but not type 1 diabetes. Diabetes 2007; 56(8): 2016-27.
[http://dx.doi.org/10.2337/db07-0197] [PMID: 17475933]

[176] Marciniak SJ, Ron D. Endoplasmic reticulum stress signaling in disease. Physiol Rev 2006; 86(4): 1133-49.
[http://dx.doi.org/10.1152/physrev.00015.2006] [PMID: 17015486]

[177] Marciniak SJ, Yun CY, Oyadomari S, et al. CHOP induces death by promoting protein synthesis and oxidation in the stressed endoplasmic reticulum. Genes Dev 2004; 18(24): 3066-77.
[http://dx.doi.org/10.1101/gad.1250704] [PMID: 15601821]

[178] Roussel BD, Kruppa AJ, Miranda E, Crowther DC, Lomas DA, Marciniak SJ. Endoplasmic reticulum dysfunction in neurological disease. Lancet Neurol 2013; 12(1): 105-18.
[http://dx.doi.org/10.1016/S1474-4422(12)70238-7] [PMID: 23237905]

[179] Brown AR, Rebus S, McKimmie CS, Robertson K, Williams A, Fazakerley JK. Gene expression profiling of the preclinical scrapie-infected hippocampus. Biochem Biophys Res Commun 2005; 334(1): 86-95.
[http://dx.doi.org/10.1016/j.bbrc.2005.06.060] [PMID: 15992767]

[180] Miyake K. Innate immune sensing of pathogens and danger signals by cell surface Toll-like receptors.

Semin Immunol 2007; 19(1): 3-10.
[http://dx.doi.org/10.1016/j.smim.2006.12.002] [PMID: 17275324]

[181] Wang C, Ha X, Li W, *et al.* Correlation of TLR4 and KLF7 in inflammation induced by obesity. Inflammation 2017; 40(1): 42-51.
[http://dx.doi.org/10.1007/s10753-016-0450-z] [PMID: 27714571]

[182] Song M, Jin J, Lim JE, *et al.* TLR4 mutation reduces microglial activation, increases Aβ deposits and exacerbates cognitive deficits in a mouse model of Alzheimer's disease. J Neuroinflammation 2011; 8: 92.
[http://dx.doi.org/10.1186/1742-2094-8-92] [PMID: 21827663]

[183] Biessels GJ, ter Laak MP, Hamers FPT, Gispen WH. Neuronal Ca2+ disregulation in diabetes mellitus. Eur J Pharmacol 2002; 447(2-3): 201-9.
[http://dx.doi.org/10.1016/S0014-2999(02)01844-7] [PMID: 12151012]

[184] Khachaturian ZS. Calcium hypothesis of Alzheimer's disease and brain aging. Ann N Y Acad Sci 1994; 747: 1-11.
[http://dx.doi.org/10.1111/j.1749-6632.1994.tb44398.x] [PMID: 7847664]

[185] Kristián T, Siesjö BK. Calcium-related damage in ischemia. Life Sci 1996; 59(5-6): 357-67.
[http://dx.doi.org/10.1016/0024-3205(96)00314-1] [PMID: 8761323]

[186] Biessels GJ, Kamal A, Urban IJ, Spruijt BM, Erkelens DW, Gispen WH. Water maze learning and hippocampal synaptic plasticity in streptozotocin-diabetic rats: effects of insulin treatment. Brain Res 1998; 800(1): 125-35.
[http://dx.doi.org/10.1016/S0006-8993(98)00510-1] [PMID: 9685609]

[187] Kostyuk E, Voitenko N, Kruglikov I, *et al.* Diabetes-induced changes in calcium homeostasis and the effects of calcium channel blockers in rat and mice nociceptive neurons. Diabetologia 2001; 44(10): 1302-9.
[http://dx.doi.org/10.1007/s001250100642] [PMID: 11692179]

[188] Grene DA, Lattimer SA, Sima AAF. Tissue spesific metabolic alterations in the pathogenesis of diabetic peripheral neuropathy. Front Diabetes 1990; 10: 83-96.
[http://dx.doi.org/10.1159/000418601]

[189] Brownlee M. Glycation products and the pathogenesis of diabetic complications. Diabetes Care 1992; 15(12): 1835-43.
[http://dx.doi.org/10.2337/diacare.15.12.1835] [PMID: 1464241]

[190] Atamna H, Frey WH II. Mechanisms of mitochondrial dysfunction and energy deficiency in Alzheimer's disease. Mitochondrion 2007; 7(5): 297-310.
[http://dx.doi.org/10.1016/j.mito.2007.06.001] [PMID: 17625988]

[191] Petersen KF, Befroy D, Dufour S, *et al.* Mitochondrial dysfunction in the elderly: possible role in insulin resistance. Science 2003; 300(5622): 1140-2.
[http://dx.doi.org/10.1126/science.1082889] [PMID: 12750520]

[192] Cadenas E, Davies KJ. Mitochondrial free radical generation, oxidative stress, and aging. Free Radic Biol Med 2000; 29(3-4): 222-30.
[http://dx.doi.org/10.1016/S0891-5849(00)00317-8] [PMID: 11035250]

[193] Villani G, Attardi G. *in vivo* control of respiration by cytochrome c oxidase in human cells. Free Radic Biol Med 2000; 29(3-4): 202-10.
[http://dx.doi.org/10.1016/S0891-5849(00)00303-8] [PMID: 11035248]

[194] Aliev G, Seyidova D, Lamb BT, *et al.* Mitochondria and vascular lesions as a central target for the development of Alzheimer's disease and Alzheimer disease-like pathology in transgenic mice. Neurol Res 2003; 25(6): 665-74.
[http://dx.doi.org/10.1179/016164103101201977] [PMID: 14503022]

[195] Hirai K, Aliev G, Nunomura A, *et al.* Mitochondrial abnormalities in Alzheimer's disease. J Neurosci

2001; 21(9): 3017-23.
[http://dx.doi.org/10.1523/JNEUROSCI.21-09-03017.2001] [PMID: 11312286]

[196] Castellani R, Hirai K, Aliev G, *et al.* Role of mitochondrial dysfunction in Alzheimer's disease. J Neurosci Res 2002; 70(3): 357-60.
[http://dx.doi.org/10.1002/jnr.10389] [PMID: 12391597]

[197] Armstrong JS. The role of the mitochondrial permeability transition in cell death. Mitochondrion 2006; 6(5): 225-34.
[http://dx.doi.org/10.1016/j.mito.2006.07.006] [PMID: 16935572]

[198] Serrano F, Klann E. Reactive oxygen species and synaptic plasticity in the aging hippocampus. Ageing Res Rev 2004; 3(4): 431-43.
[http://dx.doi.org/10.1016/j.arr.2004.05.002] [PMID: 15541710]

[199] Anandatheerthavarada HK, Biswas G, Robin MA, Avadhani NG. Mitochondrial targeting and a novel transmembrane arrest of Alzheimer's amyloid precursor protein impairs mitochondrial function in neuronal cells. J Cell Biol 2003; 161(1): 41-54.
[http://dx.doi.org/10.1083/jcb.200207030] [PMID: 12695498]

[200] Crouch PJ, Blake R, Duce JA, *et al.* Copper-dependent inhibition of human cytochrome c oxidase by a dimeric conformer of amyloid-β1-42. J Neurosci 2005; 25(3): 672-9.
[http://dx.doi.org/10.1523/JNEUROSCI.4276-04.2005] [PMID: 15659604]

[201] Blass JP, Gibson GE. The role of oxidative abnormalities in the pathophysiology of Alzheimer's disease. Rev Neurol (Paris) 1991; 147(6-7): 513-25.
[PMID: 1962057]

[202] Levy J, Zemel MB, Sowers JR. Role of cellular calcium metabolism in abnormal glucose metabolism and diabetic hypertension. Am J Med 1989; 87(6A): 7S-16S.
[http://dx.doi.org/10.1016/0002-9343(89)90489-0] [PMID: 2688414]

[203] De Felice FG, Ferreira ST. Inflammation, defective insulin signaling, and mitochondrial dysfunction as common molecular denominators connecting type 2 diabetes to Alzheimer disease. Diabetes 2014; 63(7): 2262-72.
[http://dx.doi.org/10.2337/db13-1954] [PMID: 24931033]

[204] Moreira PI, Santos MS, Seiça R, Oliveira CR. Brain mitochondrial dysfunction as a link between Alzheimer's disease and diabetes. J Neurol Sci 2007; 257(1-2): 206-14.
[http://dx.doi.org/10.1016/j.jns.2007.01.017] [PMID: 17316694]

[205] Moreira PI, Santos MS, Sena C, Seiça R, Oliveira CR. Insulin protects against amyloid β-peptide toxicity in brain mitochondria of diabetic rats. Neurobiol Dis 2005; 18(3): 628-37.
[http://dx.doi.org/10.1016/j.nbd.2004.10.017] [PMID: 15755688]

[206] Dröge W. Free radicals in the physiological control of cell function. Physiol Rev 2002; 82(1): 47-95.
[http://dx.doi.org/10.1152/physrev.00018.2001] [PMID: 11773609]

[207] Murray FE, Landsberg JP, Williams RJ, Esiri MM, Watt F. Elemental analysis of neurofibrillary tangles in Alzheimer's disease using proton-induced X-ray analysis. Ciba Found Symp 1992; 169: 201-10.
[PMID: 1490423]

[208] De Felice FG, Vieira MNN, Bomfim TR, *et al.* Protection of synapses against Alzheimer's-linked toxins: insulin signaling prevents the pathogenic binding of Abeta oligomers. Proc Natl Acad Sci USA 2009; 106(6): 1971-6.
[http://dx.doi.org/10.1073/pnas.0809158106] [PMID: 19188609]

[209] Picone P, Giacomazza D, Vetri V, *et al.* Insulin-activated Akt rescues Aβ oxidative stress-induced cell death by orchestrating molecular trafficking. Aging Cell 2011; 10(5): 832-43.
[http://dx.doi.org/10.1111/j.1474-9726.2011.00724.x] [PMID: 21624038]

[210] Decker H, Jürgensen S, Adrover MF, *et al.* N-methyl-D-aspartate receptors are required for synaptic

targeting of Alzheimer's toxic amyloid-β peptide oligomers. J Neurochem 2010; 115(6): 1520-9.
[http://dx.doi.org/10.1111/j.1471-4159.2010.07058.x] [PMID: 20950339]

[211] Yin F, Jiang T, Cadenas E. Metabolic triad in brain aging: mitochondria, insulin/IGF-1 signalling and JNK signalling. Biochem Soc Trans 2013; 41(1): 101-5.
[http://dx.doi.org/10.1042/BST20120260] [PMID: 23356266]

[212] Gerbitz KD, Gempel K, Brdiczka D. Mitochondria and diabetes. Genetic, biochemical, and clinical implications of the cellular energy circuit. Diabetes 1996; 45(2): 113-26.
[http://dx.doi.org/10.2337/diab.45.2.113] [PMID: 8549853]

[213] van den Ouweland JM, Maechler P, Wollheim CB, Attardi G, Maassen JA. Functional and morphological abnormalities of mitochondria harbouring the tRNA(Leu)(UUR) mutation in mitochondrial DNA derived from patients with maternally inherited diabetes and deafness (MIDD) and progressive kidney disease. Diabetologia 1999; 42(4): 485-92.
[http://dx.doi.org/10.1007/s001250051183] [PMID: 10230654]

[214] Weingarten MD, Lockwood AH, Hwo SY, Kirschner MW. A protein factor essential for microtubule assembly. Proc Natl Acad Sci USA 1975; 72(5): 1858-62.
[http://dx.doi.org/10.1073/pnas.72.5.1858] [PMID: 1057175]

[215] Drechsel DN, Hyman AA, Cobb MH, Kirschner MW. Modulation of the dynamic instability of tubulin assembly by the microtubule-associated protein tau. Mol Biol Cell 1992; 3(10): 1141-54.
[http://dx.doi.org/10.1091/mbc.3.10.1141] [PMID: 1421571]

[216] Binder LI, Frankfurter A, Rebhun LIJ. The distribution of tau in the mammalian central nervous system. J Cell Biol 1985; 101(4): 1371-8.
[http://dx.doi.org/10.1083/jcb.101.4.1371] [PMID: 3930508]

[217] Neve RL, Harris P, Kosik KS, Kurnit DM, Donlon TA. Identification of cDNA clones for the human microtubule-associated protein tau and chromosomal localization of the genes for tau and microtubule-associated protein 2. Brain Res 1986; 387(3): 271-80.
[PMID: 3103857]

[218] Goode BL, Feinstein SCJ. Identification of a novel microtubule binding and assembly domain in the developmentally regulated inter-repeat region of tau. J Cell Biol 1994; 124(5): 769-82.
[http://dx.doi.org/10.1083/jcb.124.5.769] [PMID: 8120098]

[219] Hirokawa N, Shiomura Y, Okabe S. Tau proteins: the molecular structure and mode of binding on microtubules. J Cell Biol 1988; 107(4): 1449-59.
[http://dx.doi.org/10.1083/jcb.107.4.1449] [PMID: 3139677]

[220] Lee G, Neve RL, Kosik KS. The microtubule binding domain of tau protein. Neuron 1989; 2(6): 1615-24.
[http://dx.doi.org/10.1016/0896-6273(89)90050-0] [PMID: 2516729]

[221] Goedert M, Spillantini MG, Jakes R, Rutherford D, Crowther RA. Multiple isoforms of human microtubule-associated protein tau: sequences and localization in neurofibrillary tangles of Alzheimer's disease. Neuron 1989; 3(4): 519-26.
[http://dx.doi.org/10.1016/0896-6273(89)90210-9] [PMID: 2484340]

[222] Kosik KS, Orecchio LD, Bakalis S, Neve RL. Developmentally regulated expression of specific tau sequences. Neuron 1989; 2(4): 1389-97.
[http://dx.doi.org/10.1016/0896-6273(89)90077-9] [PMID: 2560640]

[223] Stanford PM, Shepherd CE, Halliday GM, et al. Mutations in the tau gene that cause an increase in three repeat tau and frontotemporal dementia. Brain 2003; 126(Pt 4): 814-26.
[http://dx.doi.org/10.1093/brain/awg090] [PMID: 12615641]

[224] Gong CX, Iqbal K. Hyperphosphorylation of microtubule-associated protein tau: a promising therapeutic target for Alzheimer disease. Curr Med Chem 2008; 15(23): 2321-8.
[http://dx.doi.org/10.2174/092986708785909111] [PMID: 18855662]

[225] Yuan A, Kumar A, Peterhoff C, Duff K, Nixon RA. Axonal transport rates *in vivo* are unaffected by tau deletion or overexpression in mice. J Neurosci 2008; 28(7): 1682-7.
[http://dx.doi.org/10.1523/JNEUROSCI.5242-07.2008] [PMID: 18272688]

[226] Avila J, Lucas JJ, Perez M, Hernandez F. Role of tau protein in both physiological and pathological conditions. Physiol Rev 2004; 84(2): 361-84.
[http://dx.doi.org/10.1152/physrev.00024.2003] [PMID: 15044677]

[227] Matsuo ES, Shin RW, Billingsley ML, *et al.* Biopsy-derived adult human brain tau is phosphorylated at many of the same sites as Alzheimer's disease paired helical filament tau. Neuron 1994; 13(4): 989-1002.
[http://dx.doi.org/10.1016/0896-6273(94)90264-X] [PMID: 7946342]

[228] Gong CX, Liu F, Grundke-Iqbal I, Iqbal K. Post-translational modifications of tau protein in Alzheimer's disease. J Neural Transm (Vienna) 2005; 112(6): 813-38.
[http://dx.doi.org/10.1007/s00702-004-0221-0] [PMID: 15517432]

[229] Sengupta A, Kabat J, Novak M, Wu Q, Grundke-Iqbal I, Iqbal K. Phosphorylation of tau at both Thr 231 and Ser 262 is required for maximal inhibition of its binding to microtubules. Arch Biochem Biophys 1998; 357(2): 299-309.
[http://dx.doi.org/10.1006/abbi.1998.0813] [PMID: 9735171]

[230] Alonso AdelC, Mederlyova A, Novak M, Grundke-Iqbal I, Iqbal K. Promotion of hyperphosphorylation by frontotemporal dementia tau mutations. J Biol Chem 2004; 279(33): 34873-81.
[http://dx.doi.org/10.1074/jbc.M405131200] [PMID: 15190058]

[231] Abraha A, Ghoshal N, Gamblin TC, *et al.* C-terminal inhibition of tau assembly *in vitro* and in Alzheimer's disease. J Cell Sci 2000; 113(Pt 21): 3737-45.
[PMID: 11034902]

[232] Braak H, Braak E. Neuropathological stageing of Alzheimer-related changes. Acta Neuropathol 1991; 82(4): 239-59.
[http://dx.doi.org/10.1007/BF00308809] [PMID: 1759558]

[233] Grober E, Dickson D, Sliwinski MJ, *et al.* Memory and mental status correlates of modified Braak staging. Neurobiol Aging 1999; 20(6): 573-9.
[http://dx.doi.org/10.1016/S0197-4580(99)00063-9] [PMID: 10674422]

[234] Bennett DA, Schneider JA, Wilson RS, Bienias JL, Arnold SE. Neurofibrillary tangles mediate the association of amyloid load with clinical Alzheimer disease and level of cognitive function. Arch Neurol 2004; 61(3): 378-84.
[http://dx.doi.org/10.1001/archneur.61.3.378] [PMID: 15023815]

[235] Tanzi RE. Tangles and neurodegenerative disease--a surprising twist. N Engl J Med 2005; 353(17): 1853-5.
[http://dx.doi.org/10.1056/NEJMcibr055003] [PMID: 16251544]

[236] Santacruz K, Lewis J, Spires T, *et al.* Tau suppression in a neurodegenerative mouse model improves memory function. Science 2005; 309(5733): 476-81.
[http://dx.doi.org/10.1126/science.1113694] [PMID: 16020737]

[237] Goedert M, Spillantini MG. A century of Alzheimer's disease. Science 2006; 314(5800): 777-81.
[http://dx.doi.org/10.1126/science.1132814] [PMID: 17082447]

[238] De Felice FG, Wu D, Lambert MP, *et al.* Alzheimer's disease-type neuronal tau hyperphosphorylation induced by A β oligomers. Neurobiol Aging 2008; 29(9): 1334-47.
[http://dx.doi.org/10.1016/j.neurobiolaging.2007.02.029] [PMID: 17403556]

[239] Liu Y, Liu F, Grundke-Iqbal I, Iqbal K, Gong C-X. Brain glucose transporters, O-GlcNAcylation and phosphorylation of tau in diabetes and Alzheimer's disease. J Neurochem 2009; 111(1): 242-9.
[http://dx.doi.org/10.1111/j.1471-4159.2009.06320.x] [PMID: 19659459]

[240] Gong CX, Liu F, Grundke-Iqbal I, Iqbal K. Impaired brain glucose metabolism leads to Alzheimer neurofibrillary degeneration through a decrease in tau O-GlcNAcylation. J Alzheimers Dis 2006; 9(1): 1-12.
[http://dx.doi.org/10.3233/JAD-2006-9101] [PMID: 16627930]

[241] d'Abramo C, Ricciarelli R, Pronzato MA, Davies P. Troglitazone, a peroxisome proliferator-activated receptor-γ agonist, decreases tau phosphorylation in CHOtau4R cells. J Neurochem 2006; 98(4): 1068-77.
[http://dx.doi.org/10.1111/j.1471-4159.2006.03931.x] [PMID: 16787414]

[242] Iyer SP, Hart GW. Identification and cloning of a novel family of coiled-coil domain proteins that interact with O-GlcNAc transferase. Biochemistry 2003; 42: 2493-9.
[http://dx.doi.org/10.1021/bi020685a] [PMID: 12614143]

[243] Grünblatt E, Salkovic-Petrisic M, Osmanovic J, Riederer P, Hoyer S. Brain insulin system dysfunction in streptozotocin intracerebroventricularly treated rats generates hyperphosphorylated tau protein. J Neurochem 2007; 101(3): 757-70.
[http://dx.doi.org/10.1111/j.1471-4159.2006.04368.x] [PMID: 17448147]

[244] Lester-Coll N, Rivera EJ, Soscia SJ, Doiron K, Wands JR, de la Monte SM. Intracerebral streptozotocin model of type 3 diabetes: relevance to sporadic Alzheimer's disease. J Alzheimers Dis 2006; 9(1): 13-33.
[http://dx.doi.org/10.3233/JAD-2006-9102] [PMID: 16627931]

[245] Miklossy J, Qing H, Radenovic A, *et al.* Beta amyloid and hyperphosphorylated tau deposits in the pancreas in type 2 diabetes. Neurobiol Aging 2010; 31(9): 1503-15.
[http://dx.doi.org/10.1016/j.neurobiolaging.2008.08.019] [PMID: 18950899]

[246] Gunton JE, Kulkarni RN, Yim S, *et al.* Loss of ARNT/HIF1beta mediates altered gene expression and pancreatic-islet dysfunction in human type 2 diabetes. Cell 2005; 122(3): 337-49.
[http://dx.doi.org/10.1016/j.cell.2005.05.027] [PMID: 16096055]

[247] Freude S, Plum L, Schnitker J, *et al.* Peripheral hyperinsulinemia promotes tau phosphorylation *in vivo*. Diabetes 2005; 54(12): 3343-8.
[http://dx.doi.org/10.2337/diabetes.54.12.3343] [PMID: 16306348]

[248] Moroz N, Tong M, Longato L, Xu H, de la Monte SM. Limited Alzheimer-type neurodegeneration in experimental obesity and type 2 diabetes mellitus. J Alzheimers Dis 2008; 15(1): 29-44.
[http://dx.doi.org/10.3233/JAD-2008-15103] [PMID: 18780965]

[249] McEwen BS, Reagan LP. Glucose transporter expression in the central nervous system: relationship to synaptic function. Eur J Pharmacol 2004; 490(1-3): 13-24.
[http://dx.doi.org/10.1016/j.ejphar.2004.02.041] [PMID: 15094070]

[250] Dwyer DS, Vannucci SJ, Simpson IA. Expression, regulation, and functional role of glucose transporters (GLUTs) in brain. Int Rev Neurobiol 2002; 51: 159-88.
[http://dx.doi.org/10.1016/S0074-7742(02)51005-9] [PMID: 12420359]

[251] Liu Y, Liu F, Iqbal K, Grundke-Iqbal I, Gong CX. Decreased glucose transporters correlate to abnormal hyperphosphorylation of tau in Alzheimer disease. FEBS Lett 2008; 582(2): 359-64.
[http://dx.doi.org/10.1016/j.febslet.2007.12.035] [PMID: 18174027]

[252] Liu F, Iqbal K, Grundke-Iqbal I, Hart GW, Gong CX. O-GlcNAcylation regulates phosphorylation of tau: a mechanism involved in Alzheimer's disease. Proc Natl Acad Sci USA 2004; 101(29): 10804-9.
[http://dx.doi.org/10.1073/pnas.0400348101] [PMID: 15249677]

[253] Drewes G, Lichtenberg-Kraag B, Döring F, *et al.* Mitogen activated protein (MAP) kinase transforms tau protein into an Alzheimer-like state. EMBO J 1992; 11(6): 2131-8.
[PMID: 1376245]

[254] Goedert M, Cohen ES, Jakes R, Cohen P. p42 MAP kinase phosphorylation sites in microtubule-associated protein tau are dephosphorylated by protein phosphatase 2A1. Implications for Alzheimer's

disease [corrected]. FEBS Lett 1992; 312(1): 95-9.
[http://dx.doi.org/10.1016/0014-5793(92)81418-L] [PMID: 1330687]

[255] Litersky JM, Johnson GV, Jakes R, Goedert M, Lee M, Seubert P. Tau protein is phosphorylated by cyclic AMP-dependent protein kinase and calcium/calmodulin-dependent protein kinase II within its microtubule-binding domains at Ser-262 and Ser-356. Biochem J 1996; 316(Pt 2): 655-60.
[http://dx.doi.org/10.1042/bj3160655] [PMID: 8687413]

[256] Hoeflich KP, Luo J, Rubie EA, Tsao MS, Jin O, Woodgett JR. Requirement for glycogen synthase kinase-3beta in cell survival and NF-kappaB activation. Nature 2000; 406(6791): 86-90.
[http://dx.doi.org/10.1038/35017574] [PMID: 10894547]

[257] MacAulay K, Doble BW, Patel S, *et al.* Glycogen synthase kinase 3alpha-specific regulation of murine hepatic glycogen metabolism. Cell Metab 2007; 6(4): 329-37.
[http://dx.doi.org/10.1016/j.cmet.2007.08.013] [PMID: 17908561]

[258] Welsh GI, Wilson C, Proud CG. GSK3: a SHAGGY frog story. Trends Cell Biol 1996; 6(7): 274-9.
[http://dx.doi.org/10.1016/0962-8924(96)10023-4] [PMID: 15157454]

[259] Dudek H, Datta SR, Franke TF, *et al.* Regulation of neuronal survival by the serine-threonine protein kinase Akt. Science 1997; 275(5300): 661-5.
[http://dx.doi.org/10.1126/science.275.5300.661] [PMID: 9005851]

[260] Hoshi M, Takashima A, Noguchi K, *et al.* Regulation of mitochondrial pyruvate dehydrogenase activity by tau protein kinase I/glycogen synthase kinase 3beta in brain. Proc Natl Acad Sci USA 1996; 93(7): 2719-23.
[http://dx.doi.org/10.1073/pnas.93.7.2719] [PMID: 8610107]

[261] Hooper C, Killick R, Lovestone S. The GSK3 hypothesis of Alzheimer's disease. J Neurochem 2008; 104(6): 1433-9.
[http://dx.doi.org/10.1111/j.1471-4159.2007.05194.x] [PMID: 18088381]

[262] Takashima A. GSK-3 is essential in the pathogenesis of Alzheimer's disease. J Alzheimers Dis 2006; 9(3) (Suppl.): 309-17.
[http://dx.doi.org/10.3233/JAD-2006-9S335] [PMID: 16914869]

[263] Hochstrasser M. Origin and function of ubiquitin-like proteins. Nature 2009; 458(7237): 422-9.
[http://dx.doi.org/10.1038/nature07958] [PMID: 19325621]

[264] Fujimuro M, Sawada H, Yokosawa H. Production and characterization of monoclonal antibodies specific to multi-ubiquitin chains of polyubiquitinated proteins. FEBS Lett 1994; 349(2): 173-80.
[http://dx.doi.org/10.1016/0014-5793(94)00647-4] [PMID: 7519568]

[265] Hershko A, Ciechanover A. The ubiquitin system for protein degradation. Annu Rev Biochem 1992; 61: 761-807.
[http://dx.doi.org/10.1146/annurev.bi.61.070192.003553] [PMID: 1323239]

[266] Avram H. The Ubiquitin System-Past, Present, and Future Perspectives. Ubiquitin and the Biology of The Cell 1998; pp. 1-17.

[267] Ciechanover A, Hod Y, Hershko A. A heat-stable polypeptide component of an ATP-dependent proteolytic system from reticulocytes. Biochem Biophys Res Commun 1978; 81: 1000-105.
[PMID: 96823]

[268] Rodwell VW. Catabolism of proteins and of amino acid nitrogen.Harper's Biochemistry. 25th ed. Appleton and Lange 2000; pp. 313-22.

[269] Glickman MH, Ciechanover A. The ubiquitin-proteasome proteolytic pathway: destruction for the sake of construction. Physiol Rev 2002; 82(2): 373-428.
[http://dx.doi.org/10.1152/physrev.00027.2001] [PMID: 11917093]

[270] Mukhopadhyay D, Riezman H. Proteasome-independent functions of ubiquitin in endocytosis and signaling. Science 2007; 315(5809): 201-5.

[http://dx.doi.org/10.1126/science.1127085] [PMID: 17218518]

[271] Schnell JD, Hicke L. Non-traditional functions of ubiquitin and ubiquitin-binding proteins. J Biol Chem 2003; 278(38): 35857-60.
[http://dx.doi.org/10.1074/jbc.R300018200] [PMID: 12860974]

[272] Güney Y, Bilgihan A. Ubiquitin system. T Klin J Med Sci 2002; 22: 616-9.

[273] Ardley HC, Hung C-C, Robinson PA. The aggravating role of the ubiquitin-proteasome system in neurodegeneration. FEBS Lett 2005; 579(3): 571-6.
[http://dx.doi.org/10.1016/j.febslet.2004.12.058] [PMID: 15670810]

[274] Hershko A, Ciechanover A. The ubiquitin system. Annu Rev Biochem 1998; 67: 425-79.
[http://dx.doi.org/10.1146/annurev.biochem.67.1.425] [PMID: 9759494]

[275] Weissman AM. Themes and variations on ubiquitylation. Nat Rev Mol Cell Biol 2001; 2(3): 169-78.
[http://dx.doi.org/10.1038/35056563] [PMID: 11265246]

[276] López-Avalos MD, Duvivier-Kali VF, Xu G, Bonner-Weir S, Sharma A, Weir GC. Evidence for a role of the ubiquitin-proteasome pathway in pancreatic islets. Diabetes 2006; 55(5): 1223-31.
[http://dx.doi.org/10.2337/db05-0450] [PMID: 16644676]

[277] Mukherjee A, Morales-Scheihing D, Butler PC, Soto C. Type 2 diabetes as a protein misfolding disease. Trends Mol Med 2015; 21(7): 439-49.
[http://dx.doi.org/10.1016/j.molmed.2015.04.005] [PMID: 25998900]

[278] Gong B, Cao Z, Zheng P, et al. Ubiquitin hydrolase Uch-L1 rescues beta-amyloid-induced decreases in synaptic function and contextual memory. Cell 2006; 126(4): 775-88.
[http://dx.doi.org/10.1016/j.cell.2006.06.046] [PMID: 16923396]

[279] Choi J, Levey AI, Weintraub ST, et al. Oxidative modifications and down-regulation of ubiquitin carboxyl-terminal hydrolase L1 associated with idiopathic Parkinson's and Alzheimer's diseases. J Biol Chem 2004; 279(13): 13256-64.
[http://dx.doi.org/10.1074/jbc.M314124200] [PMID: 14722078]

[280] Chu KY, Li H, Wada K, Johnson JD. Ubiquitin C-terminal hydrolase L1 is required for pancreatic beta cell survival and function in lipotoxic conditions. Diabetologia 2012; 55(1): 128-40.
[http://dx.doi.org/10.1007/s00125-011-2323-1] [PMID: 22038515]

[281] Keller JN, Hanni KB, Markesbery WR. Impaired proteasome function in Alzheimer's disease. J Neurochem 2000; 75(1): 436-9.
[http://dx.doi.org/10.1046/j.1471-4159.2000.0750436.x] [PMID: 10854289]

[282] Lam YA, Pickart CM, Alban A, et al. Inhibition of the ubiquitin-proteasome system in Alzheimer's disease. Proc Natl Acad Sci USA 2000; 97(18): 9902-6.
[http://dx.doi.org/10.1073/pnas.170173897] [PMID: 10944193]

[283] Oddo S. The ubiquitin-proteasome system in Alzheimer's disease. J Cell Mol Med 2008; 12(2): 363-73.
[http://dx.doi.org/10.1111/j.1582-4934.2008.00276.x] [PMID: 18266959]

[284] Akarsu E, Pirim I, Capoğlu I, Deniz O, Akçay G, Unüvar N. Relationship between electroneurographic changes and serum ubiquitin levels in patients with type 2 diabetes. Diabetes Care 2001; 24(1): 100-3.
[http://dx.doi.org/10.2337/diacare.24.1.100] [PMID: 11194212]

[285] DeToma AS, Salamekh S, Ramamoorthy A, Lim MH. Misfolded proteins in Alzheimer's disease and type II diabetes. Chem Soc Rev 2012; 41(2): 608-21.
[http://dx.doi.org/10.1039/C1CS15112F] [PMID: 21818468]

[286] Glenner GG, Wong CW. Alzheimer's disease: initial report of the purification and characterization of a novel cerebrovascular amyloid protein. Biochem Biophys Res Commun 1984; 120(3): 885-90.
[http://dx.doi.org/10.1016/S0006-291X(84)80190-4] [PMID: 6375662]

[287] Glenner GG, Wong CW. Alzheimer's disease and Down's syndrome: sharing of a unique cerebrovascular amyloid fibril protein. Biochem Biophys Res Commun 1984; 122(3): 1131-5.
[http://dx.doi.org/10.1016/0006-291X(84)91209-9] [PMID: 6236805]

[288] Masters CL, Simms G, Weinman NA, Multhaup G, McDonald BL, Beyreuther K. Amyloid plaque core protein in Alzheimer disease and Down syndrome. Proc Natl Acad Sci USA 1985; 82(12): 4245--.
[http://dx.doi.org/10.1073/pnas.82.12.4245] [PMID: 3159021]

[289] Hardy JA, Higgins GA. Alzheimer's disease: the amyloid cascade hypothesis. Science 1992; 256(5054): 184-5.
[http://dx.doi.org/10.1126/science.1566067] [PMID: 1566067]

[290] Bird TD, Miller BL. Alzheimer's disease and other dementias.Harrison's Principles of Internal Medicine. 16th ed. McGraw-Hil 2005; pp. 2393-406.

[291] Selkoe DJ. Toward a comprehensive theory for Alzheimer's disease. Hypothesis: Alzheimer's disease is caused by the cerebral accumulation and cytotoxicity of amyloid β-protein. Ann N Y Acad Sci 2000; 924: 17-25.
[http://dx.doi.org/10.1111/j.1749-6632.2000.tb05554.x] [PMID: 11193794]

[292] Zheng H, Jiang M, Trumbauer ME, et al. β-Amyloid precursor protein-deficient mice show reactive gliosis and decreased locomotor activity. Cell 1995; 81(4): 525-31.
[http://dx.doi.org/10.1016/0092-8674(95)90073-X] [PMID: 7758106]

[293] Iwatsubo T, Odaka A, Suzuki N, Mizusawa H, Nukina N, Ihara Y. Visualization of A beta 42(43) and A beta 40 in senile plaques with end-specific A beta monoclonals: evidence that an initially deposited species is A beta 42(43). Neuron 1994; 13(1): 45-53.
[http://dx.doi.org/10.1016/0896-6273(94)90458-8] [PMID: 8043280]

[294] Miyakawa T, Kimura T, Hirata S, et al. Role of blood vessels in producing pathological changes in the brain with Alzheimer's disease. Ann N Y Acad Sci 2000; 903: 46-54.
[http://dx.doi.org/10.1111/j.1749-6632.2000.tb06349.x] [PMID: 10818488]

[295] Näslund J, Haroutunian V, Mohs R, et al. Correlation between elevated levels of amyloid beta-peptide in the brain and cognitive decline. JAMA 2000; 283(12): 1571-7.
[http://dx.doi.org/10.1001/jama.283.12.1571] [PMID: 10735393]

[296] Gong Y, Chang L, Viola KL, et al. Alzheimer's disease-affected brain: presence of oligomeric A beta ligands (ADDLs) suggests a molecular basis for reversible memory loss. Proc Natl Acad Sci USA 2003; 100(18): 10417-22.
[http://dx.doi.org/10.1073/pnas.1834302100] [PMID: 12925731]

[297] Georganopoulou DG, Chang L, Nam JM, et al. Nanoparticle-based detection in cerebral spinal fluid of a soluble pathogenic biomarker for Alzheimer's disease. Proc Natl Acad Sci USA 2005; 102(7): 2273-6.
[http://dx.doi.org/10.1073/pnas.0409336102] [PMID: 15695586]

[298] Klein WL, Krafft GA, Finch CE. Targeting small Abeta oligomers: the solution to an Alzheimer's disease conundrum? Trends Neurosci 2001; 24(4): 219-24.
[http://dx.doi.org/10.1016/S0166-2236(00)01749-5] [PMID: 11250006]

[299] Gandy S. The role of cerebral amyloid beta accumulation in common forms of Alzheimer disease. J Clin Invest 2005; 115(5): 1121-9.
[PMID: 15864339]

[300] Öztürk GB, Karan MA. The physiopathology of Alzheimer's disease. Klinik Gelişim 2009; pp. 36-45.

[301] Candeias E, Duarte AI, Carvalho C, et al. The impairment of insulin signaling in Alzheimer's disease. IUBMB Life 2012; 64(12): 951-7.
[http://dx.doi.org/10.1002/iub.1098] [PMID: 23129399]

[302] Klein WL. ADDLs & protofibrils--the missing links? Neurobiol Aging 2002; 23(2): 231-5.

[http://dx.doi.org/10.1016/S0197-4580(01)00312-8] [PMID: 11804707]

[303] Krafft GA, Klein WL. ADDLs and the signaling web that leads to Alzheimer's disease. Neuropharmacology 2010; 59(4-5): 230-42.
[http://dx.doi.org/10.1016/j.neuropharm.2010.07.012] [PMID: 20650286]

[304] Klein WL. Abeta toxicity in Alzheimer's disease: globular oligomers (ADDLs) as new vaccine and drug targets. Neurochem Int 2002; 41(5): 345-52.
[http://dx.doi.org/10.1016/S0197-0186(02)00050-5] [PMID: 12176077]

[305] Felice G, Lambert P, Fernandez S, Velasco T, Lacor N, Klein L. AD-type neuronal tau hyperphosphorylation induced by Ab oligomers. Neurobiol Aging 2008; 29(9): 1334-47.
[http://dx.doi.org/10.1016/j.neurobiolaging.2007.02.029] [PMID: 17403556]

[306] Klein WL. Synaptic targeting by A β oligomers (ADDLS) as a basis for memory loss in early Alzheimer's disease. Alzheimers Dement 2006; 2(1): 43-55.
[http://dx.doi.org/10.1016/j.jalz.2005.11.003] [PMID: 19595855]

[307] Ando K, Oishi M, Takeda S, *et al*. Role of phosphorylation of Alzheimer's amyloid precursor protein during neuronal differentiation. J Neurosci 1999; 19(11): 4421-7.
[http://dx.doi.org/10.1523/JNEUROSCI.19-11-04421.1999] [PMID: 10341243]

[308] Farris W, Mansourian S, Chang Y, *et al*. Insulin-degrading enzyme regulates the levels of insulin, amyloid beta-protein, and the beta-amyloid precursor protein intracellular domain *in vivo*. Proc Natl Acad Sci USA 2003; 100(7): 4162-7.
[http://dx.doi.org/10.1073/pnas.0230450100] [PMID: 12634421]

[309] Zhao W-Q, Townsend M. Insulin resistance and amyloidogenesis as common molecular foundation for type 2 diabetes and Alzheimer's disease. Biochim Biophys Acta 2009; 1792(5): 482-96.
[http://dx.doi.org/10.1016/j.bbadis.2008.10.014] [PMID: 19026743]

[310] de la Monte SM. Brain insulin resistance and deficiency as therapeutic targets in Alzheimer's disease. Curr Alzheimer Res 2012; 9(1): 35-66.
[http://dx.doi.org/10.2174/156720512799015037] [PMID: 22329651]

[311] Ling X, Martins RN, Racchi M, Craft S, Helmerhorst E. Amyloid beta antagonizes insulin promoted secretion of the amyloid beta protein precursor. J Alzheimers Dis 2002; 4(5): 369-74.
[http://dx.doi.org/10.3233/JAD-2002-4504] [PMID: 12446969]

[312] Cole AR, Astell A, Green C, Sutherland C. Molecular connexions between dementia and diabetes. Neurosci Biobehav Rev 2007; 31(7): 1046-63.
[http://dx.doi.org/10.1016/j.neubiorev.2007.04.004] [PMID: 17544131]

[313] Gasparini L, Xu H. Potential roles of insulin and IGF-1 in Alzheimer's disease. Trends Neurosci 2003; 26(8): 404-6.
[http://dx.doi.org/10.1016/S0166-2236(03)00163-2] [PMID: 12900169]

[314] Gustafson DR. Adiposity hormones and dementia. J Neurol Sci 2010; 299(1-2): 30-4.
[http://dx.doi.org/10.1016/j.jns.2010.08.036] [PMID: 20875649]

[315] Luchsinger JA. Adiposity, hyperinsulinemia, diabetes and Alzheimer's disease: an epidemiological perspective. Eur J Pharmacol 2008; 585(1): 119-29.
[http://dx.doi.org/10.1016/j.ejphar.2008.02.048] [PMID: 18384771]

[316] Luchsinger JA, Gustafson DR. Adiposity and Alzheimer's disease. Curr Opin Clin Nutr Metab Care 2009; 12(1): 15-21.
[http://dx.doi.org/10.1097/MCO.0b013e32831c8c71] [PMID: 19057182]

[317] Sims-Robinson C, Kim B, Rosko A, Feldman EL. How does diabetes accelerate Alzheimer disease pathology? Nat Rev Neurol 2010; 6(10): 551-9.
[http://dx.doi.org/10.1038/nrneurol.2010.130] [PMID: 20842183]

[318] Haan MN. Therapy Insight: type 2 diabetes mellitus and the risk of late-onset Alzheimer's disease. Nat

Clin Pract Neurol 2006; 2(3): 159-66.
[http://dx.doi.org/10.1038/ncpneuro0124] [PMID: 16932542]

[319] Campbell IW. Antidiabetic drugs present and future: will improving insulin resistance benefit cardiovascular risk in type 2 diabetes mellitus? Drugs 2000; 60(5): 1017-28.
[http://dx.doi.org/10.2165/00003495-200060050-00004] [PMID: 11129120]

[320] Mahley RW, Apolipoprotein E. Apolipoprotein E: cholesterol transport protein with expanding role in cell biology. Science 1988; 240(4852): 622-30.
[http://dx.doi.org/10.1126/science.3283935] [PMID: 3283935]

[321] Peila R, Rodriguez BL, Launer LJ. Type 2 diabetes, APOE gene, and the risk for dementia and related pathologies: The Honolulu-Asia Aging Study. Diabetes 2002; 51(4): 1256-62.
[http://dx.doi.org/10.2337/diabetes.51.4.1256] [PMID: 11916953]

[322] Li YM, Dickson DW. Enhanced binding of advanced glycation endproducts (AGE) by the ApoE4 isoform links the mechanism of plaque deposition in Alzheimer's disease. Neurosci Lett 1997; 226(3): 155-8.
[http://dx.doi.org/10.1016/S0304-3940(97)00266-8] [PMID: 9175590]

[323] Strittmatter WJ, Saunders AM, Schmechel D, *et al.* Apolipoprotein E: high-avidity binding to beta-amyloid and increased frequency of type 4 allele in late-onset familial Alzheimer disease. Proc Natl Acad Sci USA 1993; 90(5): 1977-81.
[http://dx.doi.org/10.1073/pnas.90.5.1977] [PMID: 8446617]

[324] Qiu WQ, Folstein MF. Insulin, insulin-degrading enzyme and amyloid-β peptide in Alzheimer's disease: review and hypothesis. Neurobiol Aging 2006; 27(2): 190-8.
[http://dx.doi.org/10.1016/j.neurobiolaging.2005.01.004] [PMID: 16399206]

CHAPTER 6

Pharmacological Mechanism of PPARγ Ratio in Diabetes and Obesity

José Roberto Santin, Marina Jagielski Goss and Nara Lins Meira Quintão[*]

Postgraduate Program of Pharmaceutical Science, Universidade do Vale do Itajaí, Itajaí, Santa Catarina, Brazil

Abstract: The worldwide prevalence of obesity has increased at alarming rates over the last four decades. Overweight and obesity featuring the excess of white adipose tissue are cardiovascular risk conditions consistently associated with the development of complex metabolic disorders, including insulin resistance, type 2 diabetes *mellitus* (T2DM) and coronary heart diseases. Many natural and synthetic agonists of peroxisome proliferator-activated receptors (PPARs; nuclear receptors) are used in the treatment of glucose and lipid disorders. PPARs perform different activities, mainly *via* endogenous ligands produced in the metabolic pathways of fatty acids; and therefore, they are called lipid sensors. PPAR agonists have different properties and specificities for individual PPAR receptor, different absorption/distribution profiles, and distinct gene expression profiles, which ultimately lead to different clinical outcomes. The isoform PPARγ is expressed in white and brown adipose tissue, large intestine and spleen. However, its expression is highest in adipocytes and it plays a key role in the adipogenesis regulation, energy balance and lipid biosynthesis. PPARγ has been the focus of intense research once its ligands have been described to treat T2DM. Some of them are currently prescribed as anti-diabetic drugs, such as thiazolidinedione. PPARγ activation modulates not only insulin sensitization, but also lipid metabolism, vascular tone and inflammation, all processes involved in atherogenesis. Considering the impact of this subject in the public health and the necessity of new approaches for the development of new drugs to treat metabolic diseases and to improve the quality of life, this chapter has the aim of revising important points concerning the involvement of the nuclear receptors in obesity, diabetes and discuss the real possibility of this target to become an effective and safe pharmacological therapy.

Keywords: Adipogenesis, Agonist, Diabetes, Glitazones, Hyperglycaemia, Metabolic syndrome, Nuclear receptors, Obesity, PPARγ, Peroxisome.

---

[*] **Corresponding author N.L.M. Quintão:** Postgraduate Program in Pharmaceutical Science, Universidade do Vale do Itajaí, Rua Uruguai, no 458, Bloco F6, CCS, sala 314, CEP 88302-901, Itajaí, SC, Brazil; Tel: (+55) 47 3341-7932; Fax: (+55) 47 3341-7652; E-mails: nara.quintao@univali.br; narafarmaco@yahoo.com.br

Atta-ur-Rahman (Ed.)
All rights reserved-© 2019 Bentham Science Publishers

INTRODUCTION

The world is experiencing an exponential increase in the incidence of obesity and related diseases, such as diabetes. Considering pre-school children, the World Health Organization (WHO) estimates that this figure will raise to 60 million by 2020. Obesity is a chronic condition with complex context, involving social, behavioural, environmental, cultural, metabolic and genetic factors. The morbidity degree is considerably increased when obesity is associated with important metabolic disruptors, such as diabetes, cardiovascular diseases and other metabolic disturbances.

Taking this scenario into account, the pharmaceutical industries and important research groups have concentrated their efforts on the search for new pharmacological tools to prevent or treat diseases such as obesity and diabetes. The peroxisome proliferator–activated receptor γ (PPARγ) belongs to a class of nuclear receptors and has been emerged as promisor target to control fatty acid and glucose metabolism disruption, and its agonists have been widely used (e.g pioglitazone, rosiglitazone and troglitazone). Unfortunately, despite of important effectiveness, these clinically-used PPARγ agonists present important adverse effects that compromise the treatment protocol.

The first part of this chapter introduces the PPARγ receptor and its role in physiological body functions related mainly to lipid and glucose metabolism. Therefore, major approach is given to pathological conditions, such as obesity and diabetes, and to the role of PPARγ receptor and advanced researches in the search of new pharmacological tools to modulate the receptor function. Eventually, clinical trials involving this class of drugs are presented, placing the reader within the latest advances involving PPARγ activators for the treatment of obesity and diabetes.

PHARMACOLOGY OF PPARΓ

Peroxisome proliferator-activated receptors (PPARs) are nuclear hormone receptors belonging to the superfamily of phylogenetically related protein termed as nuclear hormone factor (NR1C) containing 48 members. Issemann and Green discovered the first PPAR, in 1990 [1]. These authors built a recombinant chimeric receptor containing the domains of glucocorticoid receptor and one of their cloned receptors. After screening chemicals for their abilities to activate the created recombinant chimeric receptor, they unexpectedly found that peroxisome proliferators were able to activate a gene reporter (genes that confer characteristics easily identified and measured in organism that expresses them). The cloned receptor was concluded to belong to the steroid receptor superfamily, as shown by sequence homologies, later called mPPARα (*m* for mouse and α for the first described PPAR) [2, 3].

As described in Table **1**, PPAR superfamily is composed of 3 isotypes encoded by separate genes PPARα, PPARβ/δ and PPARγ [4]. PPARγ protein exists in three isoforms derived from the same gene expression by utilizing different promoters and 5'exons. They are called PPARγ1, PPARγ2 and PPARγ3. PPARγ1 and PPARγ3 are identical, but PPARγ2 differs by the existence of 30 extra amino acids that confers higher transcriptional activity. Other peculiarity of PPARγ2 is that under physiological conditions its expression is restricted to white and brown adipose tissue (WAT and BAT) [3, 5].

Table 1. PPAR isoforms, species and tissue origins*.

PPAR isoform	Species	Tissue Expression	Ligands	Functions
PPARα	Mouse (m) Rat (r) Human (h) *Xenopus* (x)	Muscle Liver Heart Kidney	**Natural:** Unsaturated fatty acids; LTB-4; 8-Hydroxy-eicosatetraenoic acid **Synthetic:** Fenofibrate; Clofibrate; Gemfibrozil	Lipids and lipoprotein metabolism
PPARβ	*Xenopus*	Abundantly expressed throughout the body Liver (low level)	**Natural:** Unsaturated fatty acids; Carbaprostacycline; components of VLDL **Synthetic:** GW 501516	Lipid metabolism and energy balance
PPARγ	Mouse Rat Hamister (ha) Human *Xenopus*	Adipocyte Large intestine Hepatopoietic cells Kidney Liver Muscle Pancreas Small intestine	**Natural:** Unsaturated fatty acids; 15-Hydrox--eicosatetraenoic acid; 9- and 13-Hydroxy-octadecadienoic acid; 15-deoxy Δ12,14-prostaglandin J2; PGJ$_2$ **Synthetic:** Rosiglitazone; pioglitazone; troglitazone; ciglitazone; farglitazone; S26948; INT131	Adipocyte differentiation Lipid metabolism; Glucose homeostasis; Immune cells modulation; Cell proliferation
PPARδ		Abundantly expressed throughout the body Liver (low level)	**Natural:** Unsaturated fatty acids; Carbaprostacycline; components of VLDL **Synthetic:** GW 501516	Lipid metabolism and energy balance

* [3, 6].

In general, the three-dimensional structure of PPARs consists of DNA binding domain (DBD) in the N-terminus and a ligand-binding domain (LBD) in the C-terminus. Additionally, it has an A/B domain (harbors a ligand-independent transcriptional activating function – AF-1), the hinge-region and strong C-ligand dependent transcription activating function (AF-2; responsible for the interaction with coactivators and corepressors) (Fig. **1A** and **B**). The DBD is highly conserved among PPARs and is formed by two zinc finger-like motifs that recognize the peroxisome proliferator hormone response elements (PPREs). After ligand interaction, the receptor is translocated to the nucleus and forms a heterodimer with the retinoid X receptor (RXR). Then, this heterodimer binds to a specific region of target genes called PPREs and activates transcription of various genes involved in physiological and pathological conditions. The number of coactivators and corepressors linked to the receptor will modify its function [7] (Fig. **1C**).

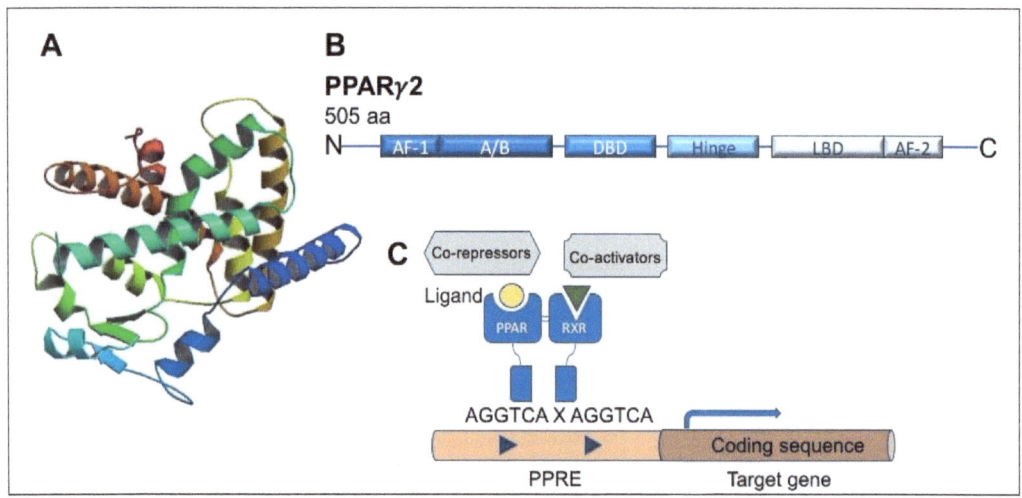

Fig. (1). PPARγ2 structure. (**A**) 3D structure; (**B**) main domains: AF-1 A/B (activation function 1 – A/B domain required for ligand-independent activation); AF-2 (activation function 2, required for ligand dependent activation, ligand dependent dimerization, co-activator recruitment and co-repressor release); DBD (DNA-binding domain); LBD (ligand-binding domain); PPAR (peroxisome proliferator-activated receptor); PPRE (PPAR responsive element); RXR (retinoid X receptor). Adapted from [8, 9].

Scientific studies about PPARs date from 1990 and have increased in number of publications since then as shown in Fig. (**2**).

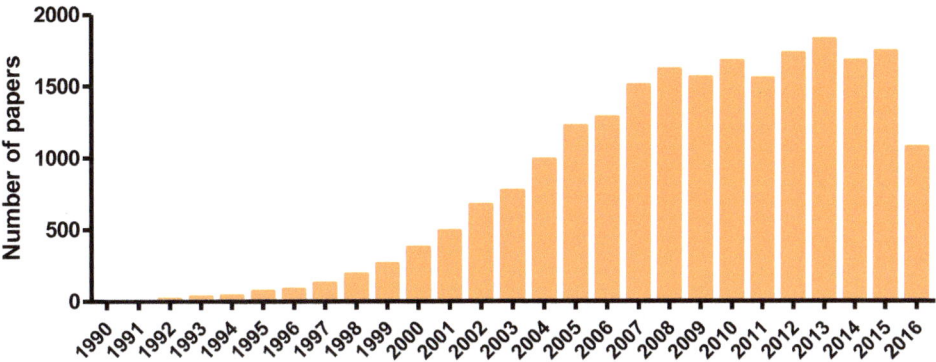

Fig. (2). Number of papers listed in PubMed using the term PPARs (date of search: 11/20/2016).

More specifically, the physiological role of PPARγ called researchers' attention more than two decades ago. It has been implicated mainly in adipogenesis, more specifically pre-adipocyte differentiation and fatty acid storage in mature adipocyte [10]. Literature data has shown the relevant role of PPARγ in the stimulation of pre-adipocytes differentiation and also in the mobilization of bone marrow-derived circulating progenitor cells to WAT and their subsequent differentiation into adipocytes. This receptor is the major regulator of lipid storage [11]. PPARγ is important for insulin sensitivity and glucose metabolism. During this process, hydrolysed triglyceride-rich lipoproteins generate fatty acids that are redirected towards adipose tissue rather than skeletal muscle, increasing glucose metabolism in the muscle. The expression of genes involved in glucose uptake and insulin signalling is also increased affecting glucose processing in the periphery [12].

The increase of publications mentioned beforehand is probably linked to the progressive increase in the worldwide metabolic diseases incidence, including obesity, diabetes and metabolic syndrome. (Fig. **3**) presents the number of papers found in PubMed when PPAR is linked to obesity (Fig. **3A**), diabetes (Fig. **3B**) or metabolic syndrome (Fig. **3C**). All terms present similar profile.

Mutations in PPARγ structure have clinical relevance for several pathological conditions, such as obesity, abdominal obesity, metabolic syndrome, atherosclerosis susceptibility, T2DM, colorectal cancer, glioblastoma and giant cell gliobastoma, gliosarcoma, prostate cancer and thyroid carcinoma [6].

Several PPARγ gene knockout studies demonstrate the fundamental role of this receptor for the glucose and lipid metabolism, keeping normal insulin sensitivity and whole-body glucose and lipid homeostasis (Table **2**).

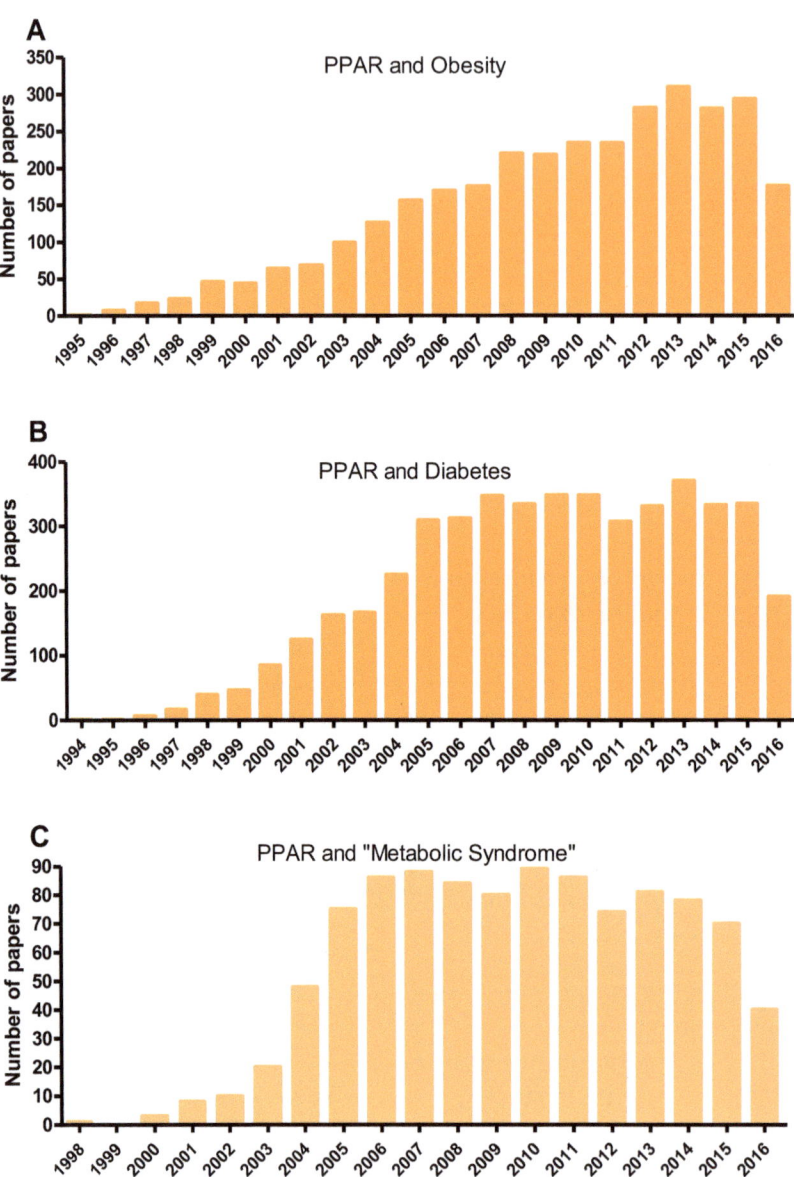

Fig. (3). Number of papers listed in PubMed using the term PPARs linked to (**A**) obesity, (**B**) diabetes or (**C**) "metabolic syndrome" (date of search: 11/20/2016).

Table 2. Transgenic mouse for PPARγ and the pathophysiological implication.

Mutation	Tissue Distribution	Phenotype
KO	White adipocyte	Retarded growth; severe lipodystrophy; hyperglycaemia
KO	Adipocyte (white and brown)	ND
KO	Macrophage	ND
KO	General	Not viable – defect in placenta formation
OE	Hepatocyte	Steatosis
KO	Muscle	Progressive insulin resistance; increase adipose tissue mass
Heterozygous	-	Reduced body size and weight; reduced insulin resistance; smaller adipocytes and fat depots.
KO	Pancreatic beta cells	Significant islet hyperplasia on chow diet; blunted expansion of beta cell mass.

KO – knockout; OE – overexpression; ND – not determined [6].

OBESITY AND DIABETES

The worldwide prevalence of obesity has progressively increased over the past decades. In fact, data obtained in 2014 shows that over 1.9 billion adults were overweight, including 600 millions of obese. According to WHO Global Health Observatory Gallery (Fig. **4**), about 39% of adults aged 18 years and above were overweight in 2014, and 13% were obese. However, not only there were more adults becoming obese, but obesity was also striking at a much younger age leading to high number of obese children and adolescents [13].

Additionally, it is important to mention that most of the world populations live in countries where obesity is evident. Consequently, obesity is nowadays the most prevalent chronic metabolic disorder and one of the most important global public-health challenges. Besides, obesity is a complex disease and one of the strongest risk factor for the development of metabolic disorders, such as T2DM [14].

Diabetes *mellitus* (DM) is one of the largest epidemic the world has faced, in both developed and developing nations. Diabetes is defined as a metabolic condition in which the body does not produce sufficient insulin to regulate blood glucose levels or the produced insulin is unable to work effectively, being classified as T1DM or T2DM, respectively [16].

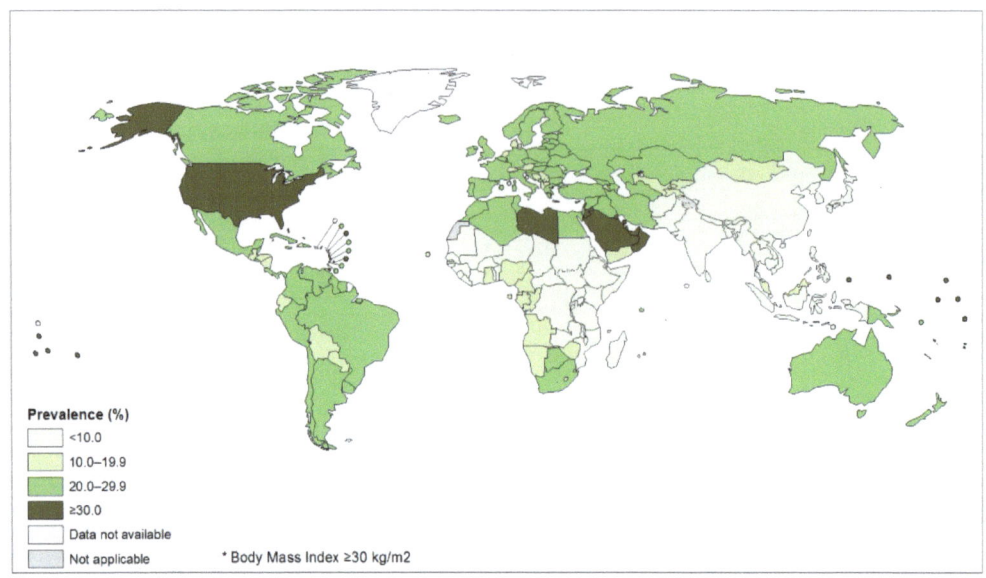

Fig. (4). Prevalence of obesity in adults (2014) [15].

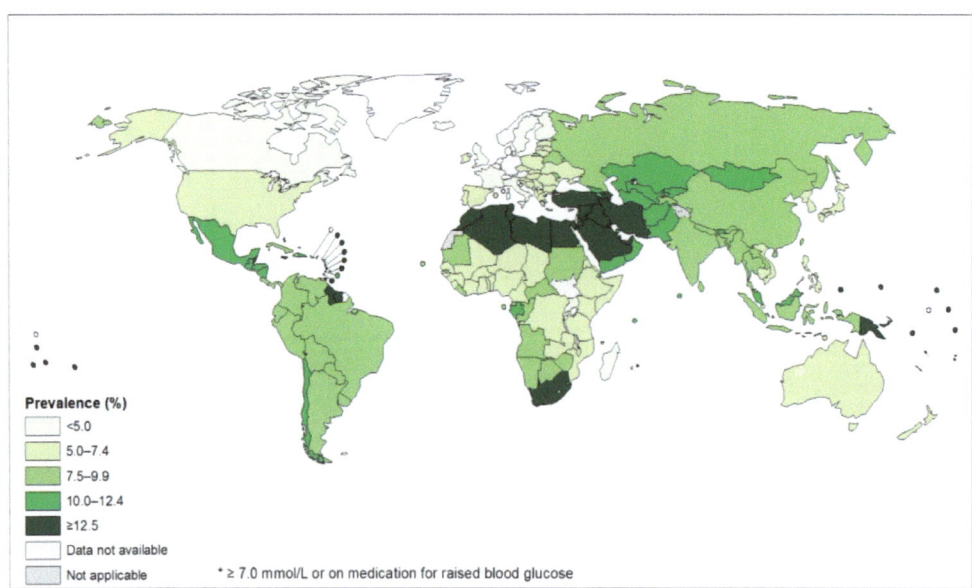

Fig. 5 cont.....

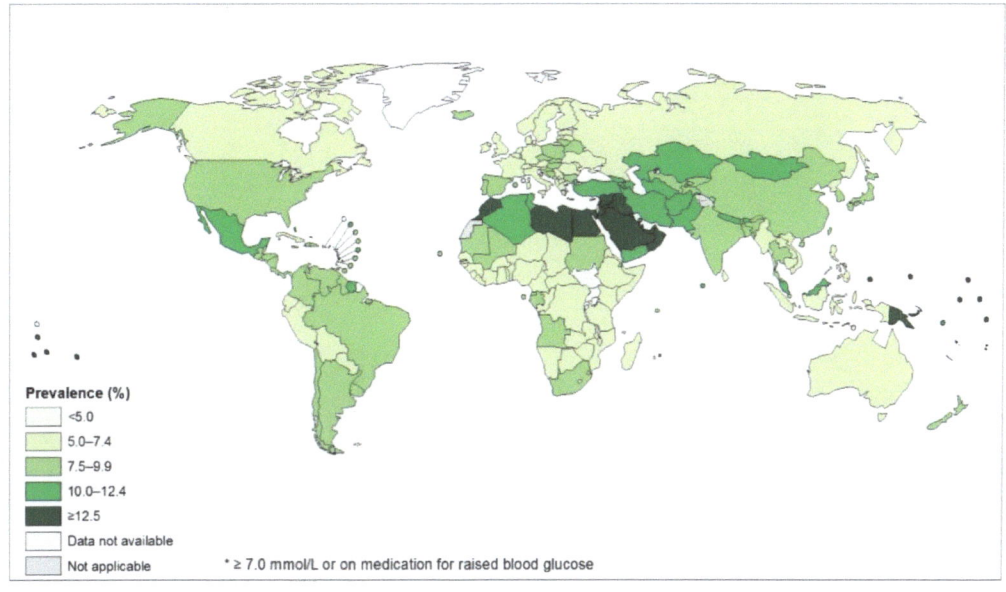

Fig. (5). Prevalence of raised fasting blood glucose, ages 18+, in Female and Male, 2014 [15].

The number of people with T2DM has more than doubled during the past 20 years, in accordance with the WHO estimative of 422 millions of people with DM (Fig. **5**). In this context, diabetes is a serious global public health issue, which has been described as the most challenging health problem in the 21st century [15].

Individuals with diabetes are two to four times more likely to develop cardiovascular disease related to the general population and have two to five-fold greater risk of dying due to these conditions. Diabetes is a significant cause of blindness in adults, non-traumatic lower limb amputations and end-stage renal disease needing transplantation or dialysis [17].

There is a close association between obesity and T2DM. The severity of T2DM is closely linked to body mass index (BMI). There is seven times greater risk of diabetes in obese people than those of healthy weight, with a three-fold increase in risk in overweight people. This close association may be visualized in Fig. (**6**), obtained from CDC - Division of Diabetes Translation, that shows the prevalence of obesity and diabetes from the United States of America (USA) in 2014. The same correlation was recently found by Public Health England, which suggests that 90% of adults with T2DM are overweight or obese (Public Health England, Adult obesity and T2DM. 2014)

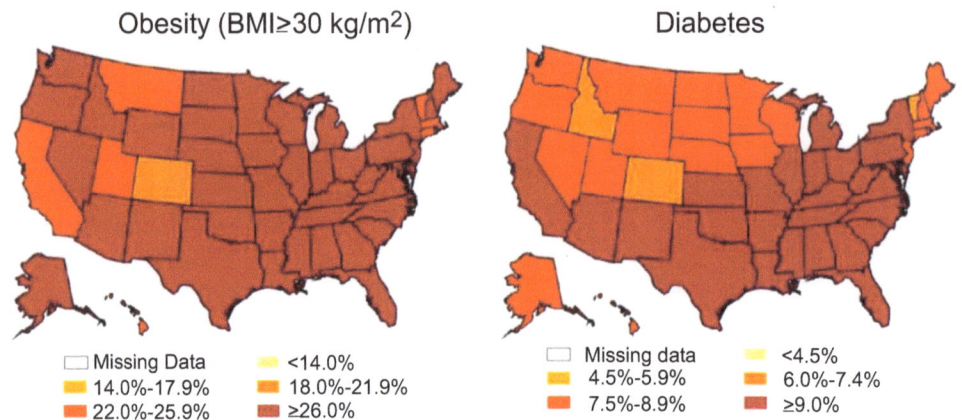

Fig. (6). Age–Adjusted Prevalence of Obesity and Diagnosed Diabetes Among US Adults in 2014. CDC's Division of Diabetes Translation. United States Diabetes Surveillance System available at http://www.cdc.gov/diabetes/data [18].

Obesity requires long-term chronic-disease model of care. Behavioural and/or lifestyle interventions, including high-quality diet, caloric restriction, reduced sedentary behaviour and increased physical activity, form the cornerstones of traditional care. In addition to lifestyle changes, a broad range of safe and effective interventions are needed to manage obesity and its effects on physical and mental function, quality of life, end-organ damage and mortality [19].

Adipose Tissue: Obesity and Diabetes

Initially, WAT was regarded as simply sink of excess calories in the fed state and a reservoir from which free fatty acids are released during fasting to fuel the organism's energy demand. However, it is known nowadays that the WAT accumulation could be the inductor of several metabolic disorders including obesity. Obesity may be defined as a result of the imbalance between the intake and spending energy. The excess of calories is initially stored in subcutaneous fat tissue. Therefore, when this storage capacity is overwhelmed, several endocrine modifications occur in the adipose tissue [19, 20].

WAT has interstate characteristics, as the extraordinary capacity to change its dimension in response to nutritional status by remodelling a series of cellular mechanism. The WAT is able to alter its size, function, inflammatory state and whole-body distribution. In addition, it modifies its extracellular matrix composition, vascularization, adipocyte size and number, oxidative stress levels, adipokine secretory profile and the inflammatory state of immune cells [13].

In a healthy patient, fuel distribution between WAT depots and other tissues mostly involves subcutaneous WAT, which is the major and the safest lipid storage organ. However, during long-term caloric excess, the subcutaneous WAT expandability occurs by hypertrophy of existing white adipocytes and increased differentiation of adipocyte progenitor and/or pre-adipocytes (hyperplasia). It is initially sufficient to scavenger triglycerides away from the liver and other tissues sensitive to triglyceride-induced lipotoxicity. The latter process embodies *de novo* adipocyte formation (adipogenesis). When subcutaneous WAT can no longer accommodate this lipid excess, ectopic fat accumulation appears leading to a lipotoxic metabolic stress, which it promotes low-grade inflammation and metabolic dysfunction in the adipose tissue [13, 19, 21].

This pro-inflammatory condition is evoked mainly by hypertrophied adipocytes and adipose tissue-resident immune cells (mainly lymphocytes and macrophages) which contribute to the increased circulating levels of pro-inflammatory cytokines. The adipose tissue from slim individual secretes anti-inflammatory adipokines such as adiponectin, transforming growth factor beta (TGF- β), interleukin 10 (IL-10), IL-4, IL-13, IL-1 receptor antagonist (IL-1Ra), and apelin. However, obese adipose tissue releases mainly pro-inflammatory cytokines, such as TNF, IL-6, leptin, vistafin, resistin, angiotensin II and monocyte chemoattractant protein 1 (MCP-1), with parallel decrease in anti-inflammatory factors, such as IL-10 and adiponectin [14].

The obesity-associated state of chronic low-grade systemic inflammation, termed "metabolic inflammation," is considered a focal point in the pathogenesis of insulin resistance (in the muscle, liver and fat), resulting in increased blood glucose levels. In this context, the anti-inflammatory adipokines released from non-obese adipose tissue mediate physiological functions, whereas the pro-inflammatory adipokines are responsible for modulating the insulin resistance by acting directly in the insulin-signalling pathway or indirectly by stimulating inflammatory pathways.

As obesity progress, MCP-1-release induces inflammatory cells recruitment, especially monocytes. Resident WAT immune cells and recruited monocyte cells become polarized to pro-inflammatory M1 "classically activated" macrophages, the opposite of anti-inflammatory M2 "alternatively activated" macrophage. The M1 macrophages are responsible for deregulating local adipose signalling and impairing the insulin sensitive. It happens primarily by the secretion of pro-inflammatory cytokines and by surrounding apoptotic fat cells that create crown-like structures that are an aggregation of single or fused macrophages around a single adipocyte in adipose tissue. These structures are typically associated with obesity, adipose tissue dysfunction and chronic inflammation [19].

It is known that several adipokines may act inducing the phosphorylation of insulin receptor substrate (IRS). In addition, the IRS phosphorylation could be promoted *via* inflammatory pathways, such as c-jun N-terminal kinase (JNK) pathway and I-κB Kinase (IKKbeta/NF-κB) pathway that disrupt the insulin signalling pathways.

Other point to be discussed is the effect of obesity in insulin-dependent glucose transport in adipose and skeletal muscle. This mechanism of transport is important because the bulk of insulin-dependent glucose transport results from the insulin-dependent recruitment of the GLUT4 (glucose transporter type 4 isoform), the facilitative glucose transporter in the cell surface of skeletal muscle and adipose tissue. This protein, in obesity condition, suffers decrease of mRNA and consequently expression repression that are evoked mainly by adipose tissue expansion. It is known that during adipose tissue expansion an inflammatory process is started. TNF is abundantly released by the adipocytes of obese animals causing deregulation of lipid and carbohydrate metabolism, as well as insulin signalling leading to insulin resistance, mainly by the decrease of GLUT4 expression. The decrease of GLUT4 expression alters the metabolism of adipose tissue in such a way that impacts whole-body glucose homeostasis. PPARγ agonists are known by upper regulating GLUT4 and it improves insulin sensitivity.

Role of PPARγ in Obesity and Insulin Resistance

PPARγ is considered the master regulator of adipogenesis, and has been extensively studied in the context of obesity. This nuclear receptor has an essential role in modulating multiple genes, including those of lipid metabolism, adipogenesis, adipokine production and inflammation. In addition, the PPARγ2 is expressed at similar levels in both WAT and BAT, and it acts as dominant regulator of adipogenesis and a powerful modulator of lipogenic pathways and insulin sensitivity [13, 22].

The PPAR gene expression could be differently modulated in obese patients when compared to non-obese. In obese patients, who present a metabolic inflammation, the PPARγ is phosphorylated by a protein induced by pro-inflammatory signals called cyclin-dependent-like kinase 5 (CDK5). This phosphorylation is activated in fat-depots, leading to deregulation of a subset of genes that are important for metabolic homeostasis, mainly adiponectin and adipsin (Fig. **7**) [13, 22].

Another important protein involved in the PPARγ gene disruption is Sirtuin-1 (SIRT1), which it hinders the adipocyte differentiation and suppress both PPARγ and NF-κB activity. In fact, data from literature reports that SIRT1-knockout mice present enhanced PPARγ expression, and consequently better insulin sensitivity.

The immune cells present in WAT express PPARγ. In fact, PPARγ has a crucial role in the macrophage functions, mainly related to anti-inflammatory activity and maintenance of lipid homeostasis. It was demonstrated that specific ablation of PPARγ gene in macrophages difficult the maturation of anti-inflammatory M2 macrophages and promotes insulin resistance. Of note, macrophage-specific PPARγ knockout mice show predisposition to diet-induced weight gain, glucose intolerance and insulin resistance (Fig. **7**) [13].

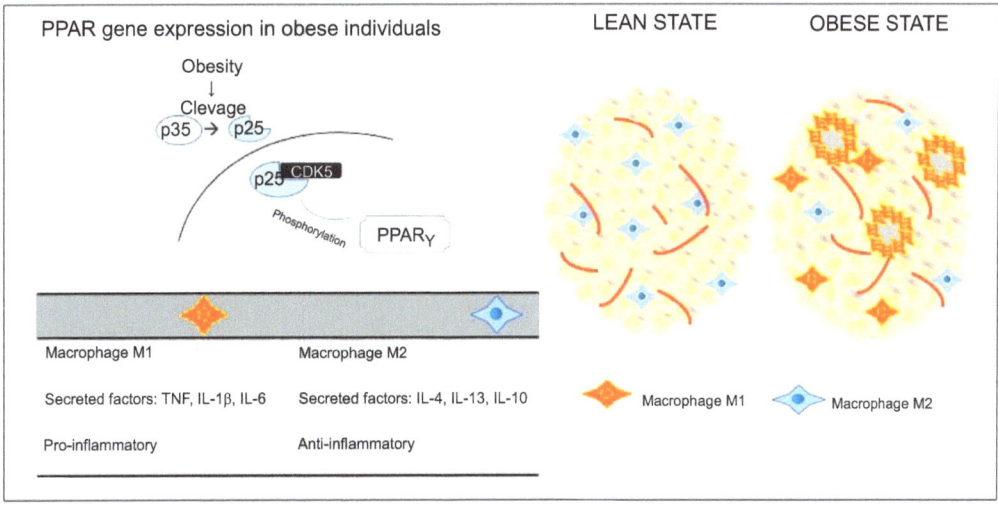

Fig. (7). In obese state MCP-1 release induces inflammatory cells recruitment, especially monocytes. Because of the chronification, the cleavage of p35 protein into p25 protein occurs. The p25 translocates to the nucleus, binds in CDK5 and, this complex p25/CDK5 phosphorylates PPARγ. This process is activated in fat-depots, leading to a deregulation of a subset of genes that are important to metabolic homeostasis, such as adiponectin and adipsin. Phosphorylated PPARγ decreases maturation of anti-inflammatory M2 macrophages and promotes insulin resistance. In this context, PPARγ agonists are pointed out as important target to treat metabolic syndrome, acting mainly in the insulin resistance.

PRE-CLINICAL RESEARCH IN OBESITY AND DIABETES X PPARΓ

There is no doubt that PPARγ agonists are important for the treatment of obesity and associated morbidities. On the other hand, the existence of significant side effects has led some drugs to limited use or even to its market discontinuity. Based on this context, new concepts regarding PPARγ-targeted drugs have appeared, including partial agonists or multi-targeted PPARs agonists, such as dual or pan agonists [23]. The use of *in silico* analysis (docking approach) and *in vitro* assays (gene reporter), have allowed researchers to design and create new molecules that present one of the characteristics mentioned above. The aim of these strategies is to optimize safety-profile, to improve glycolipid metabolism and, at the same time, to prevent side effects occurrence. The Table **3** presents some of natural or synthetic compounds already evaluated *in vitro* or *in vivo*.

Table 3. New PPARγ ligands with pre-clinical effects.

Compound	Activity	Indication	Ref.
Natural Compounds			
Pseudoginsenoside F11 (*Panax quinquefo-lium* L.)	PPARγ partial agonist	Adiponectin oligomerization and secretion, improved insulin resistance	[24]
Protopanaxatriol (ginseng)	PPARγ antagonist	Reduce obesity and serum lipid levels; improve insulin resistance, steatosis, and hyperlipidaemia in the liver.	[25]
Oregano extract	PPARγ antagonist	Body weight reduction	[26]
Commipheric acid (*Commiphora mukul*)	PPARγ/α dual agonist	Hypoglycaemic activity	[27]
Synthetic Compounds			
KY210	PPARγ partial agonist	Reduced by-effects and exerted little influence on blood and bone in normoglycaemic mice	[28]
L312	PPARγ partial agonist	Induced transcription activity of PPARγ with a less intensity, weakly promoted expression of adipogenic related genes; improved insulin resistance with reduced side-effects.	[29]
2,4,6-Trisubstitutedpyrimidine-5-carboxylic acid Compound 4	PPARγ partial agonist	Hypoglycaemic activity	[30]
MK-0533	PPARγ partial agonist	*In vitro* insulin-sensitizing Activity; *in vivo* desirable pharmacokinetic characteristics.	[31]
PAM-1616	PPARγ partial agonist	Hypoglycaemic activity; low side effects	[32]
Indenones KR-62776	PPARγ partial agonist	*In vitro* and *in vivo* PPARγ agonist activity without side effects	[33]
1,3- and 1,4- Oxybenzyl pyrrolidine acid	PPARγ/α dual agonist	Decreased the blood glucose levels and triglyceride in diabetic mice model.	[34]
Compound 20 (Oxime ether)	PPARγ/α dual agonist	Reduced the glucose level and improved fat distribution *in vivo*.	[35]
MHY908 (fibrate analog)	PPARγ/α dual agonist	Reduction in liver fat level; reduced insulin resistance; increased adiponectin levels without body weight gain.	[36]
Compound 28 (Zwitterionic compound)	PPARγ/α dual agonist	Glucose-lowering effect	[37]
Compound 29 (Zwitterionic compound)	PPARγ/α dual agonist	Glucose-lowering effect	[37]
Cevoglitazar	PPARγ/α dual agonist	Reduced glucose tolerance and decreased obesity.	[38]
Chiglitazar	PPARα/γ/δ pan-agonist	Reduced side effects	[39]
GQ-16	PPARγ partial agonist	Reduction of weight gain in mice despite increasing energy intake; reduce epididymal fat mass, reduced liver triglyceride content, increased BAT activity, increased expression of thermogenesis-related genes. Promoted insulin sensitization.	[40]

(Table 3) cont.....

Compound	Activity	Indication	Ref.
GQ-177	PPARγ partial agonist	Improvement of insulin sensitivity and lipid profile, increase of plasma adiponectin and GLUT4 mRNA in adipose tissue, no effect on body weight, food consumption, fat accumulation and bone density.	[41]

* for review read [22].

CLINICAL TRIALS

As already explained, PPARγ is associated with insulin sensitivity, glycaemic control, endothelial function, adiponectin and leptin levels and makers of inflammation. For this reason, PPARγ has been studied as potential target for the treatment of T2DM and obesity, as well as to help the management of metabolic diseases' symptoms.

Glitazones, also called thiazolidinediones (TZDs), are PPARγ ligands and are commonly employed in the diabetes management combined with metformin. Though the existence of many PPAR's agonists with high pleiotropic action and cross-talking with others signalling pathways, they seem to present important adverse effects, such as body-weight gain, fluid retention, congestive heart failure, bone fractures and increased risk of myocardial infarction [42]. The main clinical trials involving PPARγ-targeted drugs of the latest three years are listed below.

Synthetic Ligands

Troglitazone was the first TZD approved as medication for T2DM treatment in 1997. The compound showed beneficial effects in glucose levels, insulin sensitivity and free fatty acid concentration, however it was removed from the market in 2000 due to severe hepatotoxicity. For a certain period rosiglitazone had also been banned in Europe and restricted in the USA because of increased cardiovascular mortality. Rivoglitazone still under investigation [42]. Some studies in 2012 have reported the probable association between pioglitazone use and bladder cancer. Cohort studies and follow-ups were performed to assess the risk of long-term therapy in different population and it was suggested that pioglitazone alone cannot be considered the cause of increased incidence of bladder cancer in diabetic patients [43, 44].

Pioglitazone

PROactive was a 10-year follow-up, randomized and double-blind study involving 5,238 T2DM patients with pre-existing macrovascular disease. The study prospectively evaluated the effects of TZDs on cardiovascular outcomes and compared the incidence of newly diagnosed malignancies according to the originally assigned pioglitazone or placebo. Pioglitazone resulted in a non-significant 10% relative risk reduction in the primary endpoint of all-cause mortality, myocardial infarction, acute coronary syndrome, cardiac intervention, stroke, major leg amputation and leg revascularization. During this double blinded treatment, there was no difference in the cumulative incidence of malignancies between the treated groups [45].

INT131 Besylate

INT131 besylate is a selective PPARγ modulator. Its molecular structure is distinct from the TZDs and represents a new class of non-TZDs PPARγ ligand. INT131 was specifically designed to retain the insulin sensitivity and glucose-lowering actions, like the PPARγ full agonists, but mitigating or eliminating the undesirable effects. The study of this modulator is in phase II, but the primary efficacy analysis showed change in glycated haemoglobin after 24 weeks of use, in comparison to placebo, and a better control of this parameter at doses of 2 and 3 mg of INT131 in comparison to pioglitazone.

INT131 besylate was specifically designed as a selective PPARγ modulator to keep the glucose lowering after PPARγ activation without the adipogenic proprieties of full agonist. INT131 does not stimulate rodent or human preadipocytes to differentiate or accumulate lipid and has equal or greater efficacy in stimulating adiponectin, a marker of PPARγ activation in rodents and healthy subjects [46].

Aleglitazar

Aleglitazar, a PPARγ agonist, presented in a phase II (randomized dose-ranging) study, showed reduction in glycosylated haemoglobin in dose dependent manner without many cases of hypoglycaemia. It also decreased triglycerides and enhanced HDL-cholesterol. Therefore, phase III trials were initiated to investigate the effects of Aleglitazar on cardiovascular events in patients with metabolic disease. This trial was interrupted prematurely when the independent data safety monitoring board (ALECARDIO) recommended early termination due to unfavourable benefit-risk balance. Most of the safety concerns in the ALECARDIO trial (oedema/heart failure and fractures) could have been expected for a PPARγ agonist, however several smaller trials with short duration had not revelled these complications. It reinforces that favourable effects on phase II studies may not be translated into clinical efficacy and safety in phase III evaluation [47].

Saroglitazar

Saroglitazar is a dual PPARα/γ agonist, where PPARα action increases lipoprotein-lipase activation together with rising hepatic oxidation and reducing the triglycerides production in the liver. Although Saroglitazar is predominantly a PPARα agonist, it also causes activation of PPARγ. Then, it can regulate the transcription of insulin-responsive genes involved in the glucose use control added to the modulation of glucose production and transport. In both phase II and III clinical trials, Saroglitazar was effective in controlling dyslipidaemia and glycaemic parameters compared to placebo, in a T2DM population with hypertriglyceridemia not controlled with statin therapy. Besides that, it was found to be exempted from conventional adverse events of typical PPARα agonists, as well as of PPARγ agonists (such as pedal oedema, weight gain, or congestive heart failure) [48].

Lobeglitazone

Non-alcoholic fatty liver disease (NAFLD) is a condition where triglyceride accumulates in the hepatocytes of individual who have not consumed excessive amounts of alcohol. The prevalence of NAFLD ranges from 6.3% to 33%, and it is expected to rise as obesity rates increase, populations become older, and physical activity levels decrease. Moreover, there is an increased prevalence of NAFLD in T2DM patients and insulin resistance may aggravate the severity of NAFLD.

This pyrimidine derivative containing TZD is a novel PPARγ agonist, and showed, in both *in vitro* and *in vivo* studies, to be more potent than reference compounds (*i.e.* pioglitazone and rosiglitazone). ELLEGANCE evaluates the efficacy and safety of Lobeglitazone in T2DM patients with NAFLD. The data was conducted in five medical centres in Korea. The patients received doses of Lobeglitazone for 24 weeks and liver parameters, lipid serum and fasting glycaemic and insulin was measured. Lobeglitazone improved glycaemic parameters, such as the fasting levels of glucose and insulin, as well as insulin resistance, assessed by HOMAIR (which has also been ameliorated). Among lipid profile components, Lobeglitazone increased HDL-C and decreased triglycerides levels. Furthermore, all components of hepatic profile improved. Thus, data indicates that Lobeglitazone is a valid, novel therapeutic drug that could be used to treat NAFLD patients with T2DM [49].

Natural Ligands

Many natural and synthetic ligands influence the expression of PPARs. New generation drugs - PPARα/γ dual agonists presented anti-hyperlipidemic, hypotensive, anti-atherogenic, anti-inflammatory and anticoagulant effect. Precise data of the expression and function of natural PPAR agonists on glucose and lipid metabolism are still missing, mostly because the same ligand interferes with other receptors and several reports have provided conflicting results. To date, we know that PPARs have the capability of accommodating and binding a variety of natural and synthetic lipophilic acids, such as essential fatty acids, eicosanoids, phytanic acid and palmitoylethanolamide [50].

Essential Fatty Acids

Studies with cell culture, experimental animal models, as well as human clinical trials suggest that n-3 polyunsaturated fatty acids (PUFAs) improve lipid metabolism and anti-inflammatory/antioxidant capability through the regulation of genes presented in PPARγ signalling.

In 2015, researches from Iran evaluated the effect of docosahexaenoic acid (DHA)-rich fish oil supplementation on the modulation of some PPARγ responsive genes related to lipid metabolism. A randomized, double-blind, placebo-controlled clinical trial was conducted on 72 T2DM-patients that consume 2.4 g of fish oil DHA: 1,450 mg and Eicosapentaenoic Acid (EPA): 400 mg) or 2.4 g of paraffin oil for 8 weeks. Dietary intake, physical activity, anthropometric parameters and lipid profile were evaluated. Furthermore, RNA expression of certain gene was also evaluated. The results confirmed that supplementation with a moderate dose of DHA-rich fish oil reduced triglyceride levels in patients with T2DM. Moreover, CD36 expression was upregulated in hypertriglyceridemic subjects after 8 weeks of supplementation with DHA-rich fish oil. CD36, one of PPARγ target genes, is a lipid ligand responsible for recognizing and internalizing modified forms of low-density lipoprotein, including oxLDL and production of triglyceride-rich lipoproteins that are rapidly cleared from the blood.

Another research group are evaluating the effect of fish oil supplementation (3 g/day) on resting energy expenditure, skeletal muscle membrane composition and metabolism in elderly subjects. Incorporation of omega-3 into cell membranes may alter energy metabolism and the release of EPA e DHA into the cytosol which will act as ligands for PPARs regulating genes involved in the metabolism. The authors have the aim of determining the expression and content of PPARs and proteins involved in translocation of fatty acids transport (AMPK, ERK1/2, CamKII). In addition, it will be correlated with body composition, mitochondrial proteins, basal metabolic rate and fat oxidation [51].

Bofutsushosan

The obesity pharmacogenomics is a new and challenging research area, which could be a valuable tool to understand why some patients respond to pharmacological and/or behavioural interventions, while others do not. A single nucleotide polymorphism is a variation in a single base-pair in the DNA sequence of a gene, such as deletion, insertion, or substitution of a base, and represents the most common form of genetic variation. Several genetic polymorphisms in the PPARγ gene have been discovered. Among them, Pro12Ala variant is the missense mutation more prevalent in the PPARγ2 isoform involving a C→G substitution at nucleotide 34, resulting in the replacement of a proline by alanine at residue 12 of the PPARγ2 protein. Several studies have aimed to find a relationship between the Pro12Ala polymorphism and metabolic syndrome, but the results are still not conclusive. For instance, T2DM with the Pro12Ala genotype in the PPAR2 gene exhibited better therapeutic response to rosiglitazone (a TZD class) than patients with Pro12Pro genotype. Korean researches evaluated, in obese subjects, the effects of *Bofutsushosan*, a traditional Eastern Asian herbal medicinal formulation (with 18 crude drugs), in metabolic parameters and its effect on subjects with polymorphism in genes associated with obesity, including Pro12Ala polymorphism of PPARγ2. They concluded that the anti-obesity impact of *Bofutsushosan* was more pronounced in the wild type Pro/Pro genotype than in mutated Pro/Ala variant, where the function of PPARγ2 is thought to be impaired. It suggests the interaction between PPARγ and the ingredients of *Bofutsushosan*, especially Ephedra, that seems to modulate cyclic adenosine monophosphate (cAMP) signalling pathway cascade [52].

Doenjang

Korean researches investigate whether supplementation with fermented soybeans (*doenjang*) affects body fat distribution in overweight subjects with the C and T alleles of the PPARγ polymorphism. The polymorphism C1431T located in exon 6 of PPARγ is associated with susceptibility to cardiovascular diseases, leptin concentrations and body mass index (BMI). Anthropometric measures, laboratory analysis and dietary intakes composed the evaluation of the anti-obesity and antioxidant activity of *doenjang* in a 12-week, double-blind, placebo-controlled, randomized clinical trial. This study suggests that *doenjang* has anti-obesity and antioxidant effects in overweight individuals with mutant alleles of PPARγ, noticing the decrease in visceral fat area of subjects with the mutant T allele that received the supplementation compared to the placebo treatment [53].

The Table **4** shows show the clinical trials since 2014 using new molecules, medications already approved to use, natural products modulators or agonists of PPARγ.

Table 4. 3-year clinical trials involving PPARγ targeted drugs.

Compound	Chemical Structure	Type of Study/ Duration of Follow/ Year of Publication	Population	Aimed / Results	Adverse Events or Problems	Ref.
Pioglitazone		Phase IV Retrospective cohort 2016	56,337 diabetic patients from healthcare databases from Finland, the Netherlands, Sweden, and the United Kingdom	It analysed the incidence of bladder cancer. Without association.		[54]
		Phase IV Retrospective cohort 2014	113,000 Korean diabetic patients	Not associated with bladder cancer		[55]
		Phase IV Follow-up 28 days 2015	1.01 million diabetic people in TZD use. Six populations: British Columbia, Finland, Manchester, Rotterdam, Scotland and UK clinical practice research datalink.	The analysis did not support a causal effect on bladder cancer.	The authors suggested future analyses and a longer follow-up.	[44]
		Phase IV Prospective, event-, multicentre, randomized, double-bind, placebo-controlled 10 years 2016	4,873 patients with pre-existing macrovascular disease and T2DM.	It evaluated effects on cardiovascular disease. Without difference between the groups in incidence of endpoints events.		[43]
		Phase IV Prospective, multi-centre, randomized, double-bind, placebo-controlled 6 years 2014	3,600 patients with pre-existing macrovascular disease and T2DM.	Cardiovascular outcomes in T2DM. No difference was observed.		[56]

Compound	Chemical Structure	Type of Study/ Duration of Follow/ Year of Publication	Population	Aimed / Results	Adverse Events or Problems	Ref.
Pioglitazone		Randomized, double-blind, placebo-controlled 8-week 2014	166 obese subjects.		15 subjects had gastrointestinal problems such as dyspepsia and epigastric pain	[52]
Bofutsushosan		Randomized, double-blind, placebo-controlled 8-week 2015	72 T2DM patients in Iran.	The relationship between the genetic polymorphisms and the anti-obesity impact of *Bofutsushosan*.		[51]
DHA-rich fish oil				It investigated the modulation of some PPARγ-responsive genes	One participant of DHA-rich fish oil group was excluded. He died due to gastrointestinal distress.	[53]
Doenjang		Double-blind, placebo-controlled 12-week 2014	60 adults with BMI > 23kg/m² and a waist/hip ratio > 0.9 for men and > 0.85 for women with mutant alleles of PPARγ.	It evaluated the anti-obesity and antioxidant activity based on mechanisms of PPARγ		[56]
Aleglitazar		Phase III Randomized, double-blind, placebo-controlled 2015	1,581 patients with evidence of stable cardiovascular disease and glucose abnormalities as markers of residual risk.	It determined the reduction of composite outcome of cardiovascular complication.	The trial was stopped prematurely because of unfavourable benefit-risk balance. Hypoglycaemia and muscular events were recorded in patients treated with aleglitazar.	

Compound	Chemical Structure	Type of Study/ Duration of Follow/ Year of Publication	Population	Aimed / Results	Adverse Events or Problems	Ref.
Pioglitazone		Phase II randomized, double-, placebo and active-controlled 24-week 2014	300 patients with T2DM in use of sulfonylurea or sulfonylurea plus metformin on a stable dose.	It compared the efficacy of glycaemic control and side effects		[46]
INT131 besylate		Phase II Multicentre, randomized, double-blind, placebo controlled 24-weeks 2014	173 patients with T2DM South Korean	It assessed the glucose-lowering and lipid modifying effects, as well as the safety profile as a monotherapy		[57]
Lobeglitazone		Phase III prospective, single-arm, open-label clinical trial 24-week 2017	50 patients with T2DM with NAFLD	The effects on alterations in hepatic markers as well as in glycaemic and lipid profiles. Reduction of index of controlled attenuation in NAFLD, improvement of all parameters.	Treatment showed a lower increase in the average body weight of 1.4 kg	[48]
Saroglitazar		Phase III prospective, randomized, double-blind, 2-week run-in period including lifestyle modification to either a sulphonylurea, metformin, or both treatments for at least 3 months 14 sites in India wash out followed by treatment period for 24 weeks with saroglitazar or pioglitazone	122 T2DM and hypertriglyceridemia patients that received	It established therapeutic effect on triglycerides and other lipid and glucose profile with expectation of favourable safety and tolerability in T2DM patients.		[58]

CONCLUSION

Despite the strong impact of obesity and diabetes on the world public health and the consequence that the industrialized and sedentary life style could bring, these perspectives worse year by year, and new strategies are still the main point to be considered by researchers and pharmaceutical industries.

It is important to consider that, despite the knowledge of several mechanisms involved in the pathogenesis of obesity and diabetes, there are many cell signalling pathways to be elucidated to understand how the organism could adapt the metabolism in front of different nutritional situations.

The receptor addressed in this chapter, PPARγ, has been a valuable target to interfere with lipid and glucose metabolism, with important anti-hyperglicaemic effect and interference in adipocyte differentiation. However, together with the therapeutic effects, important side effects have compromised the use of these drugs, including body weight gain and heart failure. Computational chemistry allied to *in-vitro* binding or gene reporter assays are making possible the search of drugs with new concepts of activities, such as multi-targeted compounds. These new strategies will represent, in near future, strong allies with lifestyle changes to guarantee the health of the worldwide population.

CONSENT FOR PUBLICATION

Not applicable.

CONFLICT OF INTEREST

The author (editor) declares no conflict of interest, financial or otherwise.

ACKNOWLEDGMENTS

Declared none.

REFERENCES

[1] Issemann I, Green S. Activation of a member of the steroid hormone receptor superfamily by peroxisome proliferators. Nature 1990; 347(6294): 645-50.
[http://dx.doi.org/10.1038/347645a0] [PMID: 2129546]

[2] Youssef J, Badr MZ. PPARs: history and advances. Methods Mol Biol 2013; 952: 1-6.
[http://dx.doi.org/10.1007/978-1-62703-155-4_1] [PMID: 23100221]

[3] Wang L, Waltenberger B, Pferschy-Wenzig EM, *et al*. Natural product agonists of peroxisome proliferator-activated receptor gamma (PPARγ): a review. Biochem Pharmacol 2014; 92(1): 73-89.
[http://dx.doi.org/10.1016/j.bcp.2014.07.018] [PMID: 25083916]

[4] Dreyer C, Krey G, Keller H, Givel F, Helftenbein G, Wahli W. Control of the peroxisomal beta-oxidation pathway by a novel family of nuclear hormone receptors. Cell 1992; 68(5): 879-87.

[http://dx.doi.org/10.1016/0092-8674(92)90031-7] [PMID: 1312391]

[5] Janani C, Ranjitha Kumari BD. PPAR gamma gene--a review. Diabetes Metab Syndr 2015; 9(1): 46-50.
[http://dx.doi.org/10.1016/j.dsx.2014.09.015] [PMID: 25450819]

[6] Alexander SP, Cidlowski JA, Kelly E, *et al.* CGTP Collaborators. The Concise Guide to PHARMACOLOGY 2015/16: Nuclear hormone receptors. Br J Pharmacol 2015; 172(24): 5956-78.
[http://dx.doi.org/10.1111/bph.13352] [PMID: 26650443]

[7] Choi SS, Park J, Choi JH. Revisiting PPARγ as a target for the treatment of metabolic disorders. BMB Rep 2014; 47(11): 599-608.
[http://dx.doi.org/10.5483/BMBRep.2014.47.11.174] [PMID: 25154720]

[8] Murphy GJ, Holder JC. PPAR-gamma agonists: therapeutic role in diabetes, inflammation and cancer. Trends Pharmacol Sci 2000; 21(12): 469-74.
[http://dx.doi.org/10.1016/S0165-6147(00)01559-5] [PMID: 11121836]

[9] Sauer S. Ligands for the Nuclear Peroxisome Proliferator-Activated Receptor Gamma. Trends Pharmacol Sci 2015; 36(10): 688-704.
[http://dx.doi.org/10.1016/j.tips.2015.06.010] [PMID: 26435213]

[10] Tontonoz P, Hu E, Spiegelman BM. Stimulation of adipogenesis in fibroblasts by PPAR gamma 2, a lipid-activated transcription factor. Cell 1994; 79(7): 1147-56.
[http://dx.doi.org/10.1016/0092-8674(94)90006-X] [PMID: 8001151]

[11] Anghel SI, Wahli W. Fat poetry: a kingdom for PPAR gamma. Cell Res 2007; 17(6): 486-511.
[http://dx.doi.org/10.1038/cr.2007.48] [PMID: 17563755]

[12] Picard F, Auwerx J. PPAR(gamma) and glucose homeostasis. Annu Rev Nutr 2002; 22: 167-97.
[http://dx.doi.org/10.1146/annurev.nutr.22.010402.102808] [PMID: 12055342]

[13] Gross B, Pawlak M, Lefebvre P, Staels B. PPARs in obesity-induced T2DM, dyslipidaemia and NAFLD. Nat Rev Endocrinol 2017; 13(1): 36-49.
[http://dx.doi.org/10.1038/nrendo.2016.135] [PMID: 27636730]

[14] Makki K, Froguel P, Wolowczuk I. Adipose tissue in obesity-related inflammation and insulin resistance: cells, cytokines, and chemokines. ISRN Inflamm 2013; 22: 139239.
[http://dx.doi.org/10.1155/2013/139239]

[15] WHO. Global Health Observatory Map Gallery 2014. Available at: http://gamapserver.who.int/mapLibrary/app/searchResults.aspx

[16] García-Jiménez C, Gutiérrez-Salmerón M, Chocarro-Calvo A, García-Martinez JM, Castaño A, De la Vieja A. From obesity to diabetes and cancer: epidemiological links and role of therapies. Br J Cancer 2016; 114(7): 716-22.
[http://dx.doi.org/10.1038/bjc.2016.37] [PMID: 26908326]

[17] Zimmet P, Alberti KG, Magliano DJ, Bennett PH. Diabetes mellitus statistics on prevalence and mortality: facts and fallacies. Nat Rev Endocrinol 2016; 12(10): 616-22.
[http://dx.doi.org/10.1038/nrendo.2016.105] [PMID: 27388988]

[18] CDC - Nation's Health Protection Agency - Division of Diabetes Translation. Maps of Trends in Diagnosed Diabetes and Obesity 2016. Available at: http://www.cdc.gov/diabetes/data

[19] Kusminski CM, Bickel PE, Scherer PE. Targeting adipose tissue in the treatment of obesity-associated diabetes. Nat Rev Drug Discov 2016; 15(9): 639-60.
[http://dx.doi.org/10.1038/nrd.2016.75] [PMID: 27256476]

[20] Inagaki T, Sakai J, Kajimura S. Transcriptional and epigenetic control of brown and beige adipose cell fate and function. Nat Rev Mol Cell Biol 2016; 17(8): 480-95.
[http://dx.doi.org/10.1038/nrm.2016.62] [PMID: 27251423]

[21] Santos GM, Neves FdeA, Amato AA. Thermogenesis in white adipose tissue: An unfinished story

about PPARγ. Biochim Biophys Acta 2015; 1850(4): 691-5.
[http://dx.doi.org/10.1016/j.bbagen.2015.01.002] [PMID: 25583560]

[22] Yin R, Huang H, Zhang J, Zhu J, Jing H, Li Z. Dietary n-3 fatty acids attenuate cardiac allograft vasculopathy *via* activating peroxisome proliferator-activated receptor-gamma. Pediatr Transplant 2008; 12(5): 550-6.
[http://dx.doi.org/10.1111/j.1399-3046.2007.00849.x] [PMID: 18466197]

[23] Zhang J, Liu X, Xie XB, Cheng XC, Wang RL. Multitargeted bioactive ligands for PPARs discovered in the last decade. Chem Biol Drug Des 2016; 88(5): 635-63.
[http://dx.doi.org/10.1111/cbdd.12806] [PMID: 27317624]

[24] Wu G, Yi J, Liu L, Wang P, Zhang Z, Li Z. Pseudoginsenoside F11, a Novel Partial PPARγ Agonist, Promotes Adiponectin Oligomerization and Secretion in 3T3-L1 Adipocytes. PPAR Res 2013; 701017.

[25] Zhang Y, Yu L, Cai W, *et al.* Protopanaxatriol, a novel PPARγ antagonist from Panax ginseng, alleviates steatosis in mice. Sci Rep 2014; 4: 7375.
[http://dx.doi.org/10.1038/srep07375] [PMID: 25487878]

[26] Mueller M, Lukas B, Novak J, Simoncini T, Genazzani AR, Jungbauer A. Oregano: a source for peroxisome proliferator-activated receptor gamma antagonists. J Agric Food Chem 2008; 56(24): 11621-30.
[http://dx.doi.org/10.1021/jf802298w] [PMID: 19053389]

[27] Cornick CL, Strongitharm BH, Sassano G, *et al.* Identification of a novel agonist of peroxisome proliferator-activated receptors alpha and gamma that may contribute to the anti-diabetic activity of guggulipid in Lep(ob)/Lep(ob) mice. J Nutr Biochem 2009; 20(10): 806-15.
[http://dx.doi.org/10.1016/j.jnutbio.2008.07.010] [PMID: 18926687]

[28] Kubo M, Fukui M, Ito Y, *et al.* Insulin sensitization by a novel partial peroxisome proliferator-activated receptor γ agonist with protein tyrosine phosphatase 1B inhibitory activity in experimental osteoporotic rats. J Pharmacol Sci 2014; 124(2): 276-85.
[http://dx.doi.org/10.1254/jphs.13236FP] [PMID: 24553405]

[29] Xie X, Zhou X, Chen W, *et al.* L312, a novel PPARγ ligand with potent anti-diabetic activity by selective regulation. Biochim Biophys Acta 2015; 1850(1): 62-72.
[http://dx.doi.org/10.1016/j.bbagen.2014.09.027] [PMID: 25305559]

[30] Seto S, Okada K, Kiyota K, *et al.* Design, synthesis, and structure-activity relationship studies of novel 2,4,6-trisubstituted-5-pyrimidinecarboxylic acids as peroxisome proliferator-activated receptor gamma (PPARgamma) partial agonists with comparable antidiabetic efficacy to rosiglitazone. J Med Chem 2010; 53(13): 5012-24.
[http://dx.doi.org/10.1021/jm100443s] [PMID: 20527969]

[31] Acton JJ III, Black RM, Jones AB, *et al.* Benzoyl 2-methyl indoles as selective PPARgamma modulators. Bioorg Med Chem Lett 2005; 15(2): 357-62.
[http://dx.doi.org/10.1016/j.bmcl.2004.10.068] [PMID: 15603954]

[32] Kim MK, Chae YN, Choi SH, *et al.* PAM-1616, a selective peroxisome proliferator-activated receptor γ modulator with preserved anti-diabetic efficacy and reduced adverse effects. Eur J Pharmacol 2011; 650(2-3): 673-81.
[http://dx.doi.org/10.1016/j.ejphar.2010.10.044] [PMID: 20974124]

[33] Kim J, Han DC, Kim JM, *et al.* PPAR gamma partial agonist, KR-62776, inhibits adipocyte differentiation *via* activation of ERK. Cell Mol Life Sci 2009; 66(10): 1766-81.
[http://dx.doi.org/10.1007/s00018-009-9169-4] [PMID: 19347570]

[34] Zhang H, Ding CZ, Lai Z, *et al.* Synthesis and biological evaluation of novel pyrrolidine acid analogs as potent dual PPARα/γ agonists. Bioorg Med Chem Lett 2015; 25(6): 1196-205.
[http://dx.doi.org/10.1016/j.bmcl.2015.01.066] [PMID: 25686852]

[35] Oon Han H, Kim SH, Kim KH, *et al.* Design and synthesis of oxime ethers of alpha-acyl-be-

α-phenylpropanoic acids as PPAR dual agonists. Bioorg Med Chem Lett 2007; 17(4): 937-41.
[http://dx.doi.org/10.1016/j.bmcl.2006.11.050] [PMID: 17157019]

[36] Park MH, Park JY, Lee HJ, *et al.* Potent anti-diabetic effects of MHY908, a newly synthesized PPAR α/γ dual agonist in db/db mice. PLoS One 2013; 8(11): e78815.
[http://dx.doi.org/10.1371/journal.pone.0078815] [PMID: 24244369]

[37] Shibata Y, Kagechika K, Yamaguchi M, Kubo H, Usui H. Design, synthesis and evaluation of novel zwitterionic compounds as PPARα/γ dual agonists (1). Bioorg Med Chem Lett 2012; 22(23): 7075-9.
[http://dx.doi.org/10.1016/j.bmcl.2012.09.092] [PMID: 23084275]

[38] Laurent D, Gounarides JS, Gao J, Boettcher BR. Effects of cevoglitazar, a dual PPARalpha/gamma agonist, on ectopic fat deposition in fatty Zucker rats. Diabetes Obes Metab 2009; 11(6): 632-6.
[http://dx.doi.org/10.1111/j.1463-1326.2008.01017.x] [PMID: 19175377]

[39] He BK, Ning ZQ, Li ZB, *et al.* *In Vitro* and *In Vivo* Characterizations of Chiglitazar, a Newly Identified PPAR Pan-Agonist. PPAR Res 2012; 546548.

[40] Amato AA, Rajagopalan S, Lin JZ, *et al.* GQ-16, a novel peroxisome proliferator-activated receptor γ (PPARγ) ligand, promotes insulin sensitization without weight gain. J Biol Chem 2012; 287(33): 28169-79.
[http://dx.doi.org/10.1074/jbc.M111.332106] [PMID: 22584573]

[41] Silva JC, César FA, de Oliveira EM, *et al.* New PPARγ partial agonist improves obesity-induced metabolic alterations and atherosclerosis in LDLr(-/-) mice. Pharmacol Res 2016; 104: 49-60.
[http://dx.doi.org/10.1016/j.phrs.2015.12.010] [PMID: 26706782]

[42] Fröhlich E, Wahl R. Chemotherapy and chemoprevention by thiazolidinediones. Biomed Res Int 2015; 845340.
[http://dx.doi.org/10.1155/2015/845340]

[43] Balaji V, Seshiah V, Ashtalakshmi G, Ramanan SG, Janarthinakani M. A retrospective study on finding correlation of pioglitazone and incidences of bladder cancer in the Indian population. Indian J Endocrinol Metab 2014; 18(3): 425-7.
[http://dx.doi.org/10.4103/2230-8210.131223] [PMID: 24944944]

[44] Levin D, Bell S, Sund R, *et al.* Scottish Diabetes Research Network Epidemiology Group; Diabetes and Cancer Research Consortium. Pioglitazone and bladder cancer risk: a multipopulation pooled, cumulative exposure analysis. Diabetologia 2015; 58(3): 493-504.
[http://dx.doi.org/10.1007/s00125-014-3456-9] [PMID: 25481707]

[45] Erdmann E, Harding S, Lam H, Perez A. Ten-year observational follow-up of PROactive: a randomized cardiovascular outcomes trial evaluating pioglitazone in type 2 diabetes. Diabetes Obes Metab 2016; 18(3): 266-73.
[http://dx.doi.org/10.1111/dom.12608] [PMID: 26592506]

[46] DePaoli AM, Higgins LS, Henry RR, Mantzoros C, Dunn FL. INT131-007 Study Group. Can a selective PPARγ modulator improve glycemic control in patients with type 2 diabetes with fewer side effects compared with pioglitazone? Diabetes Care 2014; 37(7): 1918-23.
[http://dx.doi.org/10.2337/dc13-2480] [PMID: 24722496]

[47] Erdmann E, Califf R, Gerstein HC, *et al.* Effects of the dual peroxisome proliferator-activated receptor activator aleglitazar in patients with Type 2 Diabetes mellitus or prediabetes. Am Heart J 2015; 170(1): 117-22.
[http://dx.doi.org/10.1016/j.ahj.2015.03.021] [PMID: 26093872]

[48] Sosale A, Saboo B, Sosale B. Saroglitazar for the treatment of hypertrig-lyceridemia in patients with type 2 diabetes: current evidence. Diabetes Metab Syndr Obes 2015; 8: 189-96.
[http://dx.doi.org/10.2147/DMSO.S49592] [PMID: 25926748]

[49] Lee YH, Kim JH, Kim SR, *et al.* Lobeglitazone, a Novel Thiazolidinedione, Improves Non-Alcoholic Fatty Liver Disease in Type 2 Diabetes: Its Efficacy and Predictive Factors Related to Responsiveness.

J Korean Med Sci 2017; 32(1): 60-9.
[http://dx.doi.org/10.3346/jkms.2017.32.1.60] [PMID: 27914133]

[50] Grygiel-Górniak B. Peroxisome proliferator-activated receptors and their ligands: nutritional and clinical implications--a review. Nutr J 2014; 13: 17.
[http://dx.doi.org/10.1186/1475-2891-13-17] [PMID: 24524207]

[51] Mansoori A, Sotoudeh G, Djalali M, *et al.* Effect of DHA-rich fish oil on PPARγ target genes related to lipid metabolism in type 2 diabetes: A randomized, double-blind, placebo-controlled clinical trial. J Clin Lipidol 2015; 9(6): 770-7.
[http://dx.doi.org/10.1016/j.jacl.2015.08.007] [PMID: 26687697]

[52] Park J, Bose S, Hong SW, *et al.* Impact of GNB3-C825T, ADRB3-Trp64Arg, UCP2-3'UTR 45 bp del/ins, and PPARγ-Pro12Ala polymorphisms on Bofutsushosan response in obese subjects: a randomized, double-blind, placebo-controlled trial. J Med Food 2014; 17(5): 558-70.
[http://dx.doi.org/10.1089/jmf.2013.2836] [PMID: 24827746]

[53] Cha YS, Park Y, Lee M, *et al.* Doenjang, a Korean fermented soy food, exerts antiobesity and antioxidative activities in overweight subjects with the PPAR-γ2 C1431T polymorphism: 12-week, double-blind randomized clinical trial. J Med Food 2014; 17(1): 119-27.
[http://dx.doi.org/10.1089/jmf.2013.2877] [PMID: 24456362]

[54] Korhonen P, Heintjes EM, Williams R, *et al.* Pioglitazone use and risk of bladder cancer in patients with type 2 diabetes: retrospective cohort study using datasets from four European countries. BMJ 2016; 354: i3903.
[http://dx.doi.org/10.1136/bmj.i3903] [PMID: 27530399]

[55] Jin SM, Song SO, Jung CH, *et al.* Risk of bladder cancer among patients with diabetes treated with a 15 mg pioglitazone dose in Korea: a multi-center retrospective cohort study. J Korean Med Sci 2014; 29(2): 238-42.
[http://dx.doi.org/10.3346/jkms.2014.29.2.238] [PMID: 24550651]

[56] Erdmann E, Song E, Spanheimer R, van Troostenburg de Bruyn AR, Perez A. Observational follow-up of the PROactive study: a 6-year update. Diabetes Obes Metab 2014; 16(1): 63-74.
[http://dx.doi.org/10.1111/dom.12180] [PMID: 23859428]

[57] Kim SG, Kim DM, Woo JT, *et al.* Efficacy and safety of lobeglitazone monotherapy in patients with type 2 diabetes mellitus over 24-weeks: a multicenter, randomized, double-blind, parallel-group, placebo controlled trial. PLoS One 2014; 9(4): e92843.
[http://dx.doi.org/10.1371/journal.pone.0092843] [PMID: 24736628]

[58] Pai V, Paneerselvam A, Mukhopadhyay S, *et al.* A Multicenter, Prospective, Randomized, Double-blind Study to Evaluate the Safety and Efficacy of Saroglitazar 2 and 4 mg Compared to Pioglitazone 45 mg in Diabetic Dyslipidemia (PRESS V). J Diabetes Sci Technol 2014; 8(1): 132-41.
[http://dx.doi.org/10.1177/1932296813518680] [PMID: 24876549]

CHAPTER 7

Hydrogen Sulfide and Carbohydrate Metabolism

Zahra Bahadoran[1], Parvin Mirmiran[1] and Asghar Ghasemi[2,*]

[1] *Nutrition and Endocrine Research Center, Research Institute for Endocrine Sciences, Shahid Beheshti University of Medical Sciences, Tehran, Iran*

[2] *Endocrine Physiology Research Center, Research Institute for Endocrine Sciences, Shahid Beheshti University of Medical Sciences, Tehran, Iran*

Abstract: Hydrogen sulfide (H_2S) is an important gasotransmitter with diverse biological actions in the body and has been receiving much attention over the last two decades. It has been characterized as a key regulator of cardiovascular homoeostasis, cell growth and differentiation, mitochondrial biogenesis, adipose tissue metabolism, inflammation and liver function. H_2S-donor molecules have hence been considered as being potential therapeutic options for a variety of human diseases including hypertension, atherosclerosis, obesity, oxidative stress and chronic inflammation. It has been shown that huge amount of endogenous H_2S originates in the liver, and may be involved in the development of insulin resistance and diabetes. H_2S is considered as an important mediator of carbohydrate homeostasis. H_2S production and bioavailability are impaired during development of obesity, diabetes and its complications, highlights the potential therapeutic effects of H_2S in metabolic syndrome. This issue is however controversial due to some findings that show increased H_2S disturbs pancreatic β-cell function and may be responsible for reduced insulin secretion. H_2S also contributes to increased blood glucose levels by accelerating glycogenolysis and gluconeogenesis, effects which could intensify hyperglycemia in diabetes. Furthermore, reduced basal and insulin-stimulated glucose uptake was observed following treatment of adipocytes with H_2S; in contrast, the protective effect of H_2S on β-cell function against a high-fat diet, as well as its insulin-sensitizing properties has been reported in both *in vitro* and *in vivo* models of insulin resistance. Regarding the increasing interest in therapeutic applications of H_2S-donors in cardiometabolic disorders, its potential unexpected effects on glucose/insulin metabolism, especially in the case of diabetes, should be considered. In this review, we focus on the current knowledge available on exogenous and endogenous H_2S and carbohydrate metabolism, including both regulation of hepatic glucose production and hepatic and peripheral glucose uptake and β-cell function.

Keywords: Carbohydrate metabolism, Cystathionine γ lyase, Cystathionine β

---

[*] **Corresponding author Asghar Ghasemi:** Endocrine Physiology Research Center, Research Institute for Endocrine Sciences, Shahid Beheshti University of Medical Sciences, Tehran, Iran; Tel: +982122432500; Fax: +982122416264; Email: Ghasemi@endocrine.ac.ir

Atta-ur-Rahman (Ed.)
All rights reserved-© 2019 Bentham Science Publishers

synthase, Diabetes, Glucose, Hydrogen sulfide, Insulin, Nitric oxide, Sodium sulfide, 3-mercaptopyruvate sulfurtransferase.

INTRODUCTION

Hydrogen sulfide (H_2S) has recently become a molecule of high interest, and is now highlighted as the third gasotransmitter, following nitric oxide (NO) and carbon monoxide (CO) [1]. H_2S has been characterized as a key regulator of homoeostasis of various systems in the human body, including cardiovascular, neuronal, gastrointestinal, respiratory, renal, liver and reproductive systems [2]. It also plays an important role in regulation of cell growth and differentiation, mitochondrial biogenesis, adipose tissue metabolism, and inflammation [2, 3].

Currently, potential clinical applications of H_2S-releasing drugs have been highlighted as H_2S exerts a number of cytoprotective and anti-inflammatory effects on many organs; a number of H_2S-releasing derivatives of existing drugs have been therefore developed and extensively tested in preclinical models [4]. The well-known mechanisms documented for the protective effects of H_2S are increased glutathione (GSH) levels, decreased mitochondrial superoxide production, activation of reactive oxygen species (ROS) scavengers, modulation of mitochondrial activity and preservation of its integrity, modulation of vascular proliferation and leukocyte adhesion, modulation of cytokine production, activation of K_{ATP} and Ca^{+2}-activated K^+ channels, modulation of apoptotic signaling, and regulation of NO biosynthesis [1].

It is believed that H_2S exerts its biological properties mainly by S-sulfhydration of target proteins, a reaction in which sulfur is added to the thiol groups of reactive cysteine residues of proteins to form hydropersulfide [5]. S-sulfhydration of a protein modifies its functions, localization inside the cells, stability in cells, and resistance to oxidative stress [6]. S-sulfhydration by H_2S is involved in modification of inflammation, endoplasmic reticulum (ER) stress signaling and vascular tone [6].

Current approaches to change H_2S levels in experimental settings include modulating the expression or activity of H_2S-producing enzymes and also its exposure to sodium sulfide salts (NaSH, Na_2S), gaseous H_2S, slow-releasing H_2S donors, hybrids of H_2S donors (non-steroid anti-inflammatory hybrid drugs like S-aspirin, a H_2S-releasing form of aspirin), and cysteine analogues (S-propyl cysteine, S-allyl cysteine, S-propargyl cysteine, N-acetyl cysteine) [1, 7]. Naturally occurring H_2S compounds including garlic (*Allium sativum*), sulforaphane, erucin and iberin, are also considered as a class of H_2S-donating compounds [8].

AN OVERVIEW ON H$_2$S BIOSYNTHESIS AND METABOLISM

H$_2$S is a naturally occurring colorless gas in the body [9]. H$_2$S is slightly acidic with a pK$_{a1}$ value of 6.76 and pK$_{a2}$ >12 at 37°C [9, 10], *i.e.* only 18.5% of H$_2$S remains undissociated at physiologic pH=7.40 [9]. In aqueous solution, H$_2$S dissociates to hydrosulfide anions (HS$^-$, pK$_a$ 7.04 and S^{2-}, pK$_a$ 11.96). H$_2$S is highly lipophilic and its solubility in lipophilic solvents is 5-fold higher than in water [9] and therefore can freely penetrate cells [9]. Solubility of H$_2$S in water is 80 mM at body temperature [11]. High concentration of H$_2$S (> 700 ppm or ~ 20000 μM) is acutely lethal in both human and animals [12]. Circulating levels of H$_2$S are ~ 50 μM in both humans and animals [9, 13] while its concentrations in mammalian tissues were reported to range between 50-160 μM [9, 14]. Sulfan sulfur is the major storage pool of H$_2$S [9].

H$_2$S is produced *via* enzymatic and non-enzymatic pathways [15]. In mammalian tissues, non-enzymatic pathways are responsible for a limited amount of H$_2$S production [16]. Glucose, GSH, inorganic and organic poly-sulfides (present in garlic) and elemental sulfur are involved in non-enzymatic synthesis pathway of H$_2$S [17]; Searcy *et al.* proposed that H$_2$S can be generated from glucose either *via* glycolysis or from phosphogluconate *via* nicotinamide adenine dinucleotide phosphate (NADPH) oxidase [17]. Glucose produces H$_2$S in reaction with methionine, homocysteine or cysteine. Direct reduction of GSH and elemental sulfur, by reducing the nicotinamide adenine dinucleotide (NADH) or NADPH, produces H$_2$S equivalents of the glucose oxidation pathway [15, 17].

Enzymatic production of H$_2$S is organ-specific and is catalyzed by pyridoxal-5′-phosphate (PLP) dependent enzymes, cystathionine β synthase (CBS) and cystathionine γ lyase (CSE), and a non-PLP dependent enzyme, 3-mercaptopyruvate sulfurtransferase (3-MST) [15, 18, 19] (Table **1**).

Table 1. Enzymes involved in H$_2$S production in the body.

Enzymes	EC	Location	Distribution
Cystathionine-γ-lyase (CSE)	4.4.1.1	Cytosol	Cardiovascular system (heart, aorta, mesenteric artery, pulmonary artery, and portal vein) [9, 20, 40, 41], small intestine [9, 14, 20], stomach [9], liver [9, 20, 41], kidney [9, 20, 41], uterus [20], brain [20], pancreatic islets [20], placenta [20], penile tissue [20]
Cystathionine-β-synthetase (CBS)	4.2.1.22	Cytosol and mitochondria	Brain [14, 20, 40, 41], CNS [9], liver [9, 20], kidney [9, 20], pancreatic islets [20], ileum [20], uterus [20], placenta [20], not detectable in the blood vessels [40], penile tissue [20]

Enzymes	EC	Location	Distribution
3-Mercaptopyruvate sulfortransferase (3-MST)	2.8.1.2	Cytosol and mitochondria	Brain [9, 20] vascular endothelium [9], liver [20], kidney [20], heart [20], lung [20], thymus [20], testis [20], thoracic aorta [20]

The major source for H_2S production in mammals is L-cysteine, which is desulfhydrated during the trans-sulfuration pathway by CBS, CSE, and 3-MST [18, 19]. H_2S can also be produced from cystathionine by CSE [18, 19]. CBS also produces cystathionine from serine and homocysteine. In normal conditions, ~70% of H_2S is produced from cysteine and ~30% from homocysteine [20]. L-Cysteine, along with α-ketoglutarate (α-KG), is converted by cystine aminotransferase (CAT) into 3-mercaptopyruvate (3-MP) which can then be broken down by 3-MST to form H_2S [18, 19]. Fig. (1) summarizes biosynthesis pathways of H_2S.

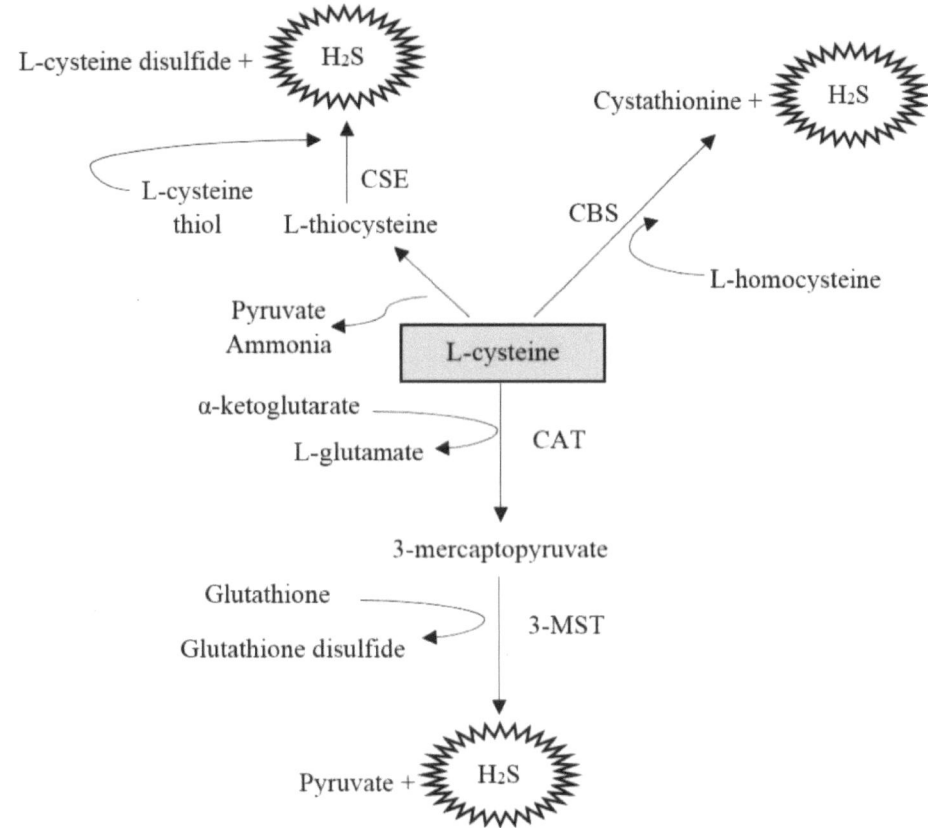

Fig. (1). Biosynthesis pathways of H_2S. CSE, cystathionine-γ-lyase; CBS, cystathionine-β-synthase; CAT, cysteine aminotransferase; 3-MST 3-mercaptopyruvate sulfurtransferase.

CBS is expressed in several organs and tissues including liver, kidney, brain, ileum, uterus, placenta and pancreatic islets [18, 19]. CBS and CSE are hemeproteins primarily localized in the cytoplasm, while 3-MST is a zinc-dependent protein, being in the cytosol and in particular in mitochondrial matrix, with an optimal pH of ~8 [15, 21]. $_{DL}$-Propargylglycine (PPG) and β-cyanol-$_L$-alanine (BCA) are characterized as the main CSE inhibitors; BCA is more potent in inhibiting CSE than PPG [22]. L-aminoethoxyvinylglycine (AVG) and aminooxyacetic acid (AOAA) have been characterized as other inhibitors of CSE; AOAA can also inhibit CBS and have a more potent inhibitory effect on CSE, compared to BCA and PPG [22].

An additional pathway for H_2S production has also been reported from D-cysteine [23], in which H_2S is produced from 3-MP provided by D-amino acid oxidase; this pathway seems to operate predominantly in the cerebellum and the kidney and is different from the L-cysteine pathway with respect to optimal pH and PLP-dependency [24]. It has been shown that D-cysteine protects cerebellar neurons from oxidative stress and in the kidney, attenuates ischemia-reperfusion injury more effectively than L-cysteine [23].

As an allosteric activator of CBS, S-adenosylmethionine (SAM) enhances CBS activity, whereas NO and CO suppress it [25, 26]. The expression of CSE has been found to be regulated by the specificity protein 1 (SP1) transcription factor [27, 28]; phosphorylation of SP1 by extracellular-signal-regulated kinase (ERK) reduces its activity [29]. It has been shown that high glucose concentrations stimulate the phosphorylation of SP1 *via* p38 mitogen-activated protein kinase (MAPK) activation, and subsequently decrease CSE promoter activity, and down-regulate CSE gene expression [28]. Accordingly, inhibition of H_2S production *via* the glucose-mediated decrease in CSE activity has been suggested to be involved in the fine control of glucose-induced insulin secretion [28]. In contrast, CSE expression through SP1 can be activated by the tumor necrosis factor-α (TNF-α) [30]. Expression of CSE can be modulated by the nuclear factor (erythroid-derived 2)-like 2 (Nrf2), an oxidative stress-sensitive transcription factor which mediates transcription of protective genes, including glutathionine S-transferase or heme oxygenase [1]. Data indicate that sumoylation and carbonylation can also cause post-translational modification of both CSE and CBS [31].

Unlike CBS and CSE, 3-MST activity needs reducing substances such as dithiothreitol, thioredoxin, dihydrolipoic acid, GSH, NADPH, NADH, and coenzyme A (CoA) [32].

Evidence indicates that hormones regulate CBS and CSE expression [33, 34]. Insulin decreases the expression and activity of CBS and CSE in the pancreas and liver [35], whereas glucagon and glucocorticoids up-regulate their expression [36].

Briefly, modulation of SP1, Nrf2, farnesoid X receptor (FXR) responsive element, and micro-RNA 21 are currently considered potential targets for modulation of the expression of CSE and CBS [37].

H_2S freely diffuses through membranes without utilizing specific transporters and exerts its biological effects on the target tissues [38]. Inactivation of H_2S mainly occurs *via* mitochondrial oxidation during three consecutive reactions; two membrane-bound sulfide:quinone oxidoreductases are involved in oxidation of sulfide to elemental sulfur [39].

H_2S METABOLISM IN DIABETES

Animal Studies

The potential role of H_2S in pathophysiology of diabetes is highly complex. Some evidence implies contribution of endogenous H_2S in the onset and progression of diabetes. Yusuf *et al*. reported a significant increase in CSE and CBS mRNAs, as well as elevated endogenous H_2S production in both the liver and pancreas of streptozotocin-induced diabetic rats, while renal activities of these enzymes remained unchanged [35]. Likewise, Jacobs *et al*. also reported increased activities of the hepatic CBS and CSE in untreated diabetic rats where insulin treatment restored elevated activities to normal [42]. However, it has been suggested that in the early stage of diabetes development, high-glucose-induced pancreatic CSE overexpression may serve as a protective mechanism, as H_2S can neutralize oxidative/nitrosative stress [43].

Exposure of pregnant rats to 20, 50, and 75 PPM (600, 1500, and 2200 µM, respectively) of H_2S from day 1 of gestation until day 21 postpartum (7 hours per day) increased serum glucose levels at day 21 postpartum [44]. Wu *et al*. showed that pancreatic CSE expression and H_2S production are higher in Zucker diabetic fatty (ZDF) rats (a diabetic model with obesity and hyperinsulinemia) than in Zucker fatty and Zucker lean rats [45]; inhibition of pancreatic H_2S production with daily intraperitoneal injections of PPG for 4 weeks in ZDF rats significantly increased serum insulin levels and reduced hyperglycemia [45]. There is a strong correlation between high H_2S levels and disrupted insulin release in pancreatic β-cells [46], suggesting that inhibition of pancreatic H_2S production may be a new therapeutic option for the management of diabetes by increasing insulin secretion [45, 46].

Conversely, Wilinski *et al*. showed that administration of metformin in diabetic rats increases H_2S concentration in the liver, kidney, heart and brain [47]. In addition, endogenous H_2S production was significantly impaired in non-obese diabetic (NOD) mice and plasma H_2S levels paralleled with the severity of diabetes [48]. Hyperglycemia produces H_2S deficiency in endothelial cells in both *in vitro* and *in vivo* conditions and administration of H_2S could be a potential therapy in hyperglycemia [49].

Overall, current literature on the pro-diabetic or anti-diabetic effects of H_2S is inconclusive and further studies are needed to elucidate the potential role of H_2S manipulation in treatment of diabetes.

Human Evidence

Plasma H_2S concentrations in humans have been reported to range between 30–100 µM [50]. Limited data are however available regarding plasma levels of H_2S and possible changes during the onset and development of type 2 diabetes. Suzuki *et al.* reported decreased plasma H_2S levels in patients with type 2 diabetes, in particular those with a history of cardiovascular disease [51]. In this study, plasma H_2S levels were negatively correlated with hemoglobin A1c (HbA1c), duration of diabetes, and systolic and diastolic blood pressures [51]. Similarly, in a study by Jain *et al.*, diabetic patients had significantly lower blood levels of H_2S compared with their age-matched normal control subjects [52]. Median plasma H_2S levels (25th-75th percentiles) in age-matched lean, overweight and type 2 diabetic patients were reported to be 38.9 (29.7-45.1), 22.0 (18.6-26.7) and 10.5 (4.8-22.0) µmol/L, respectively; in addition, waist circumference was an independent predictor of plasma H_2S (r^2= 0.42, standardized β= -0.65, P<0.001) and diabetes contributed only a further 5% to the model (r^2=0.47) [53]. In this study, plasma H_2S was negatively correlated with fasting glucose (r= -0.49, P=0.001), HbA1c (r= -0.42, P=0.006) and insulin sensitivity (as measured by insulin tolerance test, r= -0.49, P=0.004) and positively correlated with insulin resistance (calculated by the homeostatic model assessment of insulin resistance, r= 0.56, P<0.001) [53].

Regarding the association between diabetes and H_2S deficiency [52, 53], H_2S modulation has potential therapeutic effects in diabetes, a scenario similar to one has previously reported for NO, another gasotransmitter, in diabetes. Diabetes is associated with NO deficiency [54] and NO donors like nitrate and nitrite have been proposed as new therapeutic agents for managing type 2 diabetes [55 - 58].

H_2S AND GLUCOSE/INSULIN HOMEOSTASIS

H_2S and Glucose Output and Utilization in Hepatocytes

Hepatic glycogenolysis (biochemical breakdown of glycogen to glucose) and gluconeogenesis (generation of glucose from non-carbohydrate carbon substrates including lactate, glycerol, and glucogenic amino acids) contribute to hepatic glucose output [59]. In type 2 diabetes, hepatic glucose output is increased by an elevated net hepatic gluconeogenesis [60]. Chronic increased blood glucose in poorly controlled diabetic patients is mainly attributed to increased rate of gluconeogenesis [61].

H_2S which plays a crucial role in glucose metabolism and insulin signaling in hepatocytes, activates gluconeogenesis and glycogenolysis and stimulates glucose production [62]. In ZDF rats, overexpression of hepatic CBS has been reported in both prediabetic insulin-resistance and frank diabetic stages [63]. However, insulin, at physiological concentrations, inhibits CSE expression in HepG2 cells [64].

Incubation of the insulin-resistant $HepG_2$ cell model [HepG2 cells incubated with 500 nM insulin for 24 h with high-glucose DMEM (33 mM)] with NaSH (10 μM), significantly decreased glucose uptake in both basal and insulin-stimulated states, decreased glycogen storage, and enhanced gluconeogenesis and glycogenolysis. Inactivation of the AMP-activated protein kinase (AMPK) signaling pathway and suppression of glucokinase (GK) activity contribute to this effect [64]. Glycogen content of hepatocytes also reduced in a dose-dependent manner (up to 42%) via both decreased glucose uptake and increased glycogenolysis. H_2S also caused hyperglycemia by increasing gluconeogenesis, an effect on glycogenolysis is believed to be stronger than that exerted on gluconeogenesis [64]. Fig. (**2**) shows effects of H_2S on glucose metabolism in hepatocytes; as shown, H_2S-induced glucose production in hepatocytes may be a result of inhibition of Akt (a serine/threonine protein kinase) signaling which is followed by activation of phosphoenolpyruvate carboxykinase (PEPCK) [62]. Increased activity of PEPCK by H_2S is also reported to be induced by increased glucocorticoid receptor (GR) activity and decreased AMPK phosphorylation [65].

Glucose-6-phosphatase (G6Pase) and fructose-1, 6-bisphosphatase (F1-6Pase) in hepatocytes are also affected by H_2S [66]; gene expression of the rate-limiting gluconeogenic enzymes, G6Pase and F1-6Pase, is indirectly enhanced by H_2S through S-sulfhydration of the peroxisome proliferator-activated receptor gamma coactivator 1-α (PGC-1α) [65]. H_2S also directly promotes the activity of these enzymes by S-sulfhydration [67]. H_2S-induced glucose production is also mediated by S-sulfhydration and increased activity of pyruvate carboxylase (PC), a key enzyme providing fuel for gluconeogenesis [68]. H_2S-induced decreased glucose utilization in the hepatocytes, mediated by inactivation of AMPK and GK, is also another pathway known to be involved in increased hepatic glucose output [64].

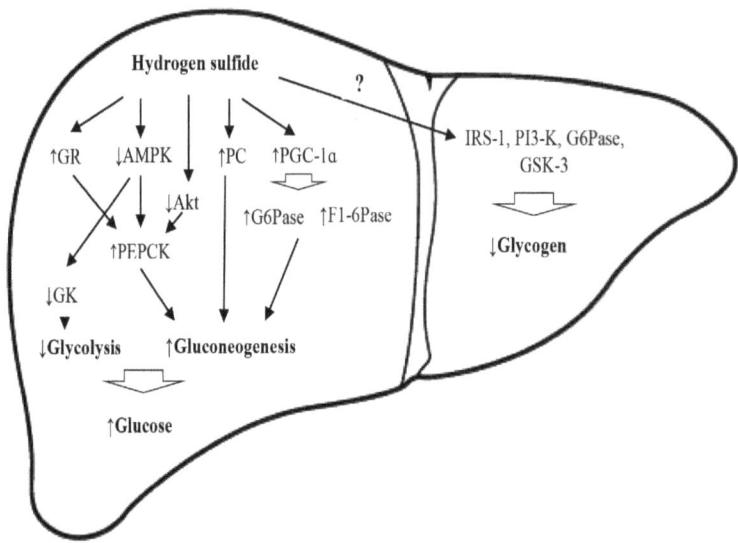

Fig. (2). H_2S-targeted pathways of glucose metabolism in the hepatocytes. Akt, serine/threonine protein kinase; G6Pase, glucose-6-phosphatase; F1-6Pase, fructose-1, 6-bisphosphatase; PGC-1α, peroxisome proliferator-activated receptor gamma coactivator 1-α; PC, pyruvate carboxylase; GK, glucokinase; PEPCK, phosphoenolpyrovate carboxykinase; AMPK, AMP-activated protein kinase

Overexpression of CSE in hepatocytes decreases glycogen content; in contrast, CSE-deficient animals have a markedly increased liver glycogen content [64]. It has been proposed that G6Pase, a key enzyme in glycogenolysis, is targeted by H_2S in this pathway [66]. Other signaling pathways including the phosphoinositide 3-kinase (PI3-K)/Akt-signaling pathway, and its upstream or downstream signaling pathways [*e.g.* insulin receptor substrate 1 (IRS-1) and glycogen synthase kinase-3 (GSK-3)] may also be involved in modifying the glycogen content of hepatocytes [64].

H_2S and Glucose Metabolism in Adipose Tissue

Both CBS and CSE contribute to adipocyte-originated H_2S, although CSE has been proposed as the main H_2S synthase in adipose tissue [69]. In rats, the rate of H_2S production in epididymal (4.76 nmol/min/protein) and brown (4.65 nmol/min/protein) fat tissues was higher than perirenal fat tissue (2.93 nmol/min/protein); in cultured cells, H_2S-generating rates in adipocytes and preadipocytes were 2.89 and 2.17 nmol/min/mg protein [69].

H_2S production in adipose tissue is affected by physiologic (age) and pathophysiologic (hyperglycemia, obesity, inflammation) factors. Expression of

CSE and H$_2$S production in adipocytes is up-regulated by aging whereas increased glucose concentration from physiological (5 mM) to hyperglycemic range (20 mM) down-regulates the CSE-H$_2$S system, in a time- and dose-dependent manner [69]. Hyperglycemia-induced generation of ROS seems to be involved in suppression of CSE in adipocytes [69]. Although CSE expression has been shown to be increased in adipose tissue macrophages (ATMs) in obese animal models, substantial reduction of endogenous H$_2$S concentrations in ATMs indicate a decreased bioavailability of H$_2$S in obesity [70]. Lipopolysaccharides (LPS)-induced inflammation in RAW264.7 cells (a mouse macrophage cell line), was associated with increased functional CSE protein, whereas H$_2$S concentration was reduced due to increased cellular demand for and consumption of H$_2$S [70]. Obesity-induced H$_2$S deficiency activates the store-operated Ca^{2+} entry (SOCE) pathway in adipose tissue macrophages and increases cytokine production [70]. Hyperglycemia- and obesity-induced decreased bioavailability of H$_2$S has accordingly been proposed to play a key role in the development of metabolic syndrome [71].

H$_2$S plays a regulatory role in adipogenesis (preadipocyte to adipocyte differentiation) [72] and adipose tissue metabolism [71]. Using the preadipose cell line 3T3-L1, Tsai *et al.* showed that both endogenous and exogenous H$_2$S promotes adipogenesis and adipose tissue maturation [72]. CBS, CSE and 3-MST up-regulated during differentiation of 3T3L1 cells and incubation of preadipocytes with H$_2$S donors, *i.e.* 50 µM GYY4137, a slow-releasing H$_2$S donor or 50 µM NaSH, induced adipocyte differentiation factors, including peroxisome proliferator-activated receptor γ (PPARγ) and CCAAT/enhancer binding protein α (CEBPα) [72].

The role of H$_2$S on insulin sensitivity and glucose uptake in adipose tissue is inconsistent; both inhibitory and stimulatory effects of H$_2$S on insulin-induced glucose uptake have been reported, and a dual role in development of insulin resistance has been illustrated for H$_2$S [69, 73 - 75].

H$_2$S increased insulin sensitivity in hyperglycemia-treated [73], and TNF-α-treated [76] adipocytes or high-fat diet (HFD)-induced diabetic mice [75]. Manna *et al.* reported two pathways related to the stimulatory effect of H$_2$S on glucose utilization in adipocytes [73]. In this study, the effects of different concentrations of L-cysteine (100, 500, and 1000 µM) and Na$_2$S (10 and 100 µM) were determined on phosphatase and tensin homolog (PTEN), phosphoinositide 3-kinase (PI3K), phosphatidylinositol-3,4,5-trisphosphate (PIP$_3$), phospho-AKT and glucose utilization levels in high glucose (25 mM)-treated 3T3L1 adipocyte cells. Treatment with H$_2$S prevented glucose-induced decrease of PI3-K, increased cellular levels of PIP3, and restored phosphorylation of IRS1, activated Akt, and insulin-regulated glucose transporter 4 (GLUT4) translocation in adipocytes, and eventually up-regulated metabolic actions of insulin and increased glucose utilization [73]. H_2S also prevented hyperglycemia-induced PTEN, a negative regulator of glucose utilization; all of the above were published in the first report demonstrating that H_2S could improve insulin signaling pathways by increasing PIP3 [73].

Cai *et al.* reported that treatment with H_2S (H_2S gas saturation buffer, 100 µmol/kg/day or GYY4137, 100 µmol/kg/day), in HFD-induced obese mice (45% of energy from fat), increased gene expression and sulfhydration of PPARγ in adipose tissue, improved the oral glucose tolerance- (OGTT) and the insulin tolerance (ITT) tests, and attenuated the development of insulin resistance [74]. H_2S also inhibited lipolysis through the protein kinase A (PKA)-perilipin/hormone-sensitive lipase pathway, which promote and sensitize insulin response in adipocytes [75].

In contrast, there is some evidence pointing towards a negative effect of H_2S on glucose uptake in the adipose tissue. Feng *et al.* reported that H_2S (from 10 to 1000 µm), dose-dependently inhibited both basal and insulin-stimulated glucose uptake in the mature adipocytes, an effect which was reversed using CSE inhibitors [69]. PPG and BCA increased adipocyte basal glucose uptake by 48 and 17%, respectively, and increased insulin-stimulated glucose uptake by 37.6 and 20.2%, respectively [69]. In this report, the PI3K pathway was involved in the inhibitory effect of H_2S on GLUT4 translocation and glucose uptake in rat adipocytes [69].

H_2S has also been identified as a mediator of insulin resistance in fructose-fed rat models; CSE protein expression was significantly up-regulated and levels of adipose-generated H_2S increased 1.9-fold in fructose-induced insulin-resistant rats as compared with controls [69]. In fructose-induced diabetes in rats, insulin stimulated glucose uptake activity of adipose tissue decreased by 67% compared to controls, a situation in which, a negative correlation was also observed between endogenous H_2S production and adipocyte glucose uptake ($R^2=0.67$, $P<0.01$). The authors accordingly suggested that adipose-generated H_2S might contribute to the pathogenesis of insulin resistance and diabetes [69].

H_2S and Glucose Metabolism in Skeletal Muscle

The expression of CSE and CBS in human skeletal muscle cells is comparable to their expression levels in the liver [77]; mouse skeletal muscles completely lack H_2S producing enzymes whereas all the three H_2S synthases (CBS, CSE, 3-MST) were expressed in the rat skeletal muscles [78]. The rate of H_2S production in the skeletal muscles of Sprague-Dawley rats was 0.17 nmol/min/mg with a H_2S

content of 2.06 nmol/mg [78].

The role of H_2S in skeletal muscle biology is still not fully understood but Veeranki *et al.* reviewed some evidence on its role in tissue specification during the embryonic stage, its therapeutic effects in skeletal muscle wasting/fibrosis, and its potential role in exercise capacity [79]. Skeletal muscle-generated H_2S decreases in rats following ischemia-reperfusion (I-R) injury and H_2S donation exerted protective effect against I-R injury and attenuated oxidative stress in skeletal muscle [78]. Anti-oxidant activity of H_2S is mediated by decreased malondialdehyde (MDA) content, reduced hydrogen peroxide and superoxide anion levels, increased superoxide dismutase activity and protein expression [78].

Direct evidence regarding potential role of H_2S on carbohydrate metabolism in skeletal muscle is limited. Xue *et al.* found that H_2S donation using NaSH increases glucose uptake in myotubes; at concentrations of 25, 50, and 100 µM, NaSH treatment for 24-h significantly promoted insulin-induced glucose uptake in myotubes by 1.54, 1.72 and 2.06-fold [80]. This effect was mediated by increased phosphorylation of insulin receptors (IRs), PI3K, and Akt [80].

Some evidence shows that H_2S can activate the Wnt/β-catenin signaling pathway [81], an effect in the pancreatic β-like cell which may partly explain the H_2S effect on insulin sensitivity in skeletal muscle cells. Activation of Wnt/β-catenin signaling in skeletal muscle cells improved insulin sensitivity by: 1) decreasing intra-myocellular lipid deposition through down-regulation of sterol regulatory element-binding proteins-1c (SREBP-1c), 2) increasing insulin effects through a differential activation of the Akt/protein kinase B (PKB) and AMPK pathways and 3) inhibiting the MAPK pathway [82].

H_2S and Insulin Release in Pancreatic β-cells

Although both CBS and CSE are produced in the pancreatic tissues, CSE plays a more important role in endogenous H_2S production in β-cells and regulation of insulin secretion under physiological and pathological conditions [83, 84]. The expression and activities of H_2S-producing enzymes as well as L-cysteine and H_2S levels in the body have been shown to be changed in insulin resistance and diabetic conditions [84, 85]. Furthermore, both exogenous and endogenous H_2S can regulate insulin secretion [83, 84, 86] and modulate circulating glucose levels [87].

In a study by Patel *et al.* acute treatment of Wistar rats with H_2S (intraperitoneal injection of 2 mg/kg NaSH) increased glucose and decreased insulin concentrations, whereas chronic intraperitoneal injection of H_2S had no effect on insulin/glucose levels [87]. The acute effects of H_2S on insulin secretion are reported to be partially blocked by glibenclamide, a sulfonylurea characterized as a classical K_{ATP} channel blocker [87].

Some evidence indicate that expression of CSE, but not CBS, increased in the pancreatic islets following glucose stimulation; a stimulatory concentration of glucose increased the levels of endogenous H_2S [88]. In contrast, glucose stimulation has been reported to decrease the H_2S-producing activity in the homogenates of INS-1E cells [86]. These contradictory effects of glucose on H_2S production have been attributed to species-specific differences in conditions for gene induction of CSE [89]. Fig. (**3**) summarizes insulin secretion pathways from pancreatic β-cells and the potential endogenous and exogenous effects of H_2S on these pathways.

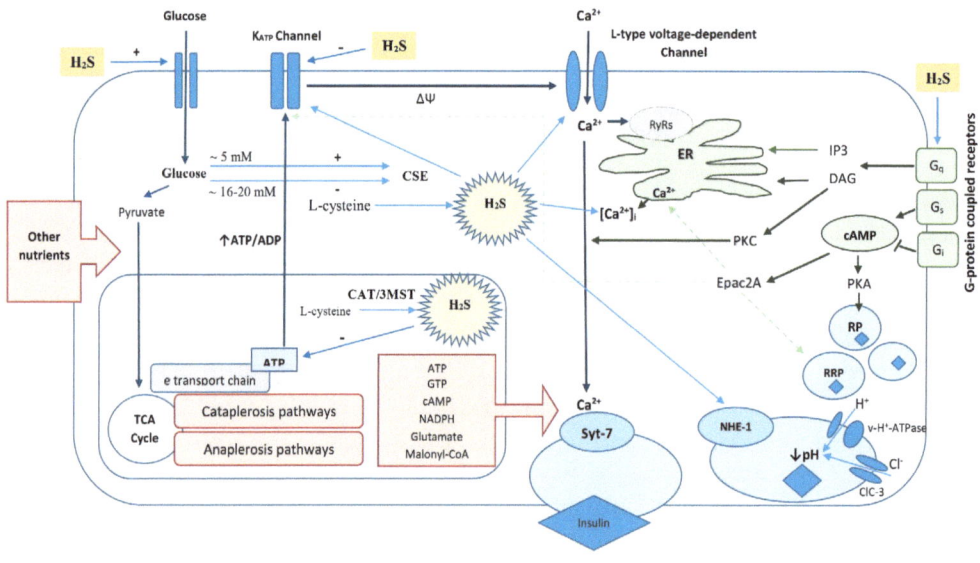

Fig. (3). Triggering and amplifying pathways of insulin secretion from pancreatic β-cell and potential endogenous and exogenous effects of H_2S on these pathways. Syt, Synaptotagmin; RyRs; Ryanodine receptors; CSE, Cystathionine γ-lyase; CAT, cysteine aminotransferase; 3-MST, 3-mercaptopyruvate sulfurtransferase; TCA cycle, tricarboxylic acid cycle; IP3, Inositol trisphosphate; DAG, diacylglycerol; PKC, protein kinase C; PKA, protein kinase; RP, releasable pool; RRP, readily releasable pool; ER, endoplasmic reticulum; Epac, exchange protein activated by cAMP; ATP, Adenosine triphosphate; GTP, Guanosine triphosphate; NADPH, Nucleotide adenine diphosphate; NHE-1, sodium-hydrogen exchanger-1.

H₂S and Triggering Pathway of Insulin Secretion

Glucose-stimulated insulin secretion is generated by triggering and amplifying pathways in the β-cell [90]. Briefly, the triggering pathway involves sequential events including entrance of glucose through GLUT1/GLUT2, glucose metabolism by oxidative glycolysis, increased ATP to ADP ratio, closing ATP-sensitive potassium (K_{ATP}) channels, membrane depolarization, opening of voltage-dependent Ca^{2+} channels (VDCC), Ca^{2+} influx and increased cytoplasmic Ca^{2+} levels, which activates the insulin exocytosis pathway [90, 91]. In Fig. (**3**), this pathway is illustrated in blue. The triggering pathway is also called the K_{ATP}-dependent glucose-induced insulin secretion [91] or the K_{ATP}-Ca^{2+} pathway [92].

The K_{ATP} channel-dependent pathway induces insulin secretion from releasable pool (RP) that is responsible for the first phase of insulin release in response to glucose metabolism [91]. Synaptotagmin 7 (Syt-7), is a Ca^{2+}-binding protein and major component of the exocytotic machinery in pancreatic β-cells; Syt-7 is co-localized with insulin granules, and has been proposed to be a major Ca^{2+} sensor for insulin granule exocytosis [93].

In the pancreatic β-cell, K_{ATP} channels act as metabolic sensors and play crucial role in coupling of metabolism to cell membrane electrical activity and insulin release; opening of these K_{ATP} channels induces β-cells hyperpolarization and suppresses insulin secretion [94].

The role of H_2S as a novel endogenous gaseous opener of K_{ATP} channel has been reported for the first time in the vascular smooth muscle cells [13]; direct confirmation of the effects of H_2S on the K_{ATP} channel in pancreatic β-cell has been reported by Yang *et al.* [86], who showed that endogenous H_2S level may be a switch for turning on/off of K_{ATP} channels at different glucose levels, in rat insulin secreting β-cells (INS-1E) [86]. Under physiological states, in the presence of low levels of extracellular glucose (~ 5 mM), a higher endogenous level of H_2S activates K_{ATP} channels, induces hyperpolarization of the INS-1E membrane cells and inhibits insulin secretion, a situation in which, K_{ATP} channels are not sensitive to exogenous H_2S [86]. *Vice versa*, in the presence of high glucose concentration (16-20 mM), a dose-dependent suppression of endogenous H_2S production is occurred which lead to reduced activity of K_{ATP} channels and increased insulin secretion as well as re-sensitizing of the cells to exogenous H_2S [86].

Significant decreased H_2S production, following partially knockdown of the CSE gene in the INS-1E cells, using the CSE-siRNA technique, also revealed that CSE may be a main enzyme for H_2S production in β-cells [86]. Interaction of H_2S with K_{ATP} channels was shown to be directly induced by functional manipulation, probably through reducing selective cysteine residues of K_{ATP} channel protein, independent of cytosolic second messengers [86]. The authors suggested that interaction between H_2S, glucose and K_{ATP} channels in β-cells may be a novel target for control of pancreatic insulin secretion in both physiological and pathophysiological conditions [86]. Sulfhydration of K_{ATP} channels has been proposed as an underlying mechanism which could explain the effect of H_2S on insulin secretion [95].

Similarly, Kaneko *et al.* demonstrated that NaSH and L-cysteine have potent inhibitory effects on insulin release *via* multiple actions on the insulin secretion machinery [84]. They showed that NaSH (10 μmol/L to 1 mmol/L) and L-cysteine (0.1-10 mmol/L) dose-dependently inhibited glucose (10 mmol/L)-induced insulin secretion in both pancreatic islets and MIN6 (a mixed cell line with other pancreatic endocrine hormones) cells, an effect not observed in non-stimulatory concentration of glucose (3 mmol/L) [84]. Both NaSH and L-cysteine also strongly inhibited the effect of tolbutamide (a K_{ATP} blocker) and α-ketoisocaproate (a mitochondrial fuel) on the K_{ATP}-insulin releasing pathway [84]. Inhibitory effects of NaSH and L-cysteine on insulin release in high K^+ conditions (30 mmol/L), when the VDCC was opened, in the presence of K_{ATP} opener diazoxide (200 μmol/L), indicated that inhibitory effect of H_2S on insulin release may occur *via* an K_{ATP}-independent pathway [84].

Another mechanism explained by Kaneko *et al*, was the inhibitory effect of H_2S on insulin release *via* suppression of glucose-induced Ca^{2+} influx, a critical pathway for secretagogue-induced insulin release [84]. Exposure of mouse pancreatic β-cells to NaSH (100 μmol/L) and L-cysteine (3 mmol/L), promptly inhibited glucose-induced $[Ca^{2+}]_i$ oscillation without changes in the mean $[Ca^{2+}]_i$ value [84].

More interestingly, this study revealed that the inhibitory effect of H_2S on insulin secretion may even be partially independent of the change of $[Ca^{2+}]_i$; in the permeabilized islets (treated with streptolysin-O; SLO), Ca^{2+}-stimulated insulin release was suppressed by NaSH (100 μmol/L) and L-cysteine (3 mmol/L). Furthermore, NaSH and L-cysteine suppressed insulin release stimulated by the copresence of guanosine 5'-0-3-thiotriphosphate (GTPγS) and Ca^{2+} from SLO-treated islets, a situation in which, $[Ca^{2+}]_i$ is not altered by either Ca^{2+} influx or mobilization, because its chelation by ethylene glycol-bis(β-aminoethyl ether)-N,N,N',N'-tetraacetic acid (EGTA) [84].

H_2S also exerts its inhibitory effect on insulin release *via* modulation of glucose metabolism in the pancreatic β-cells [84]. NaSH (100 μmol/L) and L-cysteine (3 mmol/L) dose-dependently decrease ATP production from glucose [84]. Decreased ATP/ADP ratio in the cytoplasm, activates K_{ATP} channels and inhibits insulin release. In this pathway, L-cysteine and NaSH, in particular in an immediate and more potent manner, inhibit glucose-induced mitochondrial membrane hyperpolarization, activation of the respiratory chain and H transport through the inner mitochondrial membrane, and consequently inhibit ATP production [84]. In the pancreatic β-cells, increased intracellular ATP/ADP following increasing glucose levels closes the K_{ATP} channels and decreases the hyperpolarizing outward K^+ flux, which causes depolarization of the plasma membrane, influx of extracellular Ca^{2+} through the VDCC, a sharp increase in intracellular Ca^{2+} and activation of protein motors and kinases, which then mediate exocytosis of insulin-containing vesicles [96]. It must be noted that in all above-mentioned situations, the inhibitory effect of L-cysteine on insulin release from the β-cell was much smaller than that exerted by NaSH [84].

The authors speculate that this inhibition may, at least in part, be independent of the inhibitory effect of H_2S on ATP production, because in the presence of metabolic inhibitors, like carbonylcyanide *m*-chlorophenylhydrazone and oligomycine, H_2S could not decrease Ca^{2+}-induced insulin release from permeabilized islets [84].

Considering the role of H_2S in endogenous production of NO [97], and NO-mediated insulin secretion [98] and β-cell survival [99], the authors of this study also examined the effect of NaSH on the intracellular NO levels in the pancreatic cell lines, but observed no evidence for the involvement of H_2S in the regulation of NO synthesis in the β-cell [84].

L-type VDCC, an important pancreatic β-cell channel involved in regulation of insulin secretion, is also known as another target for H_2S-induced modulation of pancreatic insulin secretion. Inhibitory effect of H_2S on L-type VDCC was shown for the first time in the isolated rat cardiomyocytes [100]. Inactivation of VDCC by H_2S in β-cells had been initially attributed to K_{ATP} channel-induced hyperpolarization of cell membrane [86, 101]. However Tang *et al.* provided a novel mechanism regarding direct inhibitory effect of H_2S on VDCC in β-cell [83], which showed that NaSH decreases L-type VDCC current density and insulin secretion from the pancreatic β-cell, in a dose-dependent manner [83].

H_2S and Amplifying Pathways of Insulin Secretion

The amplifying pathways of glucose-induced insulin secretion proposed to be K_{ATP}-independent [91], act in synergy with K_{ATP} channel-dependent pathway and contribute to the typical rising second-phase of insulin release in response to glucose [90, 91]. These pathways act mainly *via* augmentation of the Ca^{2+} effect on insulin secretion without further rise in $[Ca^{2+}]_i$ and therefore are known as Ca^{2+}-dependent pathways [90], although a "metabolic amplifying pathway" has been proposed to be a more accurate term [102]. Mitochondrial features of these pathways, highlighted in recent years, are mainly implicated by anaplerosis pathways (pyruvate carboxylase and glutamate dehydrogenase) and cataplerosis pathways (pyruvate-malate, pyruvate-citrate, pyruvate-isocitrate shuttle) [92, 103]; these pathways do not require a rise in Ca^{2+} (Fig. **3**) [103]. Cataplerosis pathways in pancreatic β-cells provide mitochondrial metabolites to the cytosol. These metabolites are involved in glucose-stimulated insulin secretion as coupling factors associated with the metabolic amplifying pathway [104], illustrated in brown in Fig. (**3**).

Glucose-induced insulin secretion is also modulated by nutrients and neuro-hormonal inputs through G protein coupled receptors (GPCRs: G_s, G_q and G_i) [102], pathways shown in green in Fig. (**3**). There is evidence showing that H_2S may activate G_q and its downstream pathways [105], that stimulate phospholipase C and production of inositol 1,4,5-trisphosphate (IP3) and diacylglycerol (DAG), which is followed by Ca^{2+} release from the ER and activation of protein kinase C (PKC), respectively [102]. Elevated intracellular Ca^{2+} and PKC activation stimulate insulin secretion *via* amplification of glucose metabolism [102].

There is also some indirect evidence regarding the possible underlying mechanisms which may explain the effect of H_2S on insulin secretion from the pancreatic β-cell. One of these hypothesized mechanisms is the role of H_2S on modification of intracellular acidification of pancreatic β-cell, *via* both direct and indirect pathways.

Weak acids (*e.g.* sulfamerazine, acetate, propionate) induce immediate and transient intracellular acidification in the β-cells, inhibit Na^+/Ca^{2+} exchange system and Ca-activated K^+ channels, induce cell depolarization and potentiate insulin release [106]. H_2S, as a weak acid, may therefore induce insulin secretion. Whether H_2S could mimic similar effects on β-cell acidification, or which H_2S concentrations would be needed to effectively change intracellular pH in the presence of pH-maintaining systems, remains to be clarified.

In vascular smooth muscle cells, NaSH (10 µM to 1 mM) decreases intracellular pH in a concentration-dependent manner; Lee *et al*. reported that H_2S donors induce intracellular acidification *via* activation of the Cl^-/HCO_3^- exchanger [107]. NaSH (10 to 200 µM) also decreases intracellular pH dose-dependently in glia cells, *via* increase in the activity of the Cl^-/HCO_3^- exchanger and inhibition of the Na^+/H^+ exchanger (NHE) [108]. In insulin secreting cells (HIT-T15), NHE has a critical role in regulation of intracellular pH [109]; the Cl^-/HCO_3^- exchanger seems to make limited contribution to maintenance of intracellular pH in the β-cell [109].

NHE-1 protects β-cells against strong acidification [110]. It has been shown that the short-splice variant of NHE-1 in the β-cell, localizes specifically to insulin-containing large dense core vesicles (LDCVs), however its function on LDCVs is still unknown [110]. Acidification of insulin granule, a critical process for its maturation and insulin exocytosis, is thought to be mediated by the ATP-dependent H^+ pump, ClC_3 CL^- channels, and NHE-1 [111, 112]. Likewise, NHA-2 has been shown to play a critical role in insulin secretion from the β-cells, through clathrin-mediated endocytosis with an indirect impact on LDCV exocytosis [113, 114].

In rat cardiac myocytes, NaSH (10-1000 μM) dose-dependently suppressed NHE-1 activity through the PI3K/Akt/protein kinase G-dependent pathway [115].

Mitochondrial glutamate acts as a messenger in glucose-induced insulin exocytosis [116]; it has been proposed that uptake of glutamate into insulin-containing secretory granules decreases pH and increases its sensitivity to the Ca^{2+}-induced exocytosis of insulin [117]. Glutamate is produced from the α-KG by glutamate dehydrogenase, through the tricarboxylic acid cycle (TCA) in the pancreatic β-cell [118], and can also be synthesized through the 3-MST/CAT pathway, the most common pathway of H_2S production in the mitochondria [32]. It can therefore be speculated that H_2S may be involved in insulin secretion through the glutamate pathway.

The potential modulatory effect of H_2S on N-methyl-D-aspartate (NMDA) receptors may also be an underlying pathway that explain its capacity to modify insulin secretion; activation of these receptors, which are membrane-bound ionotropic and glutamate receptors, reduce the amount of insulin secretion in response to stimulatory concentrations of glucose [119, 120]. In the neuronal and glial cell lines, physiological concentrations of H_2S have been shown to enhance NMDA receptor-mediated responses, to their ligands, glutamate or its synthetic agonists, which are augmented by H_2S [121, 122], an effect of H_2S, which is mediated *via* increased production of cAMP and is specifically inhibited by adenylyl cyclase-specific inhibitors [122].

Bcl-2-associated death promoter (BAD), is another pathway speculated to be involved in the potential inhibitory effect of H_2S on insulin secretion from the pancreatic β-cell; H_2S has been found to phosphorylate and inactivate BAD in pancreatic β-like cells [81]. BAD contributes to formation and activation of the mitochondrion glucokinase-containing complex, regulation of glucose-driven respiration and maintenance of glucose homeostasis. It also plays a central role in the physiological regulation of insulin secretion and β cell mass [123]. Lack of BAD in β-cells impairs increased ATP/ADP levels and the insulin secretion pathway in response to glucose. Moreover, glucose-induced changes in mitochondrial membrane hyperpolarization were significantly reduced in Bad $^{-/-}$ beta cells [123]. Average $[Ca^{2+}]i$ response to stimulatory concentrations of glucose was significantly lower under these conditions [123].

Considering the observed inhibitory effect of H_2S on insulin secretion from the pancreatic β-cells, low H_2S levels found in the plasma of type 2 diabetic patients could be a compensatory response [95], a mechanism which may contribute to the development of hyper-insulinemia. This state is required to maintain normal glucose levels in an insulin-resistant state [95]. Furthermore, the inhibitory effect of hyperglycemia-induced H_2S overproduction on insulin release, is thought to prevent β-cells exhaustion, and protecting them against apoptotic cell death in chronic exposure to high-glucose [89].

H_2S and Insulin Synthesis

Beyond insulin release, H_2S could potentially contribute in insulin synthesis in pancreatic β-cells. In a novel investigation, Matei *et al.* reported that treatment of pancreatic β-like cells, derived from dental pulp stem cells (DPSCs), with H_2S (1 ng/mL) resulted in increased GLUT2, intracellular C-peptide, and insulin levels in both 5.5-mM and 25.0-mM glucose-treated cells; the authors hence suggested that H_2S may stimulate insulin production in both non-glucotoxic and glucotoxic conditions [81].

Effects of H_2S on Pancreatic β-cell Differentiation and Survival

Recently, much attention is being focused on the potential effects of H_2S on cell survival/death, especially in pancreatic β-cells [88, 124]. Endogenous H_2S is known to protect pancreatic β-cells of mice from apoptosis, induced by oxidative stress or glucotoxicity [88]. Kenako *et al.* suggested that CSE-released H_2S may act as an "intrinsic brake" against glucose-induced apoptotic death in β-cells [88]. In their study, CSE expression was markedly increased by both low and high-glucose treatment (5 and 20 mM) although CBS expression remained relatively constant [88]. It has been speculated that this glucose-induced expression of CSE may be involved in the prevention of β-cell cytotoxicity [88]. Since glucose-induced β-cell apoptosis is mediated by the persistent Ca^{2+} influx [125], and L-cysteine and NaSH exert an inhibitory effect on Ca^{2+} oscillation in the pancreatic β-cell [84], the authors speculated that increased CSE expression may suppress glucose-induced apoptotic death *via* its effect on the intracellular Ca^{2+} dynamics. Furthermore, incubation of MIN6 β-cells with L-cysteine (3 mM) or NaSH (100 μM) suppressed high glucose-induced DNA-laddering, increased GSH content and improved the secretory responsiveness following stimulation by glucose, leading the authors to conclude that this pathway may be considered as a self-defense against β-cell exhaustion in hyperglycemic condition, if glucose-induced CSE expression could be confirmed *in vivo* [88]. H_2S donation was also able to protect β-cells against apoptosis induced by fatty acid, cytokines and hydrogen peroxide, and promote cell proliferation and survival *via* phosphorylation and activation of Akt signaling [124]. Taniguchi *et al.* reported that NaSH (0.1 μmol/L) suppressed DNA fragmentation and islet cell death (measured by caspase-3 and -7), in MIN6 cells cultured with palmitate (5 mmol/L), cytokines mixture (1000 u/mL TNF-α, 1000 u/mL interferon-γ, 50 u/mL interleukin-1β), and hydrogen peroxide (10 μmol/L) [124].

Endogenous H_2S production has also been shown to have a protective role against HFD-induced glucotoxicity in pancreatic β-cells [126]. In a report by Okamoto *et al.*, 8 weeks HFD (60% of calorie from fat) resulted in hyperglycemia, impaired insulin secretion and glucose intolerance, increased thioredoxin binding protein-2 (TBP-2) expression and elevated DNA fragmentation of the islet cells in CSE knockout (CSE-KO), compared to wild-type (WT) mice [126]. Administration of NaSH (0.1 mM) reduced TBP-2 gene levels in isolated islets from CSE-KO mice [126]. Considering the role of TBP-2 in β-cell apoptosis, it has been speculated that H_2S may protect β-cells from glucotoxicity-induced apoptosis by suppressing TBP-2 expression levels [126]. The authors eventually conclude that CSE-produced H_2S protects β-cells from glucotoxicity *via* regulation of TBP-2 expression levels, thereby preventing the onset and development of type 2 diabetes [126].

Conversely, a part of inhibitory effect of H_2S on insulin production has been attributed to its potential effect to induce β-cell apoptosis and reduction of β-cell mass, events suggested to be contribute to development of diabetes [127]. For the first time, Yang *et al.* reported that overexpression of CSE (induced by adenovirus) in INS-1E cells, significantly decreased cell viability by induction of apoptosis [127]. H_2S donation at physiological concentrations (100 μM) also reduce cell viability and induce apoptosis of INS-1E cells [127]. Lack of changes in cellular GSH levels, following H_2S treatment indicate that the observed apoptotic properties of H_2S in INS-1E cells were independent of redox status [127]. In this study, treatment with exogenous H_2S or overexpression of CSE in INS-1E cells, increased expression of binding immunoglobulin protein (BiP), CCAAT/enhancer-binding protein homologous protein (CHOP), and sterol regulatory element-binding transcription factor 1 (SREBF1), as indicators of ER stress [127]; ER stress plays a key role in pancreatic β-cell apoptosis and development of diabetes [128].

These contradictory effects of H_2S on β-cell apoptosis have been attributed to different types of insulin-secreting cells (*e.g.* INS-1E cells *vs.* mouse pancreatic β-cells) used in different studies and their sensitivity to H_2S [126]. Potential differences between exogenous *vs.* endogenous H_2S on pancreatic β-cell survival, may also be another contributing factor [65]. Differences in the dose and duration of exposure could also be a factor determining the overall effect of H_2S on the islet function [95].

Some evidence indicates that H_2S may contribute in pancreatic β-cell differentiation. Okada *et al.* demonstrated that H_2S exposure enhances DPSC cells differentiation into pancreatic β-like cells and activates the core components of the PI3K/Akt pathway, a mediator of survival, proliferation and the insulin response [129]. H_2S, at a concentration of 1 ng/ml, also effectively inhibited the adverse effects of glucotoxicity in DPSC-derived pancreatic β-like cells [81]. Matei *et al.* reported that H_2S did not significantly influence apoptosis in DPSC-derived pancreatic β-like cells; they observed that both apoptotic and anti-apoptotic pathways were activated in the H_2S-exposed cells, in a balanced manner [81]. H_2S exposure induced genes associated with PTEN-dependent cell-cycle arrest and apoptosis simultaneously with genes associated with BAD phosphorylation and anti-apoptotic pathways [81].

H_2S AND DIABETES COMPLICATIONS

Despite an overview of current data implying contribution of H_2S in the development of early stages of diabetes, H_2S exerts several protective roles in the prevention and improvement of diabetes complications [2]; H_2S protects pancreatic β-cells against hyperglycemia-induced apoptotic pathways, and exerts anti-oxidative properties and anti-atherogenic effects [43]. Administration of H_2S in diabetic models prevented endothelial dysfunction, retinopathy, cardiomyopathy, and nephropathy [49, 130 - 132].

Effects of H_2S on Diabetic Cardiovascular Complications

Xie *et al.* showed that daily systemic administration of the slow-releasing H_2S donor (GYY4137) decreased atherosclerotic lesion size in streptozotocin-induced diabetic mice, fed a high-fat diet, independent of any change in circulating blood glucose or cholesterol [133]. This protective effect was attributed to inhibition of oxidative stress and activation of Nrf2 signaling pathway *via* Kelch ECH-Associating Protein 1 (Keap1) sulfhydration at Cys151 [133]. It has been proposed that restoration of H_2S in diabetic conditions, using administration of H_2S donor molecules, gene delivery of H_2S-generating enzymes, or dietary sulfate supplementation, may activate the Nrf2–heme oxygenase-1 signaling pathway [134]. H_2S restoration can also inhibit development of atherosclerosis by blocking diabetes-induced oxidative and inflammatory stress in endothelial cells, decrease ROS levels and prevent foam cell formation by macrophages [134]. In a model of streptozotocin-induced diabetic rat, H_2S supplementation (100-300 μM) returned plasma H_2S to normal levels, and improved the endothelium-dependent relaxant responses of the thoracic aorta *ex vivo*, without affecting the degree of hyperglycemia [49].

H_2S attenuates the development of diabetic cardiomyopathy; in a study by Zhou *et al.* 16 weeks administration of NaSH (14 μmol/kg body weight/day) improved left ventricular function and prevented cardiac hypertrophy and myocardial fibrosis in diabetic rats [132]. H_2S also attenuated hyperglycemia-induced inflammation, oxidative stress and apoptosis in the cardiac tissue, favorable effects which were accompanied by activation of Nrf2/antioxidant response-element (ARE) signaling pathway and up-regulated expression of antioxidant proteins, including hem oxygenase-1 and NADPH: quinone oxidoreductase 1 [132]. In a streptozotocin model of diabetic cardiomyopathy, intraperitoneal injection of NaSH (14 μmol/kg body weight/day) reduced myocardial hypertrophy, improved the histological picture of the diabetic hearts, and reduced the degree of fibrosis, effects which were attributed to considerable reductions in the up-regulated matrix metalloproteinase 2 and transforming growth factor (TGF)-β1 in the hearts of diabetic animals [135]. In this study, H_2S donating improved antioxidant status including elevated GSH levels and reduced levels of myocardial hydroxyproline [135].

Effects of H_2S on Renal Complications of Diabetic Models

An 8-week administration of H_2S (NaSH at a dose of 100 μmol/kg) improved renal tissue fibrosis in diabetic rats by up-regulation of superoxide dismutase, inhibition of autophagy-related proteins, and down-regulation of serine/threonine kinase, TGF-β1 and NF-κB protein, as key mediators of diabetic nephropathy [136]. H_2S donating also decreased 24 h proteinuria in diabetic rats [136].

H_2S donation using 12-week administration of 14 μmol/kg body weight/day NaSH in diabetic rats improved renal function and attenuated glomerular basement membrane thickening, mesangial matrix deposition, and renal interstitial fibrosis [130]. Moreover, H_2S reduced high glucose-induced oxidative stress through activation of the Nrf2 antioxidant pathway and alleviation of inflammatory pathways by inhibiting NF-κB signaling [130]. Another beneficial effect of H_2S was its capacity to attenuate high glucose-induced mesangial cell proliferation by suppression of the MAPK signaling pathway; H_2S administration also inhibited the renin-angiotensin system in the diabetic kidney [130].

Likewise, in a study by Yuan *et al.*, treatment of streptozotocin-diabetic rats with intraperitoneal NaSH (50 μmol/kg body weight/day) reduced high blood urea nitrogen levels, and attenuated renal collagen and TGF-β1 expression, without affecting the hyperglycemia levels [137].

H_2S and Other Diabetes Complications

Impairment of cognitive function, an undesirable complication of diabetes, could be a result of hyperglycemia, vascular disease, hypoglycemia, and insulin resistance [138]. Hippocampal ER stress is substantially involved in diabetic cognitive impairment, and H_2S exerts its therapeutic effect on diabetes-associated cognitive dysfunction, through inhibition of hippocampal ER stress [139].

Data shows that down-regulation of CSE may contribute in the pathogenesis of diabetic impaired wound healing. Zhao *et al*. recently reported that exogenous administration of H_2S improved diabetic impaired wound healing by attenuating inflammation and increasing angiogenesis, an effect related to decreased neutrophil and macrophage infiltration, reduced production of TNF-α and interleukin-6, decreased metalloproteinase (MMP)-9, as well as increased collagen deposition in the tissue [140]. In a wound healing model of diabetic rats, using 2% sodium bisulfide ointment on wounds significantly increased levels of vascular endothelial growth factor (VEGF) and intercellular cell adhesion molecule-1 (ICAM-1) [141]. Treatment with sodium bisulfide also reduced coagulation activity, induce superoxide dismutase (SOD) and heme oxygenase-1 protein expression, and decreased TNF-α protein expression in diabetic rats [141].

CONCLUSION AND FUTURE PERSPECTIVE

The role of H_2S in glucose and insulin homeostasis seems to be complex. In different stages of diabetes, H_2S plays opposite roles. Some evidence indicates that H_2S overproduction or its exogenous donation may contribute to development of insulin resistance and the onset of diabetes, whereas others believe that overproduction of H_2S, especially in tissues like pancreas, may act as an intrinsic brake to reduce cellular damage and exhaustion. H_2S also has paradoxical functions in regulation of glucose/insulin metabolism in hepatocytes, adipocytes and skeletal muscle. Regarding exogenous administration of H_2S in animal models, overall, its multiple functions against diabetes-induced vascular diseases, cardiomyopathy, nephropathy, neuropathy and other complications have been documented. These effects seem to be mainly mediated by anti-oxidant and anti-inflammatory properties and inhibiting the expression of pro-fibrotic mediators. Effect of H_2S donation on glycemic control is still unclear and future studies are needed to clarify therapeutic targets for H_2S regarding glucose and insulin homeostasis.

CONSENT FOR PUBLICATION

Not applicable.

CONFLICT OF INTEREST

The authors confirm that this chapter contents have no conflict of interest.

ACKNOWLEDGEMENT

We wish to acknowledge Ms. Niloofar Shiva for critical editing of English grammar and syntax of the manuscript.

REFERENCES

[1]   Bos EM, van Goor H, Joles JA, Whiteman M, Leuvenink HG. Hydrogen sulfide: physiological properties and therapeutic potential in ischaemia. Br J Pharmacol 2015; 172(6): 1479-93.
[http://dx.doi.org/10.1111/bph.12869] [PMID: 25091411]

[2]   Wallace JL, Wang R. Hydrogen sulfide-based therapeutics: exploiting a unique but ubiquitous gasotransmitter. Nat Rev Drug Discov 2015; 14(5): 329-45.
[http://dx.doi.org/10.1038/nrd4433] [PMID: 25849904]

[3]   Wallace JL, Blackler RW, Chan MV, *et al.* Anti-inflammatory and cytoprotective actions of hydrogen sulfide: translation to therapeutics. Antioxid Redox Signal 2015; 22(5): 398-410.
[http://dx.doi.org/10.1089/ars.2014.5901] [PMID: 24635322]

[4]   Gemici B, Elsheikh W, Feitosa KB, Costa SK, Muscara MN, Wallace JL. 2015.

[5]   Mustafa AK, Gadalla MM, Sen N, *et al.* H2S signals through protein S-sulfhydration. Sci Signal 2009; 2(96): ra72.
[http://dx.doi.org/10.1126/scisignal.2000464] [PMID: 19903941]

[6]   Paul BD, Snyder SH. H_2S signalling through protein sulfhydration and beyond. Nat Rev Mol Cell Biol 2012; 13(8): 499-507.
[http://dx.doi.org/10.1038/nrm3391] [PMID: 22781905]

[7]   Pircher J, Fochler F, Czermak T, *et al.* Hydrogen sulfide-releasing aspirin derivative ACS14 exerts strong antithrombotic effects *in vitro* and *in vivo*. Arterioscler Thromb Vasc Biol 2012; 32(12): 2884-91.
[http://dx.doi.org/10.1161/ATVBAHA.112.300627] [PMID: 23023375]

[8]   Kashfi K, Olson KR. Biology and therapeutic potential of hydrogen sulfide and hydrogen sulfide-releasing chimeras. Biochem Pharmacol 2013; 85(5): 689-703.
[http://dx.doi.org/10.1016/j.bcp.2012.10.019] [PMID: 23103569]

[9]   Liu YH, Yan CD, Bian JS. Hydrogen sulfide: a novel signaling molecule in the vascular system. J Cardiovasc Pharmacol 2011; 58(6): 560-9.
[http://dx.doi.org/10.1097/FJC.0b013e31820eb7a1] [PMID: 21283022]

[10] Szabo C. Hydrogen sulfide, an enhancer of vascular nitric oxide signaling: mechanisms and implications. Am J Physiol Cell Physiol 2017; 312(1): C3-C15.
[http://dx.doi.org/10.1152/ajpcell.00282.2016] [PMID: 27784679]

[11] Kanagy NL, Szabo C, Papapetropoulos A. Vascular biology of hydrogen sulfide. Am J Physiol Cell Physiol 2017; 312(5): C537-49.
[http://dx.doi.org/10.1152/ajpcell.00329.2016] [PMID: 28148499]

[12] Khan AA, Schuler MM, Prior MG, *et al.* Effects of hydrogen sulfide exposure on lung mitochondrial respiratory chain enzymes in rats. Toxicol Appl Pharmacol 1990; 103(3): 482-90.
[http://dx.doi.org/10.1016/0041-008X(90)90321-K] [PMID: 2160136]

[13] Zhao W, Zhang J, Lu Y, Wang R. The vasorelaxant effect of H(2)S as a novel endogenous gaseous K(ATP) channel opener. EMBO J 2001; 20(21): 6008-16.
[http://dx.doi.org/10.1093/emboj/20.21.6008] [PMID: 11689441]

[14] Kubo S, Doe I, Kurokawa Y, Nishikawa H, Kawabata A. Direct inhibition of endothelial nitric oxide synthase by hydrogen sulfide: contribution to dual modulation of vascular tension. Toxicology 2007; 232(1-2): 138-46.
[http://dx.doi.org/10.1016/j.tox.2006.12.023] [PMID: 17276573]

[15] Kolluru GK, Shen X, Bir SC, Kevil CG. Hydrogen sulfide chemical biology: pathophysiological roles and detection. Nitric Oxide 2013; 35: 5-20.
[http://dx.doi.org/10.1016/j.niox.2013.07.002]

[16] Singh SB, Lin HC. Hydrogen Sulfide in Physiology and Diseases of the Digestive Tract. Microorganisms 2015; 3(4): 866-89.
[http://dx.doi.org/10.3390/microorganisms3040866] [PMID: 27682122]

[17] Searcy DG, Lee SH. Sulfur reduction by human erythrocytes. J Exp Zool 1998; 282(3): 310-22.
[http://dx.doi.org/10.1002/(SICI)1097-010X(19981015)282:3<310::AID-JEZ4>3.0.CO;2-P] [PMID: 9755482]

[18] Predmore BL, Lefer DJ, Gojon G. Hydrogen sulfide in biochemistry and medicine. Antioxid Redox Signal 2012; 17(1): 119-40.
[http://dx.doi.org/10.1089/ars.2012.4612] [PMID: 22432697]

[19] Kamoun P. Endogenous production of hydrogen sulfide in mammals. Amino Acids 2004; 26(3): 243-54.
[http://dx.doi.org/10.1007/s00726-004-0072-x] [PMID: 15221504]

[20] Kimura H. Hydrogen sulfide: its production, release and functions. Amino Acids 2011; 41(1): 113-21.
[http://dx.doi.org/10.1007/s00726-010-0510-x] [PMID: 20191298]

[21] Nagahara N, Ito T, Kitamura H, Nishino T. Tissue and subcellular distribution of mercaptopyruvate sulfurtransferase in the rat: confocal laser fluorescence and immunoelectron microscopic studies combined with biochemical analysis. Histochem Cell Biol 1998; 110(3): 243-50.
[http://dx.doi.org/10.1007/s004180050286] [PMID: 9749958]

[22] Asimakopoulou A, Panopoulos P, Chasapis CT, *et al.* Selectivity of commonly used pharmacological inhibitors for cystathionine β synthase (CBS) and cystathionine γ lyase (CSE). Br J Pharmacol 2013; 169(4): 922-32.
[http://dx.doi.org/10.1111/bph.12171] [PMID: 23488457]

[23] Shibuya N, Kimura H. Production of hydrogen sulfide from d-cysteine and its therapeutic potential. Front Endocrinol (Lausanne) 2013; 4: 87.
[http://dx.doi.org/10.3389/fendo.2013.00087] [PMID: 23882260]

[24] Shibuya N, Koike S, Tanaka M, *et al.* P33 A novel pathway for the production of hydrogen sulfide from d-cysteine in mammalian cells. Nitric Oxide 2014; 39: S26.
[http://dx.doi.org/10.1016/j.niox.2014.03.083]

[25] Taoka S, Banerjee R. Characterization of NO binding to human cystathionine beta-synthase: possible implications of the effects of CO and NO binding to the human enzyme. J Inorg Biochem 2001; 87(4): 245-51.
[http://dx.doi.org/10.1016/S0162-0134(01)00335-X] [PMID: 11744062]

[26] Miles EW, Kraus JP. Cystathionine β-synthase: structure, function, regulation, and location of homocystinuria-causing mutations. J Biol Chem 2004; 279(29): 29871-4.
[http://dx.doi.org/10.1074/jbc.R400005200] [PMID: 15087459]

[27] Yang G, Pei Y, Teng H, Cao Q, Wang R. Specificity protein-1 as a critical regulator of human cystathionine gamma-lyase in smooth muscle cells. J Biol Chem 2011; 286(30): 26450-60.
[http://dx.doi.org/10.1074/jbc.M111.266643] [PMID: 21659522]

[28] Zhang L, Yang G, Tang G, Wu L, Wang R. Rat pancreatic level of cystathionine γ-lyase is regulated by glucose level *via* specificity protein 1 (SP1) phosphorylation. Diabetologia 2011; 54(10): 2615-25.
[http://dx.doi.org/10.1007/s00125-011-2187-4] [PMID: 21618058]

[29] Wu N, Siow YL, O K. Ischemia/reperfusion reduces transcription factor Sp1-mediated cystathionine beta-synthase expression in the kidney. J Biol Chem 2010; 285(24): 18225-33.
[http://dx.doi.org/10.1074/jbc.M110.132142] [PMID: 20392694]

[30] Sen N, Paul BD, Gadalla MM, *et al*. Hydrogen sulfide-linked sulfhydration of NF-κB mediates its antiapoptotic actions. Mol Cell 2012; 45(1): 13-24.
[http://dx.doi.org/10.1016/j.molcel.2011.10.021] [PMID: 22244329]

[31] Carballal S, Cuevasanta E, Marmisolle I, *et al*. Kinetics of reversible reductive carbonylation of heme in human cystathionine β-synthase. Biochemistry 2013; 52(26): 4553-62.
[http://dx.doi.org/10.1021/bi4004556] [PMID: 23790103]

[32] Kimura H. The physiological role of hydrogen sulfide and beyond. Nitric Oxide 2014; 41: 4-10.
[http://dx.doi.org/10.1016/j.niox.2014.01.002]

[33] Zhu XY, Gu H, Ni X. Hydrogen sulfide in the endocrine and reproductive systems. Expert Rev Clin Pharmacol 2011; 4(1): 75-82.
[http://dx.doi.org/10.1586/ecp.10.125] [PMID: 22115350]

[34] Schalinske KL. Interrelationship between diabetes and homocysteine metabolism: hormonal regulation of cystathionine beta-synthase. Nutr Rev 2003; 61(4): 136-8.
[http://dx.doi.org/10.1301/nr.2003.apr.136-138] [PMID: 12795447]

[35] Yusuf M, Kwong Huat BT, Hsu A, Whiteman M, Bhatia M, Moore PK. Streptozotocin-induced diabetes in the rat is associated with enhanced tissue hydrogen sulfide biosynthesis. Biochem Biophys Res Commun 2005; 333(4): 1146-52.
[http://dx.doi.org/10.1016/j.bbrc.2005.06.021] [PMID: 15967410]

[36] Ratnam S, Maclean KN, Jacobs RL, Brosnan ME, Kraus JP, Brosnan JT. Hormonal regulation of cystathionine beta-synthase expression in liver. J Biol Chem 2002; 277(45): 42912-8.
[http://dx.doi.org/10.1074/jbc.M206588200] [PMID: 12198128]

[37] Renga B, Mencarelli A, Migliorati M, Distrutti E, Fiorucci S. Bile-acid-activated farnesoid X receptor regulates hydrogen sulfide production and hepatic microcirculation. World J Gastroenterol 2009; 15(17): 2097-108.
[http://dx.doi.org/10.3748/wjg.15.2097] [PMID: 19418582]

[38] Wang R. Two's company, three's a crowd: can H2S be the third endogenous gaseous transmitter? FASEB J 2002; 16(13): 1792-8.
[http://dx.doi.org/10.1096/fj.02-0211hyp] [PMID: 12409322]

[39] Hildebrandt TM, Grieshaber MK. Three enzymatic activities catalyze the oxidation of sulfide to thiosulfate in mammalian and invertebrate mitochondria. FEBS J 2008; 275(13): 3352-61.
[http://dx.doi.org/10.1111/j.1742-4658.2008.06482.x] [PMID: 18494801]

[40] Kiss L, Deitch EA, Szabó C. Hydrogen sulfide decreases adenosine triphosphate levels in aortic rings and leads to vasorelaxation *via* metabolic inhibition. Life Sci 2008; 83(17-18): 589-94.
[http://dx.doi.org/10.1016/j.lfs.2008.08.006] [PMID: 18790700]

[41] Shen X, Pattillo CB, Pardue S, Bir SC, Wang R, Kevil CG. Measurement of plasma hydrogen sulfide *in vivo* and *in vitro*. Free Radic Biol Med 2011; 50(9): 1021-31.
[http://dx.doi.org/10.1016/j.freeradbiomed.2011.01.025] [PMID: 21276849]

[42] Jacobs RL, House JD, Brosnan ME, Brosnan JT. Effects of streptozotocin-induced diabetes and of

insulin treatment on homocysteine metabolism in the rat. Diabetes 1998; 47(12): 1967-70.
[http://dx.doi.org/10.2337/diabetes.47.12.1967] [PMID: 9836532]

[43] Szabo C. Roles of hydrogen sulfide in the pathogenesis of diabetes mellitus and its complications. Antioxid Redox Signal 2012; 17(1): 68-80.
[http://dx.doi.org/10.1089/ars.2011.4451] [PMID: 22149162]

[44] Hayden LJ, Goeden H, Roth SH. Exposure to low levels of hydrogen sulfide elevates circulating glucose in maternal rats. J Toxicol Environ Health 1990; 31(1): 45-52.
[http://dx.doi.org/10.1080/15287399009531436] [PMID: 2213921]

[45] Wu L, Yang W, Jia X, et al. Pancreatic islet overproduction of H2S and suppressed insulin release in Zucker diabetic rats. Lab Invest 2009; 89(1): 59-67.
[http://dx.doi.org/10.1038/labinvest.2008.109] [PMID: 19002107]

[46] Desai KM, Chang T, Untereiner A, Wu L. Hydrogen sulfide and the metabolic syndrome. Expert Rev Clin Pharmacol 2011; 4(1): 63-73.
[http://dx.doi.org/10.1586/ecp.10.133] [PMID: 22115349]

[47] Wiliński B, Wiliński J, Somogyi E, Piotrowska J, Opoka W. Metformin raises hydrogen sulfide tissue concentrations in various mouse organs. Pharmacol Rep 2013; 65(3): 737-42.
[http://dx.doi.org/10.1016/S1734-1140(13)71053-3] [PMID: 23950598]

[48] Brancaleone V, Roviezzo F, Vellecco V, De Gruttola L, Bucci M, Cirino G. Biosynthesis of H2S is impaired in non-obese diabetic (NOD) mice. Br J Pharmacol 2008; 155(5): 673-80.
[http://dx.doi.org/10.1038/bjp.2008.296] [PMID: 18641671]

[49] Suzuki K, Olah G, Modis K, et al. Hydrogen sulfide replacement therapy protects the vascular endothelium in hyperglycemia by preserving mitochondrial function. Proc Natl Acad Sci USA 2011; 108(33): 13829-34.
[http://dx.doi.org/10.1073/pnas.1105121108] [PMID: 21808008]

[50] Olson KR. Is hydrogen sulfide a circulating "gasotransmitter" in vertebrate blood? Biochim Biophys Acta 2009; 1787(7): 856-63.
[http://dx.doi.org/10.1016/j.bbabio.2009.03.019] [PMID: 19361483]

[51] Suzuki K, Sagara M, Aoki C, Tanaka S, Aso Y. Clinical Implication of Plasma Hydrogen Sulfide Levels in Japanese Patients with Type 2 Diabetes. Intern Med 2017; 56(1): 17-21.
[http://dx.doi.org/10.2169/internalmedicine.56.7403] [PMID: 28049995]

[52] Jain SK, Bull R, Rains JL, et al. Low levels of hydrogen sulfide in the blood of diabetes patients and streptozotocin-treated rats causes vascular inflammation? Antioxid Redox Signal 2010; 12(11): 1333--.
[http://dx.doi.org/10.1089/ars.2009.2956] [PMID: 20092409]

[53] Whiteman M, Gooding KM, Whatmore JL, et al. Adiposity is a major determinant of plasma levels of the novel vasodilator hydrogen sulphide. Diabetologia 2010; 53(8): 1722-6.
[http://dx.doi.org/10.1007/s00125-010-1761-5] [PMID: 20414636]

[54] Ghasemi A, Jeddi S. Anti-obesity and anti-diabetic effects of nitrate and nitrite. Nitric Oxide 2017; 70: 9-24.
[http://dx.doi.org/10.1016/j.niox.2017.08.003]

[55] Bahadoran Z, Ghasemi A, Mirmiran P, Azizi F, Hadaegh F. Beneficial effects of inorganic nitrate/nitrite in type 2 diabetes and its complications. Nutr Metab (Lond) 2015; 12: 16.
[http://dx.doi.org/10.1186/s12986-015-0013-6] [PMID: 25991919]

[56] Ghasemi A, Zahediasl S. Potential therapeutic effects of nitrate/nitrite and type 2 diabetes mellitus. Int J Endocrinol Metab 2013; 11(2): 63-4.
[http://dx.doi.org/10.5812/ijem.9103] [PMID: 23825974]

[57] Gheibi S, Bakhtiarzadeh F, Jeddi S, Farrokhfall K, Zardooz H, Ghasemi A. Nitrite increases glucose-stimulated insulin secretion and islet insulin content in obese type 2 diabetic male rats. Nitric Oxide 2017; 64: 39-51.

[http://dx.doi.org/10.1016/j.niox.2017.01.003]

[58] Khalifi S, Rahimipour A, Jeddi S, Ghanbari M, Kazerouni F, Ghasemi A. Dietary nitrate improves glucose tolerance and lipid profile in an animal model of hyperglycemia. Nitric Oxide 2015; 44: 24--.
[http://dx.doi.org/10.1016/j.niox.2014.11.011]

[59] Petersen MC, Vatner DF, Shulman GI. Regulation of hepatic glucose metabolism in health and disease. Nat Rev Endocrinol 2017; 13(10): 572-87.
[http://dx.doi.org/10.1038/nrendo.2017.80] [PMID: 28731034]

[60] Magnusson I, Rothman DL, Katz LD, Shulman RG, Shulman GI. Increased rate of gluconeogenesis in type II diabetes mellitus. A 13C nuclear magnetic resonance study. J Clin Invest 1992; 90(4): 1323-7.
[http://dx.doi.org/10.1172/JCI115997] [PMID: 1401068]

[61] Boden G. Gluconeogenesis and glycogenolysis in health and diabetes. Journal of Investigative Medicine 2004; 52(6): 375-8.

[62] Mani S, Cao W, Wu L, Wang R. Hydrogen sulfide and the liver. Nitric Oxide 2014; 41: 62-70.
[http://dx.doi.org/10.1016/j.niox.2014.02.006]

[63] Wijekoon EP, Hall B, Ratnam S, Brosnan ME, Zeisel SH, Brosnan JT. Homocysteine metabolism in ZDF (type 2) diabetic rats. Diabetes 2005; 54(11): 3245-51.
[http://dx.doi.org/10.2337/diabetes.54.11.3245] [PMID: 16249451]

[64] Zhang L, Yang G, Untereiner A, Ju Y, Wu L, Wang R. Hydrogen sulfide impairs glucose utilization and increases gluconeogenesis in hepatocytes. Endocrinology 2013; 154(1): 114-26.
[http://dx.doi.org/10.1210/en.2012-1658] [PMID: 23183179]

[65] Untereiner A, Wu L. Hydrogen sulfide and glucose homeostasis - a tale of sweet & the stink. Antioxid Redox Signal 2017.
[PMID: 28699407]

[66] Yang G. H2S and glucose metabolism, how does the stink regulate the sweet? Immunoendocrinology (Houst) 2015; 3.

[67] Yang G. Protein S-sulfhydration as a major sources of H2S bioactivity. Receptors Clin Investig 2014; 1(4)

[68] Ju Y, Untereiner A, Wu L, Yang GH. 2 S-induced S-sulfhydration of pyruvate carboxylase contributes to gluconeogenesis in liver cells. Biochimica et Biophysica Acta (BBA)-. General Subjects 2015; 1850(11): 2293-303.
[http://dx.doi.org/10.1016/j.bbagen.2015.08.003] [PMID: 26272431]

[69] Feng X, Chen Y, Zhao J, Tang C, Jiang Z, Geng B. Hydrogen sulfide from adipose tissue is a novel insulin resistance regulator. Biochem Biophys Res Commun 2009; 380(1): 153-9.
[http://dx.doi.org/10.1016/j.bbrc.2009.01.059] [PMID: 19166813]

[70] Velmurugan GV, Huang H, Sun H, et al. Depletion of H2S during obesity enhances store-operated Ca2+ entry in adipose tissue macrophages to increase cytokine production. Sci Signal 2015; 8(407): ra128.
[http://dx.doi.org/10.1126/scisignal.aac7135] [PMID: 26671149]

[71] Bełtowski J, Jamroz-Wiśniewska A. Hydrogen Sulfide in the Adipose Tissue-Physiology, Pathology and a Target for Pharmacotherapy. Molecules 2016; 22(1): E63.
[http://dx.doi.org/10.3390/molecules22010063] [PMID: 28042862]

[72] Tsai CY, Peh MT, Feng W, Dymock BW, Moore PK. Hydrogen sulfide promotes adipogenesis in 3T3L1 cells. PLoS One 2015; 10(3): e0119511.
[http://dx.doi.org/10.1371/journal.pone.0119511] [PMID: 25822632]

[73] Manna P, Jain SK. Hydrogen sulfide and L-cysteine increase phosphatidylinositol 3,4,5-trisphosphate (PIP3) and glucose utilization by inhibiting phosphatase and tensin homolog (PTEN) protein and activating phosphoinositide 3-kinase (PI3K)/serine/threonine protein kinase (AKT)/protein kinase Cζ/λ

(PKCζ/λ) in 3T3l1 adipocytes. J Biol Chem 2011; 286(46): 39848-59.
[http://dx.doi.org/10.1074/jbc.M111.270884] [PMID: 21953448]

[74] Cai J, Shi X, Wang H, *et al.* Cystathionine γ lyase-hydrogen sulfide increases peroxisome proliferator-activated receptor γ activity by sulfhydration at C139 site thereby promoting glucose uptake and lipid storage in adipocytes. Biochim Biophys Acta 2016; 1861(5): 419-29.
[http://dx.doi.org/10.1016/j.bbalip.2016.03.001] [PMID: 26946260]

[75] Geng B, Cai B, Liao F, *et al.* Increase or decrease hydrogen sulfide exert opposite lipolysis, but reduce global insulin resistance in high fatty diet induced obese mice. PLoS One 2013; 8(9): e73892.
[http://dx.doi.org/10.1371/journal.pone.0073892] [PMID: 24058499]

[76] Huang CY, Yao WF, Wu WG, Lu YL, Wan H, Wang W. Endogenous CSE/H2 S system mediates TNF-α-induced insulin resistance in 3T3-L1 adipocytes. Cell Biochem Funct 2013; 31(6): 468-75.
[http://dx.doi.org/10.1002/cbf.2920] [PMID: 23080424]

[77] Chen NC, Yang F, Capecci LM, *et al.* Regulation of homocysteine metabolism and methylation in human and mouse tissues. FASEB J 2010; 24(8): 2804-17.
[http://dx.doi.org/10.1096/fj.09-143651] [PMID: 20305127]

[78] Du JT, Li W, Yang JY, Tang CS, Li Q, Jin HF. Hydrogen sulfide is endogenously generated in rat skeletal muscle and exerts a protective effect against oxidative stress. Chin Med J (Engl) 2013; 126(5): 930-6.
[PMID: 23489804]

[79] Veeranki S, Tyagi SC. Role of hydrogen sulfide in skeletal muscle biology and metabolism. Nitric Oxide 2015; 46: 66-71.
[http://dx.doi.org/10.1016/j.niox.2014.11.012]

[80] Xue R, Hao DD, Sun JP, *et al.* Hydrogen sulfide treatment promotes glucose uptake by increasing insulin receptor sensitivity and ameliorates kidney lesions in type 2 diabetes. Antioxid Redox Signal 2013; 19(1): 5-23.
[http://dx.doi.org/10.1089/ars.2012.5024] [PMID: 23293908]

[81] Matei IV, Ii H, Yaegaki K. Hydrogen sulfide enhances pancreatic β-cell differentiation from human tooth under normal and glucotoxic conditions. Regen Med 2016; 12(2): 125-41.

[82] Abiola M, Favier M, Christodoulou-Vafeiadou E, Pichard AL, Martelly I, Guillet-Deniau I. Activation of Wnt/beta-catenin signaling increases insulin sensitivity through a reciprocal regulation of Wnt10b and SREBP-1c in skeletal muscle cells. PLoS One 2009; 4(12): e8509.
[http://dx.doi.org/10.1371/journal.pone.0008509] [PMID: 20041157]

[83] Tang G, Zhang L, Yang G, Wu L, Wang R. Hydrogen sulfide-induced inhibition of L-type Ca2+ channels and insulin secretion in mouse pancreatic beta cells. Diabetologia 2013; 56(3): 533-41.
[http://dx.doi.org/10.1007/s00125-012-2806-8] [PMID: 23275972]

[84] Kaneko Y, Kimura Y, Kimura H, Niki I. L-cysteine inhibits insulin release from the pancreatic beta-cell: possible involvement of metabolic production of hydrogen sulfide, a novel gasotransmitter. Diabetes 2006; 55(5): 1391-7.
[http://dx.doi.org/10.2337/db05-1082] [PMID: 16644696]

[85] Okamoto M, Ishizaki T, Kimura T. Protective effect of hydrogen sulfide on pancreatic beta-cells. Nitric Oxide 2015; 46: 32-6.
[http://dx.doi.org/10.1016/j.niox.2014.11.007]

[86] Yang W, Yang G, Jia X, Wu L, Wang R. Activation of KATP channels by H2S in rat insulin-secreting cells and the underlying mechanisms. J Physiol 2005; 569(Pt 2): 519-31.
[http://dx.doi.org/10.1113/jphysiol.2005.097642] [PMID: 16179362]

[87] Patel M, Shah G. Possible role of hydrogen sulfide in insulin secretion and in development of insulin resistance. J Young Pharm 2010; 2(2): 148-51.
[http://dx.doi.org/10.4103/0975-1483.63156] [PMID: 21264117]

[88] Kaneko Y, Kimura T, Taniguchi S, *et al.* Glucose-induced production of hydrogen sulfide may protect the pancreatic beta-cells from apoptotic cell death by high glucose. FEBS Lett 2009; 583(2): 377-82.
[http://dx.doi.org/10.1016/j.febslet.2008.12.026] [PMID: 19100738]

[89] Taniguchi S, Niki I. Significance of hydrogen sulfide production in the pancreatic β-cell. J Pharmacol Sci 2011; 116(1): 1-5.
[http://dx.doi.org/10.1254/jphs.11R01CP] [PMID: 21512302]

[90] Henquin JC. Triggering and amplifying pathways of regulation of insulin secretion by glucose. Diabetes 2000; 49(11): 1751-60.
[http://dx.doi.org/10.2337/diabetes.49.11.1751] [PMID: 11078440]

[91] Straub SG, Sharp GW. Glucose-stimulated signaling pathways in biphasic insulin secretion. Diabetes Metab Res Rev 2002; 18(6): 451-63.
[http://dx.doi.org/10.1002/dmrr.329] [PMID: 12469359]

[92] Prentki M, Matschinsky FM, Madiraju SR. Metabolic signaling in fuel-induced insulin secretion. Cell Metab 2013; 18(2): 162-85.
[http://dx.doi.org/10.1016/j.cmet.2013.05.018] [PMID: 23791483]

[93] Gauthier BR, Wollheim CB. Synaptotagmins bind calcium to release insulin. Am J Physiol Endocrinol Metab 2008; 295(6): E1279-86.
[http://dx.doi.org/10.1152/ajpendo.90568.2008] [PMID: 18713958]

[94] Aguilar-Bryan L, Bryan J. Molecular biology of adenosine triphosphate-sensitive potassium channels. Endocr Rev 1999; 20(2): 101-35.
[PMID: 10204114]

[95] Carter RN, Morton NM. Cysteine and hydrogen sulphide in the regulation of metabolism: insights from genetics and pharmacology. J Pathol 2016; 238(2): 321-32.
[http://dx.doi.org/10.1002/path.4659] [PMID: 26467985]

[96] Fridlyand LE, Philipson LH. Glucose sensing in the pancreatic beta cell: a computational systems analysis. Theor Biol Med Model 2010; 7: 15.
[http://dx.doi.org/10.1186/1742-4682-7-15] [PMID: 20497556]

[97] Predmore BL, Julian D, Cardounel AJ. Hydrogen sulfide increases nitric oxide production from endothelial cells by an akt-dependent mechanism. Front Physiol 2011; 2: 104.
[http://dx.doi.org/10.3389/fphys.2011.00104] [PMID: 22194727]

[98] Eckersten D, Henningsson R. Nitric oxide (NO)--production and regulation of insulin secretion in islets of freely fed and fasted mice. Regul Pept 2012; 174(1-3): 32-7.
[http://dx.doi.org/10.1016/j.regpep.2011.11.006] [PMID: 22120830]

[99] Bedoya FJ, Salguero-Aranda C, Cahuana GM, Tapia-Limonchi R, Soria B, Tejedo JR. Regulation of pancreatic β-cell survival by nitric oxide: clinical relevance. Islets 2012; 4(2): 108-18.
[http://dx.doi.org/10.4161/isl.19822] [PMID: 22614339]

[100] Sun YG, Cao YX, Wang WW, Ma SF, Yao T, Zhu YC. Hydrogen sulphide is an inhibitor of L-type calcium channels and mechanical contraction in rat cardiomyocytes. Cardiovasc Res 2008; 79(4): 632-41.
[http://dx.doi.org/10.1093/cvr/cvn140] [PMID: 18524810]

[101] Tang G, Wu L, Wang R. Interaction of hydrogen sulfide with ion channels. Clin Exp Pharmacol Physiol 2010; 37(7): 753-63.
[http://dx.doi.org/10.1111/j.1440-1681.2010.05351.x] [PMID: 20636621]

[102] DeFronzo RA, Ferrannini E, Alberti KGMM, Zimmet P, Alberti G. International Textbook of Diabetes Mellitus, 2 Volume Set. John Wiley & Sons 2015.
[http://dx.doi.org/10.1002/9781118387658]

[103] Farfari S, Schulz V, Corkey B, Prentki M. Glucose-regulated anaplerosis and cataplerosis in pancreatic

beta-cells: possible implication of a pyruvate/citrate shuttle in insulin secretion. Diabetes 2000; 49(5): 718-26.
[http://dx.doi.org/10.2337/diabetes.49.5.718] [PMID: 10905479]

[104] Jitrapakdee S, Wutthisathapornchai A, Wallace JC, MacDonald MJ. Regulation of insulin secretion: role of mitochondrial signalling. Diabetologia 2010; 53(6): 1019-32.
[http://dx.doi.org/10.1007/s00125-010-1685-0] [PMID: 20225132]

[105] Moustafa A, Habara Y. Hydrogen sulfide regulates Ca(2+) homeostasis mediated by concomitantly produced nitric oxide *via* a novel synergistic pathway in exocrine pancreas. Antioxid Redox Signal 2014; 20(5): 747-58.
[http://dx.doi.org/10.1089/ars.2012.5108] [PMID: 24138560]

[106] Lynch A, Best L. Cytosolic pH and pancreatic beta-cell function. Biochem Pharmacol 1990; 40(3): 411-6.
[http://dx.doi.org/10.1016/0006-2952(90)90537-U] [PMID: 2166513]

[107] Lee SW, Cheng Y, Moore PK, Bian JS. Hydrogen sulphide regulates intracellular pH in vascular smooth muscle cells. Biochem Biophys Res Commun 2007; 358(4): 1142-7.
[http://dx.doi.org/10.1016/j.bbrc.2007.05.063] [PMID: 17531202]

[108] Lu M, Choo CH, Hu LF, Tan BH, Hu G, Bian JS. Hydrogen sulfide regulates intracellular pH in rat primary cultured glia cells. Neurosci Res 2010; 66(1): 92-8.
[http://dx.doi.org/10.1016/j.neures.2009.09.1713] [PMID: 19818370]

[109] Stiernet P, Nenquin M, Moulin P, Jonas JC, Henquin JC. Glucose-induced cytosolic pH changes in beta-cells and insulin secretion are not causally related: studies in islets lacking the Na+/H+ exchangeR NHE1. J Biol Chem 2007; 282(34): 24538-46.
[http://dx.doi.org/10.1074/jbc.M702862200] [PMID: 17599909]

[110] Moulin P, Guiot Y, Jonas JC, Rahier J, Devuyst O, Henquin JC. Identification and subcellular localization of the Na+/H+ exchanger and a novel related protein in the endocrine pancreas and adrenal medulla. J Mol Endocrinol 2007; 38(3): 409-22.
[http://dx.doi.org/10.1677/jme.1.02164] [PMID: 17339404]

[111] Rorsman P, Renström E. Insulin granule dynamics in pancreatic beta cells. Diabetologia 2003; 46(8): 1029-45.
[http://dx.doi.org/10.1007/s00125-003-1153-1] [PMID: 12879249]

[112] Hou JC, Min L, Pessin JE. Insulin granule biogenesis, trafficking and exocytosis. Vitam Horm 2009; 80: 473-506.
[http://dx.doi.org/10.1016/S0083-6729(08)00616-X] [PMID: 19251047]

[113] Deisl C, Albano G, Fuster DG. Role of Na/H exchange in insulin secretion by islet cells. Curr Opin Nephrol Hypertens 2014; 23(4): 406-10.
[http://dx.doi.org/10.1097/01.mnh.0000447013.36475.96] [PMID: 24840298]

[114] Deisl C, Simonin A, Anderegg M, *et al.* Sodium/hydrogen exchanger NHA2 is critical for insulin secretion in β-cells. Proc Natl Acad Sci USA 2013; 110(24): 10004-9.
[http://dx.doi.org/10.1073/pnas.1220009110] [PMID: 23720317]

[115] Hu L-F, Li Y, Neo KL, *et al.* Hydrogen sulfide regulates Na+/H+ exchanger activity *via* stimulation of phosphoinositide 3-kinase/Akt and protein kinase G pathways. J Pharmacol Exp Ther 2011; 339(2): 726-35.
[http://dx.doi.org/10.1124/jpet.111.184754] [PMID: 21865440]

[116] Maechler P, Wollheim CB. Mitochondrial glutamate acts as a messenger in glucose-induced insulin exocytosis. Nature 1999; 402(6762): 685-9.
[http://dx.doi.org/10.1038/45280] [PMID: 10604477]

[117] Maechler P, Wollheim CB. Mitochondrial signals in glucose-stimulated insulin secretion in the beta cell. J Physiol 2000; 529(Pt 1): 49-56.

[http://dx.doi.org/10.1111/j.1469-7793.2000.00049.x] [PMID: 11080250]

[118] Gylfe E, Hellman B. Role of glucose as a regulator and precursor of amino acids in the pancreatic beta-cells. Endocrinology 1974; 94(4): 1150-6.
[http://dx.doi.org/10.1210/endo-94-4-1150] [PMID: 4206544]

[119] Marquard J, Otter S, Welters A, *et al.* Characterization of pancreatic NMDA receptors as possible drug targets for diabetes treatment. Nat Med 2015; 21(4): 363-72.
[http://dx.doi.org/10.1038/nm.3822]

[120] Scholz O, Welters A, Lammert E. Role of NMDA Receptors in Pancreatic Islets.The NMDA Receptors. Cham: Springer International Publishing 2017; pp. 121-34.
[http://dx.doi.org/10.1007/978-3-319-49795-2_7]

[121] Abe K, Kimura H. The possible role of hydrogen sulfide as an endogenous neuromodulator. J Neurosci 1996; 16(3): 1066-71.
[http://dx.doi.org/10.1523/JNEUROSCI.16-03-01066.1996] [PMID: 8558235]

[122] Kimura H. Hydrogen sulfide induces cyclic AMP and modulates the NMDA receptor. Biochem Biophys Res Commun 2000; 267(1): 129-33.
[http://dx.doi.org/10.1006/bbrc.1999.1915] [PMID: 10623586]

[123] Danial NN, Walensky LD, Zhang CY, *et al.* Dual role of proapoptotic BAD in insulin secretion and beta cell survival. Nat Med 2008; 14(2): 144-53.
[http://dx.doi.org/10.1038/nm1717] [PMID: 18223655]

[124] Taniguchi S, Kang L, Kimura T, Niki I. Hydrogen sulphide protects mouse pancreatic β-cells from cell death induced by oxidative stress, but not by endoplasmic reticulum stress. Br J Pharmacol 2011; 162(5): 1171-8.
[http://dx.doi.org/10.1111/j.1476-5381.2010.01119.x] [PMID: 21091646]

[125] Efanova IB, Zaitsev SV, Zhivotovsky B, *et al.* Glucose and tolbutamide induce apoptosis in pancreatic beta-cells. A process dependent on intracellular Ca2+ concentration. J Biol Chem 1998; 273(50): 33501-7.
[http://dx.doi.org/10.1074/jbc.273.50.33501] [PMID: 9837930]

[126] Okamoto M, Yamaoka M, Takei M, *et al.* Endogenous hydrogen sulfide protects pancreatic beta-cells from a high-fat diet-induced glucotoxicity and prevents the development of type 2 diabetes. Biochem Biophys Res Commun 2013; 442(3-4): 227-33.
[http://dx.doi.org/10.1016/j.bbrc.2013.11.023] [PMID: 24246677]

[127] Yang G, Yang W, Wu L, Wang R. H2S, endoplasmic reticulum stress, and apoptosis of insulin-secreting beta cells. J Biol Chem 2007; 282(22): 16567-76.
[http://dx.doi.org/10.1074/jbc.M700605200] [PMID: 17430888]

[128] Oyadomari S, Araki E, Mori M. Endoplasmic reticulum stress-mediated apoptosis in pancreatic beta-cells. Apoptosis 2002; 7(4): 335-45.
[http://dx.doi.org/10.1023/A:1016175429877] [PMID: 12101393]

[129] Okada M, Imai T, Yaegaki K, Ishkitiev N, Tanaka T. Regeneration of insulin-producing pancreatic cells using a volatile bioactive compound and human teeth. J Breath Res 2014; 8(4): 046004.
[http://dx.doi.org/10.1088/1752-7155/8/4/046004] [PMID: 25358383]

[130] Zhou X, Feng Y, Zhan Z, Chen J. Hydrogen sulfide alleviates diabetic nephropathy in a streptozotocin-induced diabetic rat model. J Biol Chem 2014; 289(42): 28827-34.
[http://dx.doi.org/10.1074/jbc.M114.596593] [PMID: 25164822]

[131] Si YF, Wang J, Guan J, Zhou L, Sheng Y, Zhao J. Treatment with hydrogen sulfide alleviates streptozotocin-induced diabetic retinopathy in rats. Br J Pharmacol 2013; 169(3): 619-31.
[http://dx.doi.org/10.1111/bph.12163] [PMID: 23488985]

[132] Zhou X, An G, Lu X. Hydrogen sulfide attenuates the development of diabetic cardiomyopathy. Clin Sci (Lond) 2015; 128(5): 325-35.

[http://dx.doi.org/10.1042/CS20140460] [PMID: 25394291]

[133] Xie L, Gu Y, Wen M, *et al.* Hydrogen Sulfide Induces Keap1 S-sulfhydration and Suppresses Diabetes-Accelerated Atherosclerosis *via* Nrf2 Activation. Diabetes 2016; 65(10): 3171-84.
[http://dx.doi.org/10.2337/db16-0020] [PMID: 27335232]

[134] Durante W. Hydrogen Sulfide Therapy in Diabetes-Accelerated Atherosclerosis: A Whiff of Success. Diabetes 2016; 65(10): 2832-4.
[http://dx.doi.org/10.2337/dbi16-0042] [PMID: 27659227]

[135] El-Seweidy MM, Sadik NA, Shaker OG. Role of sulfurous mineral water and sodium hydrosulfide as potent inhibitors of fibrosis in the heart of diabetic rats. Arch Biochem Biophys 2011; 506(1): 48-57.
[http://dx.doi.org/10.1016/j.abb.2010.10.014] [PMID: 20965145]

[136] Li L, Xiao T, Li F, *et al.* Hydrogen sulfide reduced renal tissue fibrosis by regulating autophagy in diabetic rats. Mol Med Rep 2017; 16(2): 1715-22.
[http://dx.doi.org/10.3892/mmr.2017.6813] [PMID: 28656209]

[137] Yuan P, Xue H, Zhou L, *et al.* Rescue of mesangial cells from high glucose-induced over-proliferation and extracellular matrix secretion by hydrogen sulfide. Nephrol Dial Transplant 2011; 26(7): 2119-26.
[http://dx.doi.org/10.1093/ndt/gfq749] [PMID: 21208996]

[138] Kodl CT, Seaquist ER. Cognitive dysfunction and diabetes mellitus. Endocr Rev 2008; 29(4): 494--.
[http://dx.doi.org/10.1210/er.2007-0034] [PMID: 18436709]

[139] Zou W, Yuan J, Tang ZJ, *et al.* Hydrogen sulfide ameliorates cognitive dysfunction in streptozotocin-induced diabetic rats: involving suppression in hippocampal endoplasmic reticulum stress. Oncotarget 2017; 8(38): 64203-16.
[http://dx.doi.org/10.18632/oncotarget.19448] [PMID: 28969063]

[140] Zhao H, Lu S, Chai J, *et al.* Hydrogen sulfide improves diabetic wound healing in ob/ob mice *via* attenuating inflammation. J Diabetes Complications 2017; 31(9): 1363-9.
[http://dx.doi.org/10.1016/j.jdiacomp.2017.06.011] [PMID: 28720320]

[141] Wang G, Li W, Chen Q, Jiang Y, Lu X, Zhao X. Hydrogen sulfide accelerates wound healing in diabetic rats. Int J Clin Exp Pathol 2015; 8(5): 5097-104.
[PMID: 26191204]

SUBJECT INDEX

A

Acid 47, 115, 116, 117, 141, 142, 143, 146
  Ferulic 117
  nucleic 141, 142, 143, 146
  oleanolic 47
  Salvianolic 115, 116
Aconitum carmichaelii 119, 120, 124
Activation D2, 20, 23, 25, 30, 31, 32, 33, 34, 38, 135, 144, 146, 147, 150, 155, 159, 163, 165, 227, 230, 233, 237, 241, 242, 243, 245, 246, 247
  acid-dependent D2 33, 34
  bile acid 31, 32
  glucose aldose reductase 146
Activity 9, 23, 47, 62, 113, 114, 152, 165, 166, 167, 170, 199, 212, 213, 217, 219, 221, 227, 230, 233, 237, 238, 242
  antioxidant 217, 219
  γ-secretase 170
Adipocytes 16, 60, 150, 151, 153, 155, 199, 201, 203, 205, 210, 226, 234, 235, 236, 248
Adipogenesis 199, 203, 209, 210, 235
Adipokines, anti-inflammatory 209
Adiponectin 209, 210, 211, 213, 214
Adipose tissue 2, 15, 140, 153, 154, 203, 208, 209, 210, 213, 226, 227, 234, 235, 236
  expansion 210
  macrophages (ATMs) 235
  metabolism 210, 226, 227, 235
Adiposity 135, 174, 175
Adipsin 210, 211
AD-type neurodegeneration 135
  obesity 21, 207
Advanced glycation 135, 136, 141, 143, 146, 176
  products 136, 141, 143, 146
Agents 62, 63, 64, 75, 77, 84
  antidiabetic 62, 63, 64, 75, 77
AGEs formation 142, 143
Agonists, receptor dual 42

Albumin excretion, urinary 111, 113, 126
Aldose reductase (AR) 113
Alogliptin 41, 45, 46
  adding 45, 46
Alzheimer's disease 135, 136, 139, 144, 148, 158, 162, 167, 169, 171, 176
American diabetes association (ADA) 44
Amino acid residues 168, 169
AMP-activated protein kinase 233, 234
Amplifying pathways of insulin secretion 238, 241
Amyloid 136, 137, 158, 162, 163, 168, 169, 170, 171, 173
  β-derived diffusible ligands (ADDLs) 170, 171, 173
  beta deposition 136
  precursor protein (APP) 137, 158, 162, 163, 168, 169, 170, 173
Angiotensin converting enzyme inhibitor (ACEI) 113, 117, 124
Anti-diabetic 40, 59, 199, 232
  agents 59
  drugs 40, 199
  effects 232
Antidiabetics 81, 87
  oral 87
Anti-hyperglycemic agents 59
Apolipoprotein 136, 175
Apoptosis 112, 138, 154, 157, 165, 167, 244, 245, 246, 247
Astragaloside-IV (ASIV) 113, 114
Astragalus membranaceus 112, 119, 121, 123
Astrocytes 143, 144, 145, 164, 175
Atherosclerosis 141, 142, 226, 246
Authorities, regulatory 58, 61, 63, 86
Axonal transport 160, 161

B

Bariatric surgery 2, 16, 20, 30, 35, 36, 37, 48
Beta 29, 137, 140, 168, 171, 173
  amyloid 137, 168, 171, 173

cell apoptosis 29, 140
pancreatic 29, 31, 32, 205
Bile acid(s) (BA) 20, 21, 22, 23, 24, 25, 26, 27, 30, 31, 32, 33, 34, 35, 36, 38, 39, 40, 48, 49, 50
 -binding resins (BABRs) 33, 34, 35, 40, 48
 circulating conjugated 39
 interactions 20, 21
 levels 38, 39, 48
 transports 23, 24
 sequestrants 20, 33, 40, 48, 49
 synthesis 21, 22, 33
Blood brain barrier (BBB) 148, 152, 163, 164, 173, 174
Blood glucose levels 27, 47, 113, 205, 212
Blood pressure, systolic 1, 4, 5, 8, 10, 13, 15, 67, 68
Blood urea nitrogen (BUN) 110, 112, 113, 114, 115, 117, 118
Body mass index (BMI) 14, 21, 37, 42, 74, 77, 174, 175, 207, 217, 219
Bofutsushosan 216, 217, 219
Brain insulin 135, 159, 172
 resistance 135, 172
 signalling 159
Brown adipose tissue (BAT) 25, 26, 27, 33, 35, 199, 201, 210
Bushen tongluo formula (BTF) 122, 123
Buyang huanwu decoction (BHD) 108, 120, 121

C

Calcium dysregulation 136
Canagliflozin 62, 63, 68, 69, 70, 71, 73, 74
Cancer, bladder 213, 218
Carbaprostacycline 201
Carbohydrate-responsible element binding protein (ChREBP) 31, 32
Cardiomyopathy, diabetic 247
Cardiovascular diseases 1, 2, 43, 46, 110, 135, 200, 207, 217, 218, 232
Carthamus tinctorius 120, 121
Cataplerosis pathways in pancreatic β-cells 242

Cells 16, 20, 21, 25, 26, 27, 28, 29, 31, 32, 37, 38, 110, 135, 137, 138, 139, 142, 143, 145, 146, 147, 148, 150, 151, 152, 155, 156, 157, 159, 161, 165, 168, 169, 170, 171, 172, 173, 226, 227, 235, 237, 239, 240, 242, 243, 244, 245, 246
 β-islet 171, 172
 death 137, 138, 139, 156, 159, 161
 differentiation 150, 151
 growth 226, 227
 glucose-treated 244
 insulin secreting 242
 neuronal 145, 148, 152, 165, 168
 pancreatic β-like 237, 243, 244, 246
 receptors 20, 21
 renal 110
Central nervous system (CNS) 136, 228
Ceramides 136, 149, 154
Cerebrospinal fluid 148, 169
Chenodeoxycholic acid 22, 23, 47, 49
Cholestyramine 34
Choline 137, 149, 152, 229, 238
 acetyl transferase (ChAT) 152
 acetyltransferase (CAT) 137, 149, 229, 238
Circulation 22, 23, 24, 26, 33
 enterohepatic 22, 23, 33
 portal 22, 23, 24, 26
Codonopsis pilosula 119, 120
Colesevelam 34, 35, 49
Colestimide 34, 35
Combination 7, 8, 10, 67, 68, 116, 149
 of danshen Root 116
 of naltrexone and bupropion 10
 of phentermine 7, 8
 of sphingosine 149
 of topiramate and phentermine 7
 of oral hypoglycaemic agents 67, 68
C-reactive protein (CRP) 116, 122, 153
Credibility 82, 83, 84, 86
Crisis, global 58, 59
Crptotanshinone 115
Cystathionine 226, 228, 229
 -γ-lyase 228, 229
Cytosol 157, 166, 216, 228, 229, 230, 242

D

Dental pulp stem cells (DPSCs) 244
Deposition 135, 155, 168, 171
DHA-rich fish oil 216, 219
Diabetes 2, 69, 70, 75, 141, 145, 207, 208, 236
  etiology 145
  fructose-induced 236
  high plasma 141
  ketoacidosis 69, 70, 75
  prevention program (DPP) 2
  translation 207, 208
Diabetic 59, 109, 111, 141, 246
  albuminuri 111
  cardiovascular complications 246
  complications 141
  glomerulosclerosis 109
  ketoacidosis 59
Diastolic blood pressure (DBP) 4, 5, 6, 8, 9, 10, 13, 14, 15, 67, 68, 232
Dicarbonyl compounds 142, 146
Diet-induced obesity 33, 35, 154
Diseases 20, 21, 30, 42, 50, 58, 59, 60, 74, 75, 77, 88, 135, 136, 138, 139, 141, 145, 154, 159, 166, 168, 169, 174, 175, 199, 200, 213, 214, 215, 218, 248
  diabetes-induced vascular 248
  metabolic 199, 213, 214
  non-alcoholic fatty liver 20, 21, 30, 42, 50, 215
  preexisting macrovascular 218
  vascular 141, 174, 248
Disorders 21, 141, 148, 152
  obesity-related 21
Distal ileum 22
DM diabetes mellitus 125
DN 108, 109, 111, 112, 118, 119, 120
  development of 108, 109, 111
  management of 112, 118, 119, 120
  pathogenesis of 109
DNA binding domain (DBD) 202
DPSC-derived pancreatic β-like cells 246
Drugs 1, 3, 71, 77, 154
  antidiabetic 77, 154
  new glucose-lowering 71
Dysfunction 136, 155, 158, 248
  diabetes-associated cognitive 248
Dyslipidemia 30, 174, 175
Dysregulation 156, 157

E

Efficacy of SGLT2 inhibitors 63, 64, 67
Empagliflozin 59, 62, 63, 64, 65, 66, 67, 68, 69, 70, 73, 74, 77
Endogenous H_2S production 231, 236, 237, 239, 245
Endoplasmic reticulum 125, 136, 154, 227, 238
  stress 136
Endothelial cells 26, 112, 143, 157, 164, 232, 246
Energy 3, 11, 13,, 25, 27, 33, 34, 39, 40, 47, 48, 49, 135, 138, 140, 145, 149, 157, 199, 201, 216
  balance 11, 199, 201
  expenditure 3, 11, 13, 25, 27, 33, 34, 39, 47, 48, 49
  homeostasis 26, 27
  metabolism 40, 48, 135, 138, 140, 145, 149, 157, 216
Enterocytes 23, 24, 32
Enteroendocrine 20, 31, 38
Etiopathogenesis 138

F

Farnesoid X receptor (FXR) 20, 23, 24, 31, 32, 33, 34, 47, 231
Fasting 6, 8, 12, 13, 15, 29, 30, 43, 45, 63, 67, 68
  glucose levels 12, 15, 29, 30, 45
  insulin levels 8, 12, 13, 15
  plasma glucose (FPG) 6, 29, 43, 45, 63, 67, 68
Fibrosis 110, 113, 114, 247
  renal 110, 113
Flavin adenin dinucleotide (FAD) 145

Formation 110, 135, 136, 138, 140, 142, 144, 145, 146, 147, 149, 157, 161, 162, 169, 170, 171, 174, 205, 243
   free radical 140, 145, 170
   placenta 205
Fundamental diabetes complications in AGEs 143
FXR 23, 24, 31, 32
   intestinal 23, 24
   activation 31, 32

G

Gamma secretase 137, 169
Gastric bypass 20, 35, 36, 37, 38, 48, 50
   surgery 35, 36, 37, 38, 39, 48
Genes 145, 147, 149, 176, 200, 201, 202, 203, 210, 211, 212, 215, 216, 217, 233, 246
   expression 145, 201, 203, 233
   insulin receptor substrate 149
   insulin-responsive 215
Glipizide monotherapy 46
Globesity 135
Glomerular filtration rate (GFR) 68, 109, 125
GLP-1 Receptor 21, 23, 25, 27, 28, 29, 31, 33, 35, 37, 39, 41, 42, 43, 45, 47, 49
   agonist liraglutide on liver fat content 42
GLP-1 receptor agonists 40, 42, 43, 44, 46, 47
   injectable 44
Glucagon secretion 29, 30
Glucocorticoid receptor (GR) 200, 233
Glucokinase 233, 234
Gluconeogenesis 33, 34, 61, 226, 232, 233
Glucose metabolism 21, 33, 34, 35, 36, 37, 38, 42, 135, 136, 144, 162, 200, 203, 221, 228, 238, 233, 234, 239, 240, 242
   in adipose tissue 234
   oxidation pathway 228
   restoring brain 162
Glucose production 38, 60, 61, 226
   endogenous 61
   hepatic 38, 60, 226
Glucose tolerance 4, 30, 31, 32, 33, 35, 37, 42, 49, 159, 165
   impaired 4, 32, 59

   improved 31, 35, 37, 42, 49, 165
   normal 4, 37
Glucose transport 62, 164, 210
   insulin-dependent 210
Glucosuria 59, 62
Glucotoxicity 244, 245, 246
Glycation 142, 143, 146, 147, 175
Glycogenolysis 61, 233, 234
Glycosylated haemoglobin 59, 60, 62, 67, 68, 214
Growth factors 108, 110, 122, 125, 135, 149, 150, 151, 165, 247
   connective tissue 122, 125
   transforming 108, 110, 247

H

H_2S 226, 231, 234, 237, 238, 240, 241, 242, 244, 246
   adipose-generated 236
   endogenous 226, 231, 237, 238, 244, 246
   on insulin secretion 237, 240, 242
HDL-cholesterol 1, 8, 12, 13
Healing 248
   diabetic impaired wound 248
   improved diabetic impaired wound 248
Health technology assessment (HTAs) 85, 86, 89
Hepatocytes 23, 24, 26, 42, 215, 232, 233, 234, 248
   glycogen content of 233, 234
High-fat diet (HFD) 31, 32, 47, 158, 168, 226, 235, 245, 246
Hippocampal neurons 147, 162, 171, 173
Homocysteine 228, 229
Hydrogen peroxide 140, 245
Hyperglycemia 140, 141, 142, 145, 147, 148, 175, 226, 232, 233, 234, 245, 247, 248
Hyperinsulinemia 35, 231
Hypertension 2, 5, 21, 42, 62, 135, 174, 175, 226
Hypertriglyceridemia 141, 215
Hypoglycemia 4, 29, 40, 41, 43, 45, 248
   low incidence of 45

I

IGF-1 receptors 152, 163
Impaired insulin mechanisms 152
Impairment, diabetic cognitive 248
Inactive thyroxine 26, 27
Incretinmimetics 65, 66
Inflammation 25, 108, 109, 110, 115, 117, 120, 143, 150, 152, 154, 155, 199, 209, 210, 213, 226, 227, 234
   inhibiting 115, 117
   metabolic 209, 210
Inflammatory 108, 109, 112, 114, 115, 117, 118, 124, 143, 149, 152, 153, 154, 209, 211
   cells recruitment 209, 211
   cytokines 108, 109, 112, 114, 115, 117, 118, 124, 143
   mediators 152, 153, 154
Inhibitory effect(s) of 240, 241
   NaSH and L-cysteine on insulin release 240
Insulin 11, 150, 151, 152
   bonding of 150, 151, 152
Insulin like Growth Factors (IGFs) 125, 148, 150, 151
Insulin-mediated glucose disposal 135
Insulin receptors (IRs) 148, 149, 150, 151, 152, 163, 171, 172, 173, 210, 237
Insulin release 26, 28, 31, 32, 36, 41, 237, 239, 240, 241, 242, 244
Insulin resistance 1, 27, 30, 32, 36, 37, 38, 42, 60, 135, 136, 139, 149, 150, 152, 153, 154, 155, 159, 171, 172, 174, 175, 199, 205, 209, 210, 211, 212, 215, 226, 232, 235, 236, 237, 248
   decreased 36, 37
   development of 235, 236, 248
   neuronal 172
   peripheral 135, 153, 159
   reduced 205, 212
Insulin secretion 13, 29, 30, 32, 35, 38, 60, 135, 141, 159, 237, 238, 239, 240, 241, 242, 243, 244
   augmented glucose-induced 32
   pancreatic 240, 241
   regulation of 237, 241
Insulin sensitivity 6, 26, 32, 33, 34, 35, 38, 40, 42, 162, 165, 203, 210, 213, 214, 232, 235, 237
   improved 34, 35, 38, 237
   increasing brain 162
Insulin signaling 135, 154, 158, 233
   neuronal 154
Intracellular acidification 242
Intracellular signalization 144

J

JNK activation and inhibition of insulin/IGF-1 159

K

KATP 31, 239, 240, 241
   channels 31, 239, 240, 241
   -dependent glucose-induced insulin secretion 239
Kinases 137, 146, 150, 155, 241, 243

L

Laparoscopic adjustable gastric banding (LAGB) 39, 40
Large dense core vesicles (LDCVs) 243
Leonurus heterophyllus 119, 120, 124
Ligand-binding domain (LBD) 202
Ligustrazine 116
Lipid metabolism 25, 147, 175, 199, 203, 201, 210, 215, 216
   and energy balance 201
Lipid peroxydation 141, 145
Liraglutide 13, 14, 15, 16, 41, 42, 43, 44
   and placebo groups 14, 15
   effect of 13, 14, 42
Liver 23, 27, 150, 155
   cells 27, 150, 155
   X receptor (LXR) 23
Lixisenatide 41, 43
Lobeglitazone 215, 220
Long-term depression (LTD) 156

Lumen, intestinal 23, 24, 33
Luseogliflozin 59, 62, 63, 64, 68, 70, 71, 74
Lycium barbarum 118, 119, 120

M

Macrophages 25, 110, 140, 150, 153, 155, 209, 211, 246
   anti-inflammatory M2 211
Mechanism of PPARγ Ratio 201, 203, 205, 207, 209, 211, 213, 215, 217, 219, 221
Membranes, basolateral 24, 62
Mesangial cells (MCs) 110, 111, 125
Metabolic 8, 20, 25, 39, 40, 42, 47, 49, 50, 71, 109, 139, 143, 149, 150, 151, 153, 155, 159, 199, 203, 204, 205, 208, 210, 211, 217, 226, 235, 241, 242
   amplifying pathway 241, 242
   changes 39, 143
   disorders 25, 47, 109, 153, 159, 205, 208
   effects 25, 71, 151
   homeostasis 210, 211
   syndrome 8, 20, 40, 42, 49, 50, 139, 149, 150, 153, 155, 199, 203, 204, 211, 217, 226, 235
Metformin 40, 41, 43, 44, 45, 48, 63, 64, 65, 66, 67, 68, 77, 81, 213, 220, 231
Microalbuminuria 110, 111
Microtubule-associated protein (MAPs) 160, 161
Microvascular complications 59, 60
Mitochondrial 136, 157, 158, 159, 226, 227
   biogenesis 226, 227
   dysfunction 136, 157, 158, 159
Mitogen-activated protein kinase (MAPK) 118, 125, 144, 165, 230
Monocyte chemoattractant protein (MCP) 108, 111, 125, 209
Monocytes 25, 26, 209, 211
Monotherapy 4, 10, 11, 40, 41, 44, 45, 47, 63, 64, 65, 66, 67, 68, 220
Morphometric variables 111
Muscarinic receptors 137
Muscle cells 140, 143, 145, 237
   skeletal 237
   smooth 140, 143, 145
Mutation tissue distribution phenotype 205
Myocardial infarction 5, 15, 43, 44, 59, 213, 214
   nonfatal 15, 43, 44

N

Naltrexone 1, 10, 11, 12
Neocortex 162, 170
Nephropathy 141, 142, 246, 248
Nervous system 3, 9, 25, 135, 142, 145, 147, 153, 175
Neurodegeneration 136, 152, 160, 161, 163, 173
Neuronal 16, 135, 136, 138, 160, 165, 168, 173
   homeostasis 135, 136
   loss 138, 165, 168, 173
   pathways 16, 160
Neuropathological changes 156, 168
Neuropathy, diabetic 146, 168
Nicotinamide dinucleotide (NADH) 145, 228, 230
Non-alcoholic 20, 21, 30, 40, 42, 50, 215, 220
   fatty liver disease (NAFLD) 20, 21, 30, 40, 42, 50, 215, 220
   steatohepatitis 20, 42
Nonenzymatic glycation 140, 141, 142, 145, 146
Nuclear receptors 20, 23, 50, 199, 200, 210

O

Obesity 32, 203
   abdominal 203
   genetic 32
Obesity epidemic 20, 21
   growing 21
Observational studies 44, 46
Oligomers 158, 159, 169, 170
Oral 42, 48, 67, 68
   glucose tolerance test 42, 48
   hypoglycaemic agents 67, 68
Oxidase, cytochrome 138, 158

Oxidative stress 26, 42, 109, 113, 135, 136, 138, 140, 141, 142, 145, 157, 158, 159, 175, 226, 227, 230, 244, 246, 247

P

Paeonia lactiflora 118, 119, 121
Paired helical filaments (PHF) 160
Panax notoginseng saponins (PNS) 117, 126
Pancreatic β-cells 231, 237, 238, 239, 240, 241, 242, 243, 244, 245, 246
   differentiation 244, 246
Pancreatitis 14, 44, 46
   increased risk of 44, 46
Parkinson's disease (PD) 168
Pathways 209, 228, 229, 238, 239, 241, 246
   amplifying 238, 239, 241
   anti-apoptotic 246
   biosynthesis 229
   insulin-signalling 209
   non-enzymatic 228
   triggering 239
Perineal hygiene 59
Permeabilized islets 240, 241
Peroxisome 173, 199, 200, 201, 202, 203, 204, 211, 215, 216, 233, 234, 235
   proliferator-activated receptors (PPARs) 199, 200, 201, 202, 203, 204, 211, 215, 216
Pharmaceutical and devices agency (PMDA) 63, 67, 70
Phosphoinositide 3-kinase 165, 234, 235
Phosphorylation 137, 149, 161, 173, 210, 245
Pioglitazone 65, 66, 67, 68, 200, 201, 213, 214, 215, 218, 220
   reduction 67, 68
Plaques 135, 137, 138, 161, 169, 170
   amyloid 135, 137, 138, 161, 169
   solid 170
Postprandial glucose levels 34, 45
PPARα agonists 215
Preadipocytes 153, 234, 235
Prediabetic insulin-resistance 233
Prevalence of obesity in adults 206
Pro-inflammatory cytokines 155, 209
Proteasome 136, 167
Protein kinase 109, 126, 146, 147, 149, 151, 154, 159, 165, 237, 238, 242
   B (PKB) 149, 151, 165, 237
   C (PKC) 109, 126, 146, 147, 151, 154, 159, 238, 242
Protein levels 31, 113, 115
Protein phosphatases 154, 166
Proteinuria 108, 113, 116, 121, 122, 247
Proximal tubule 61, 62
Pyruvate carboxylase (PC) 233, 234, 242

R

Randomised controlled trials (RCTs) 58, 69, 73, 77, 80, 81, 82, 83, 84, 89, 113, 115, 116, 117, 121, 122, 124, 126
Reactive oxygen species (ROS) 110, 111, 126, 145, 156, 158, 227, 235
Reductive glucoses 142, 143
Regulating inflammatory cytokines 108, 124
Rehmannia glutinosa 120, 123
Releasable pool (RP) 238, 239
Renal function 59, 70, 108, 113, 116, 117, 247
Reno-protective effects 112, 113, 116, 117, 118
Retinoid X receptor (RXR) 202
Rosiglitazone 71, 162, 200, 201, 215, 217

S

Scavenger receptors 143, 144
Serum 38, 112, 113, 153, 167, 168, 173
   creatinine 112, 113
   glucose levels 38, 153
   insulin levels 173
   ubiquitin levels 167, 168
Small heterodimer partner (SHP) 23, 34
Soluble tumor necrosis factor receptors 126
Sphingosine 149, 154
Store-operated Ca^{2+} entry (SOCE) 235
Streptolysin 148
Stroke, nonfatal 15, 43, 44
Superoxide dismutase 114, 140, 247, 248
Synapses 136, 139, 170, 171

Synaptic plasticity 148, 156, 157
Synaptotagmin 238, 239
Systolic blood pressure (SBP) 1, 4, 5, 6, 8, 9, 10, 13, 14, 15, 67, 68

T

Target proteins 167, 227
Tau 136, 137, 160, 161, 162, 163, 164, 165, 166, 171
  hyperphosphorylation 136, 137, 160, 161, 162, 163, 164, 165, 166, 171
  isoforms 160
  phosphorylation 160, 161, 163, 164, 165
Thiazolidinediones 40, 41, 44, 63, 199, 213
Tissues 149, 151, 153, 158, 228
  fatty 149, 151
  mammalian 228
  penile 228
  peripheral 153, 158
Tofogliflozin 59, 62, 63, 64, 67, 68, 70, 71, 74
Toll Like Receptor (TLRs) 126, 150, 155
Topiramate 1, 5, 6, 7, 8, 9
  effects of 5, 6
Traditional chinese medicine (TCM) 108, 109, 111, 112, 113, 115, 116, 117, 118, 119, 121, 122, 123, 124, 125, 126
Transforming growth factor (TGF) 108, 110, 126, 209, 247
Transporters 24, 61, 231
Transport proteins 4, 20, 21
Triglycerides 1, 8, 12, 13, 15, 212, 215, 220
Tripterygium 118
  glycosides 118
  wilfordii extract 118
Triptolide 118
Troglitazone 71, 200, 201, 213

Tumor necrosis factor (TNF) 108, 110, 126, 146, 150, 153, 209, 210, 230

U

Ubiquitin/proteosome system (UPS) 166, 167, 168
Unsaturated fatty acids 201
Urinary albumin 111, 113, 114, 115, 117, 126
  excretion (UAE) 111, 113, 114, 115, 117, 126
  excretion rate (UAER) 117, 126
Urinary tract infections 46, 69, 70

V

Vascular 123, 126, 146, 248
  endothelial growth factor (VEGF) 123, 126, 248
Vertical sleeve gastrectomy (VSG) 39, 49
Voltage-dependent Ca^{2+} channels (VDCC) 239, 240, 241

W

Weight loss 1, 2, 3, 4, 5, 6, 7, 8, 9, 10, 11, 12, 13, 14, 16, 30, 35, 36, 37, 40, 41, 42, 44, 62, 65, 68
  meaningful 2, 7, 9, 12, 14
  showed robust 5, 7
  small amounts of 1, 2
Weight reduction 5, 59

Z

Zucker diabetic fatty (ZDF) 39, 231

www.ingramcontent.com/pod-product-compliance
Lightning Source LLC
Chambersburg PA
CBHW051144220526
45473CB00003B/646